Tomorrow's World by Raymond Baxter James Burke Edited by Michael Latham

Research by Richard Collin/Alan Dobson/
Louise Panton/Robert Symes Schutzmann/
John Parry Tomorrow's World Reporter/
Designed by Haydon James Young

British Broadcasting Corporation

To Jennifer, Graham, Caroline, and Sarah–and all their fellow citizens of tomorrow's world.

Published by the
British Broadcasting Corporation
35 Marylebone High Street
London WIM 4AA

SBN 563 10162 8

First published 1970
Reprinted 1971

Printed in England by
Sir Joseph Causton & Sons Ltd
Eastleigh Hampshire

Contents

The authors wish to record their appreciation of the many research workers, scientists, engineers, and doctors who helped in the preparation of this book. Together with universities, hospitals, research establishments and firms all over the world, they generously offered considerable expertise, private research material and illustrations without which this book, or indeed the television programme whose name it bears, could not exist.

From Nuclear Physics to Blue Cheese

Within a month of the first edition of *Tomorrow's World* in July 1965 viewers were writing to find out where they could obtain 'the book of the programme'. Since then, literally thousands of people from schoolboys to university professors have urged us to publish in more permanent form the reports which we present each week on BBC–1.

We decided to do better than that. In our first *Tomorrow's World* Book we have not only updated some of the most significant reports that have appeared in the programme but have also gone into issues and developments that it has been possible to touch on only briefly in our television survey of the critical and fast-changing world of medicine, science and technology.

The trouble is that it is a world of enormous proportions and from the outset it became only too clear that however hard we tried we could not hope to distil into these pages every single current achievement of major importance. So finally we made a selection that appealed to us, in the hope that it will appeal to you, too. Like our television programme, it is an extremely wide-ranging look at the current scene. From surgical techniques to edible petroleum; from blue cheese to nuclear physics; from the newest aircraft to the latest school desks–it is all here between these covers.

But before you delve farther into this book, a word of warning. There are aspects of life in the future that may not appeal to you, that may even shock you. For it is becoming clear that tomorrow's world could turn out to be an exceedingly dangerous place for some of us. Raymond Baxter and James Burke are your guides to the incredible world of tomorrow. Discover with them what is in store for you and the generations that will follow in the exciting decades that lie ahead.

Michael Latham
October 1970

An inflatable 'experience' in Holland

A Day in your Tomorrow

Home Computers/Time Off/
Plastic Homes/In Your Garden/
Off to Work/Laser Television/
Colour TV Cassettes/Robots/
Hot Paint/Space Heating/
Automated Hypnosis/
The Holiday Computer/
Inflatables/Disposables/
Tomorrow—like today

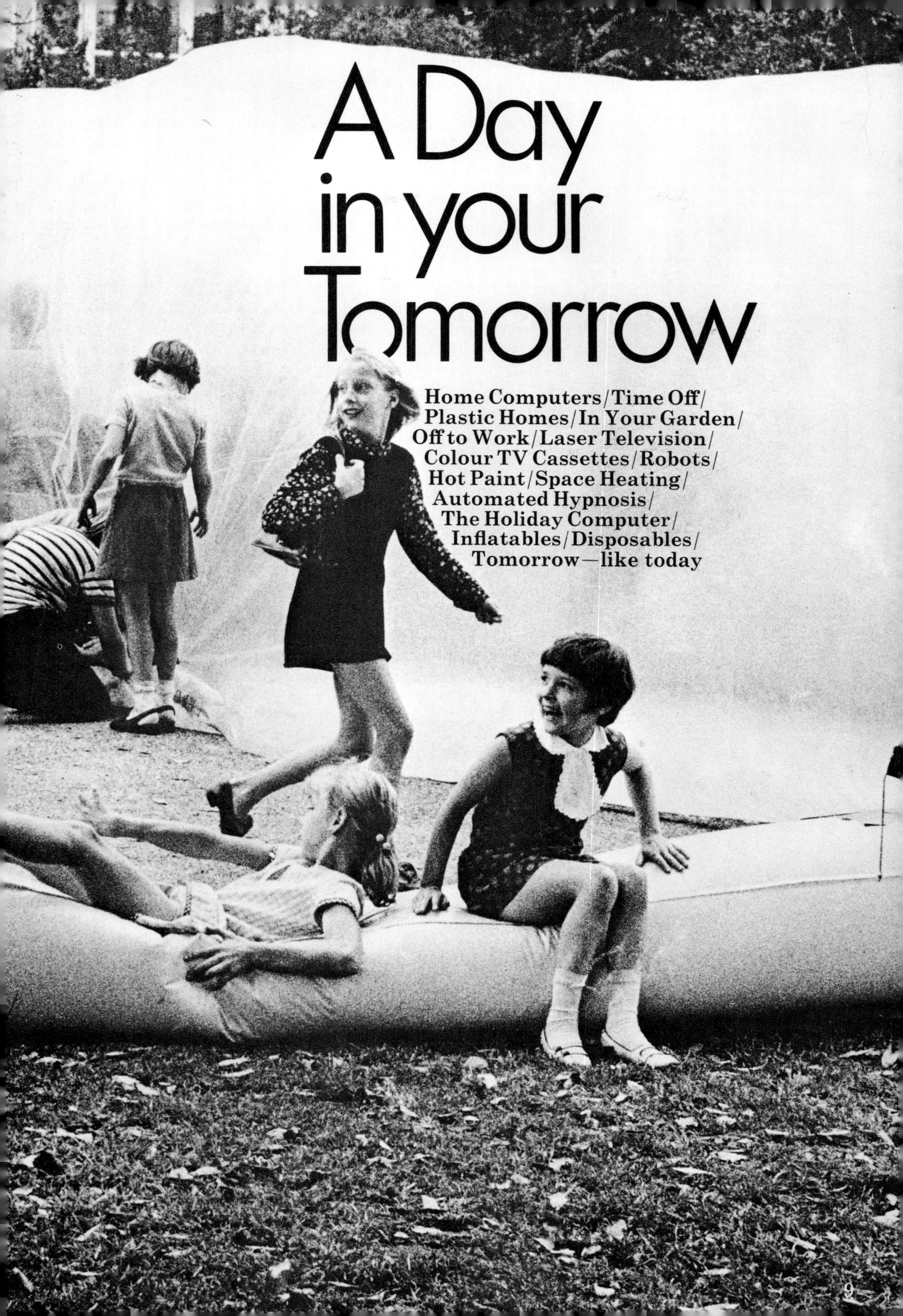

Home Computers

There is little doubt that the 1970s and 1980s will be the years in which the effects of computers will begin to be felt at every level of living. They will take over the running of our lives from the moment we wake up to the moment we drop off to sleep. Sleeping, indeed, will probably be about the only thing we shall be able to get on with by ourselves–although there is no reason why a diagnostic computer should not one day decide how much sleep we need on any given night.

The computer gourmet. Choose your own main course–the computer chooses everything else and cooks it. A bargain at £5000!

Computer Gourmet/Already computers in the home are becoming a practical and, for some, a financial possibility. For just under £5000 the American millionaires' favourite little shop in Dallas, Texas–Nieman Marcus–is offering the ultimate in kitchen gadgetry, a computer gourmet. All you have to do is choose your own main course (beef, chicken, lamb, etc.) and the computer does the rest. It selects the whole meal from the soup to the after-dinner mints, with as many courses as you want, and gives you all the relevant recipes. It also advises which wines you should drink with the different courses. If you feel confident enough to disagree with any of its choices, the computer will supply a selection of 'suitable' alternatives. The cooking is already being automated. You still have to choose your own guests, though.

Obviously, the price of a gadget like this puts it outside the price-range most people can afford, but Honeywell, the makers, believe that in the next five to six years the price of computers that can be used in the home will come down to the cost of the average family car. And that, they say, is only the beginning.

All the big computer companies have their own ideas about how the computer-controlled household should be run. Their plans may sound rather far-fetched, but they are all based on technology that is available now.

Only ten years ago a computer with the comparatively small capacity of the 'gourmet' would have cost in the region of £30,000. Today, at less than £5000, it is probably a hundred times faster and ten times more reliable. And while there will undoubtedly be a levelling-off of this rate of drop in costs, it is not yet in sight.

It is unlikely that there will be many fundamental technological breakthroughs in the computers of the 70s, but what will certainly happen is a tremendous boom in their applications. The planners see the ideal home controlled by a small computer linked to all the systems in the house. In the kitchen it will control the cooking operations, handle the household accounts, and automatically order any of the basic groceries that run out. A closed-circuit TV screen, linked to a number of stores, could help the housewife to do all her shopping without leaving home. The camera scans the shelves in the various stores and she simply chooses what she wants through her computer. The computer of course will compare prices and advise her on best buys. Everything will be credited to her account at a central bank which will give her a statement at the touch of a button.

The computer will control the home environment. Feed

in the daily schedule of your choice, and interior and exterior lights will be turned on or off, brightened or dimmed, curtains and blinds opened and closed automatically. Temperature, humidity, freshness, even odour, will all be taken care of.

Sweet Smell/There is a new odour control gadget available now that can be fitted into a ducted warm air heating system, or merely fixed on the wall with its own blower unit. It has two actions. First, there is a liquid chemical in it which combines with any smells in the air and neutralises them. Then the blower wafts the scent of your choice into the neutralised air. Floral, mint, pine smells–*you* choose. At the moment it is manually controlled, but it is the ideal system for a simple computer to take over and automatically operate.

In the study, the computer takes over the role of personal and business file, linked if necessary to the husband's business computer at his office, or to huge data banks to help the children with their homework. It could also be linked to the school's computer to operate as a teaching machine.

The computer's video screen could also be linked through to a central news bureau. At your request all the most interesting headlines of the day would be flashed on the screen and you simply key in the stories you want to read in detail. Any of them you would like to keep permanently could be printed out on a copying machine.

Computer Doctor/Eventually even the family doctor is likely to off-load some of his work on to a diagnostic computer. If your child, for instance, developed a temperature and the doctor were not available, you would be put through to the diagnostic computer service. The computer would refer to the child's previous medical history and the present symptoms. If the information were insufficient for a diagnosis, the computer would put up a questionnaire on your video screen and take you through it step by step before prescribing a course of action. The doctor would be informed as soon as he was available, and he would visit the child himself and prescribe any further treatment if necessary.

How much you make use of domestic computers in the future will be up to you, but as with most labour-saving devices, once the home adapts to them they're difficult things to do without. Just imagine waking up in the morning. The computer would call you with a tape of your favourite music–anything you like from 'Teddy Bears' Picnic' to a Bach fugue. It could even be a recording of an old-fashioned alarm clock–and, in time, no doubt that will be part of the genuine late-20th-century home style. Then while you sip your coffee, which came to the boil as you opened your sleepy eyes, the bath or shower is run to exactly the right temperature. And in the kitchen the bacon starts cooking.

After all that, you'll probably still have to sprint the last fifty yards to catch that commuter train to the office. There are some things that even a computer will never be able to do for you.

Time Off

With various forms of automation taking more and more of the hard work out of living, the leisure planners are having a field day with schemes to help us enjoy all the spare time promised in the future. Inevitably there has been talk about providing more playing fields, sports stadiums, gymnasiums, theatres, art galleries, further-education centres and libraries. Exciting as many of these projects may ultimately be, some designers have been looking for completely new approaches to the leisure problem.

Fun Palace/Londoner Keith Albarn came up with an idea that he called a Fun Palace. The Fun Palace is a collection of brilliantly coloured plastic domes held together by plastic bolts; the interior is a series of womb-like chambers linked by tubular corridors. The first one was opened in Girvan, a small Ayrshire seaside resort. The idea was that people going through it would enjoy a 'total experience'. They would be bombarded with changing sounds, light patterns, different wall-textures, moving floors; complete involvement was the key to the idea.

After analysis of the first year's operational results the town council at Girvan commissioned two 23-year-old Glasgow designers, John Ballantyne, a physicist from Strathclyde University, and Denis Barns, who had trained at the Glasgow School of Art, to redesign the interiors completely. They have come up with some ideas that should leave no one in any doubt about whether they are having an 'experience' or not.

The first room is painted throughout–ceiling, walls and floor–with aluminium paint. Sonic switches, which react to the sound of voices or footsteps, turn on stroboscopic lights and kaleidoscopic light patterns. Once you have

The 'light room'–activated by sonic switches, it changes colour as you move or speak

blinked your way through this section, you move along a tube into the inflatable room. As you stand wondering what is going to happen, the walls and ceiling (looking like giant plastic bags) start squeezing in towards you, pushing you on into the next chamber like toothpaste from a tube.

The next room is in darkness until you start speaking, and again sonic switches turn on a variety of lights, leading you to a kind of computer keyboard. By playing with the unlabelled switches and buttons on the panel, you create your own sound and light show, a sort of instant, and infinitely variable, 'Son et Lumière'. From there you fight your way through a room filled with ceiling-high polythene bags packed with rubber, and once you've squeezed and battered a path through, you find yourself in a plastic-foam maze in which you become totally disorientated. Strobe lighting is flashing all the time, and weird electronic music comes from six speakers ringed around the maze but only from one speaker at a time. And it is impossible to orient yourself by locating only one speaker.

If you ever find your way out, the ultimate experience is the Flame Room. This is painted in red and green flame patterns, with flashing red and green lights changing two or three times a second. When the red light is on, the red paint appears to be white and the green seems to be black. With the green light, everything is reversed–the green appears to be white and the red black. So with the lights changing so quickly, the changing flame pattern is highly dramatic.

The idea is that you come out of the 'experience' feeling like a new person. It should be an intriguing place to spend a quiet Sunday afternoon.

Plastic Homes

Living Pod/The use of plastics for building is fascinating architects all over the country. In Nottingham, 32-year-old David Greene has come up with a plastic home for two that he calls a Living Pod. Looking rather like an egg on stilts, it's a highly functional design that probably weighs about two tons. Its shell of thin fibre-glass gets its strength, like the egg, from its curved surfaces. The legs are adjustable hydraulically so that it can be set down on any sort of irregular surface. Like a trailer home, it could be plugged into whatever services are needed–electricity, water, telephone, computer link. All service pipes and cables, including an air-conditioning plant, are left quite blatantly on the outside, without any attempt to cover or disguise them. 'It's more honest that way,' says David Greene, 'it's simply making a visual statement'. He does not see the Living Pod as the sort of home that every couple would want to have. It is simply another design to increase consumer choice.

A view of the plastic house in Epsom

About the only conventional thing about it is that inside you still have to go upstairs to bed, or at least to an upper level. The entrance is like an igloo tunnel, interior walls and seats are inflatable, plastic covered windows are collapsible, desk, cooker, fridge are portable. The only thing you cannot move, deflate, or fold away, is the lavatory. And that has its own electrostatic sewage disposal unit. The Pod has no price tag on it yet.

Plasticottage/If you like your home to be on more conventional lines, it is still possible to have it built and furnished almost entirely with plastic. BP Plastics put together a house in Epsom to show just how effective their ideas can be. It has been built to show the advantages of lightness, strength, low cost, minimum maintenance, hygiene and colour possibilities of plastics. The average three-bedroomed house might have as much as 200 lb. of plastics in it. This house has two and a half tons of it in components like roof tiles, windows, walls, ceiling covers, floor tiles, wallpaper, kitchen fittings, bath, shower, wash basins, plumbing, tables, chairs and endless gadgetry. And it's all available *now* if you want it.

In Your Garden

If you are going to live with all that plastic around you, the ultimate is to get yourself a complete lawn made of plastic grass, also available now. Almost the only thing you have to do with it is hose it down occasionally. But if you really balk at the thought of plastic grass and insist on sticking to the genuine article, there is help on the way in the shape of a completely automatic lawn-mower. All you have to do is switch it on, then settle back in your armchair and relax in the sun.

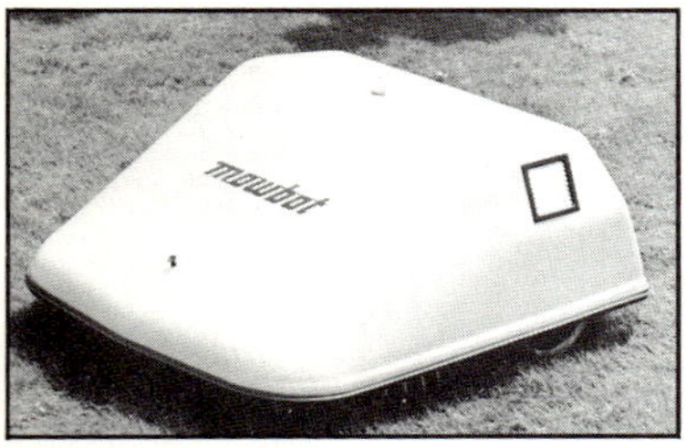

The Mowbot automatic lawn-mower. With the lid off, the rotary blade can be seen between the large and small wheels

Mowbot/An American company from New York has come up with the idea. It's called the Mowbot and costs about £400. You sink a thin cable all round the outer edge of the lawn and plug the two ends into a small transmitter. This sends an electrical signal round the cable. The mower, a battery-driven, rotary model, has a series of sensors built into it, and as it cuts its way across the grass the sensors pick up the signal and turn the mower away from the buried cable back into the middle of the lawn. So it cuts its way backwards and forwards through the grass, in completely random directions, until the whole lawn is finished.

If anything goes wrong with the system, there is no need to worry that your favourite rose bush will be chewed up. If the guiding sensors fail to pick up the signal from the cable, and the Mowbot goes trundling on over it, three more special sensors get the signal and simply cut off all the electrics immediately. And if *they* fail . . . well, there's always plastic grass.

Off to Work

In the summer of 1970 the first stage of a one-million-pound prestige office-block was completed at County Gates, near Poole in Dorset. As in an ever-increasing number of offices, the interior is spacious, airy and open-plan. But the exterior is dramatically different with its bold use of plastic. Designed by Patrick Holden and Associates, specialists in office-landscaping, the building is fully air-conditioned and artificially lit so that it is independent of external weather conditions. But in a building with so much open planning, the problem of air-conditioning it

effectively meant finding room for the huge ducts to bring in fresh air and carry out the stale. The economic thing to do was to hang the ducts vertically on the outside walls, and, since the building was to be artificially lit and the problem of window area did not therefore arise, a decision could be made on purely aesthetic grounds. But there is nothing aesthetic about great clumps of air-conditioning ducts stuck on the side of a building.

The architects decided to use giant moulded plastic covers made of glass-reinforced polyester. They needed 39 units, measuring 34 feet long, 9 feet wide and an average of 2 feet deep, but approaching 5 feet at each end. Once they had found a plastics specialist who was prepared to attempt the huge mouldings, they then had to go to the Physics department at Bristol University for advice on covering the surface with a stone aggregate to give an attractive finish. No one in the building trade seemed to know much about the problem. Eventually they used fine stone for the aggregate facing from English China Clay's quarries near Plymouth, bound to the surface of the plastic with a special gel. The end result was a non-corrosive, easily maintained, weatherproof moulding that would just need washing down with soapy water once a year to keep it clean.

And the cost: twenty-five shillings a square foot. The architect's original idea of concrete cladding for the ducts would have cost £8 a square foot.

Office Complex/In a new building going up in New York–the Port of New York Authority's £250 million World Trade Centre–the problem again was environmental

Moulding one of the giant duct covers for a new office-block in Dorset. *Below* is an artist's impression of the completed office-block with the ducts in place

control, but on a very much bigger scale. It has twin towers, each 1350 feet high, making it the world's tallest building. There are 110 stories with 10 million square feet of office space.

Controlling temperatures, humidities, pressures and power consumptions from one central point in a gigantic structure is an extremely complex task. £3 million worth of computers were brought in to deal with the problem and all its variables so that the building's 200 mechanical systems could be operated at maximum efficiency. Among the computer's jobs will be to make analyses of trends, efficiencies and operating profiles, taking into consideration not only the conditions inside the World Trade Centre but also those outside the buildings at four different elevations. Monitoring temperature, humidity, wind direction and speed at four different points is necessary because the height of the building is such that it could be sunny and hot on the 110th floor, but cold and foggy at ground level.

One man will be able to check, control and correlate the air-handling systems, pumping a million cubic feet of conditioned air through the building every minute simply by using push-buttons on a small console. He will have instant fingertip control of the largest chillers ever built–seven 7000-ton giant plants cooled by the Hudson River.

The six-building complex should be completed by the end of 1973–ready to house 50,000 office workers and 80,000 visitors a day.

Office Comfort/Although there is something basically absurd about people commuting as far as forty and fifty miles a day to a place where they can walk down a corridor to talk to a colleague who has travelled the same sort of distance, the system is unlikely to change much in this century. According to research, people apparently like to leave home and arrive somewhere else for work. So multi-million-pound office-blocks will continue to be built. Hopefully, the choice of sites will lessen rather than increase the crowding in cities.

But even with the development of video-telephones, computer communications and a variety of other electronic communications systems, there is, in business, a continuing stress placed on the value of personal contact. So the trend will be for offices to become more comfortable for the people working in them–more automated, more glossy, more efficient, more streamlined.

A group of designers in London called Complex D Development have formulated a complete programme of office environment design. The basis of their thinking is that, as computers, data transmission systems and audio-visual equipment become routine within the office environment, we must think carefully about how human beings react to them. Some studies have indicated that people who get dispirited by the overwhelming presence of machines can lose their efficiency and affect that of their colleagues. Some of the designs Complex D have produced are strictly for the super-executive.

At a cost of something approaching £1000, the centre-

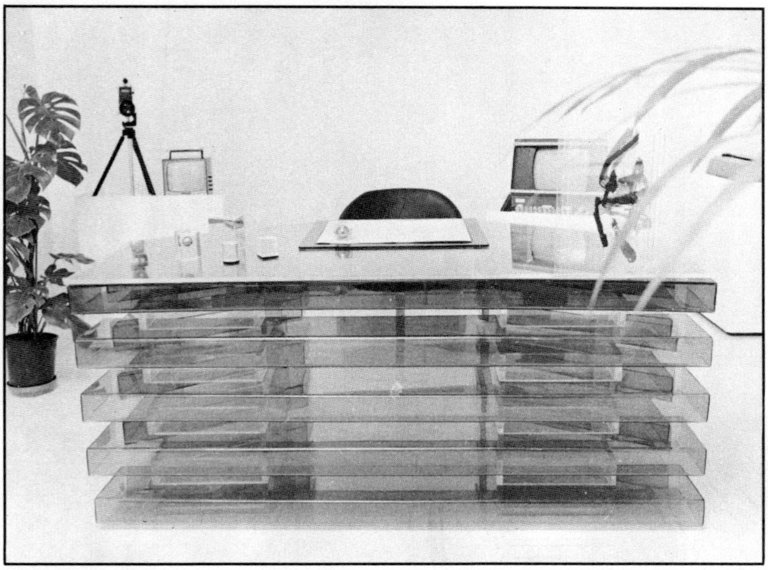

The Complex D transparent desk—with accessories

piece of their office is an acrylic desk. It is made to order in clear or tinted shades, with no drawers unless you specifically want them–because in their sort of super-office, there is no need to clutter yourself with bits of paper and common-or-garden files. All equipment is on mobile trolleys around you and you bring them to your side at the touch of a button. There's a video-phone, telephone, dictaphone, intercom, computer-terminal, attached video-screen, and a reproducing machine for sending typewritten or handwritten data, photographs, pictures and drawings over by telephone, using the principle of the electronic scanner.

Laser Television

Something else that could well find a place in office-blocks of the future, or in home projection rooms or cinemas, is a laser TV system. Present systems for projecting television on to giant screens are not as efficient as they might be. Pictures tend to lose their sharpness and colour clarity when the ordinary cathode ray tube picture is projected and enlarged.

At Expo 70 in Osaka the Japanese electronics giant, Hitachi, has come up with an alternative, a laser TV picture. The pictures are projected on to a screen measuring 12 feet by 9 feet and are startlingly brilliant, clear and sharp. In this method the ordinary TV signal is boosted and passed into three gas laser beams, coloured red, blue and green–the basic TV colour ingredients. The beams go through a set of special mirrors where they are combined into a single, multicoloured laser beam and relayed to a second mirror system.

This consists of a horizontal scanner with 16 mirror faces, spinning at a constant 60,000 rpm. A vertical scanner with 24 mirror faces spins at 150 rpm and tilts rapidly back and forth at the same time. The whole system is synchronised so that the laser-projected video-signals are literally splattered across the screen to produce a larger and brighter image than a cathode ray tube could reproduce.

Lasers produce the TV image for Expo '70 in Osaka, Japan

Its practical use is still probably five to ten years away because a number of problems remain to be solved. At the moment, because lasers use such high power, their lifetime is limited to a few hundred hours; with lower power, the lifespan could be increased to as much as a thousand hours. And the optical display devices need to be more efficient over a long term. But there is no doubt that these problems will be overcome and that laser TV will find a place in classrooms, cinemas, traffic-control and air traffic-control centres, offices and exhibition halls of the future.

Colour TV Cassettes

Development of television in the home is also booming, particularly with the use of portable, and comparatively cheap, video-recording systems. At the moment there are a number of systems which allow you to record programmes of your choice on video-tape or film, and play them back through a conventional television set. An American company, RCA, have produced a system which will probably be the first domestic consumer-product to use lasers.

They will produce full-colour programmes in cassette form that can be played into any domestic TV set, using a special laser TV tape player. They use lasers because the tape is made of vinyl–the sort of clear plastic the butcher wraps the weekend joint in–and the programme is recorded as a series of holograms. The material for the tapes will cost only a tenth of the cost of conventional film and it is scratch-proof, dust-proof and virtually indestructible under normal conditions.

At present the holographic tapes are not made directly

from life but from films, video-tapes, slides or photographs. The conversion process is quite complicated. A colour-TV programme is recorded on conventional film by means of an electron-beam recorder. This film, known as the colour encoded master, is then developed and converted by a laser to a series of holograms recorded on a plastic tape coated with photoresist–a material that softens to varying degrees depending on the intensity of the light striking it. The tape is then developed in a chemical solution that eats away the portions of the photoresist which have been softened by the laser beam. The result is a relief map of photoresist whose hills and valleys represent the original colour TV programme in encoded form. This is the hologram master.

The master is then plated with a thick coating of nickel and stripped away, leaving a nickel tape with the holograms impressed on it like a series of engravings. This is called the nickel master. Finally, by feeding the nickel master through a set of rollers along with a transparent vinyl tape of similar dimensions, the holographic engravings on the master are impressed on the smooth surface of the vinyl as holographic reliefs. And there you have it.

The nickel master can be used to press thousands of tapes without any loss of quality. RCA reckon they will soon be ready to go into production at a price in America of less than £200 for the complete playback unit.

Meanwhile, in Germany A.E.G-Telefunken have produced a disc and a record-player that you can plug into the back of your television set. They have managed to record on to the disc both sound *and* picture, and all you have to do is to put the record on the player, like any conventional machine, and switch on. The trouble is that the record spins at about 2000 rpm and at the moment lasts for only a couple of minutes.

Robots

The American engineer Buckminster Fuller once described the human body in mechanical terms: 'a self-balancing, twenty-eight-jointed biped, an electro-chemical reduction plant integral with segregated stowages of special energy extracts in storage batteries, for subsequent actuation of thousands of hydraulic and pneumatic pumps, with motors attached . . .' and so on and so on for more than a page. And we who operate this mechanism do so almost without thinking. But to make a machine that walks, for instance, is an extremely complicated problem. That is just what Professor Meredith Thring and his team at the Department of Mechanical Engineering, Queen Mary College, London, have done.

Called the Centipede Mark IV, it is powered by 24-volt aircraft batteries which drive two motors, each of which runs a tread. The tread itself consists of solid, rigid legs fitted on to rigid sections of chain. Each link of the chain is joined by a sort of axle that fits the notches of the driving wheel run by the motor. It will walk across the floor, climb stairs and is easily controlled by the person standing on it. Professor Thring is planning his next model with slightly shorter, sprung feet and with rubber tips to help it climb stairs more easily.

Two of Professor Thring's centipede robots. *Bottom* Uses a 2-tread step-mode for climbing; the other (with control bar) uses a single tread of stronger, rubber-tipped legs

Professor Thring intends the Centipede to act as a robot-base for carrying other robots around. It could one day be the means of propulsion for something that he believes no housewife will want to be without–the Robot Housemaid–which will clear tables, carry things round the house, make beds, even perhaps to go on short errands to borrow a cup of sugar from next door!

Hot Paint

One of the brilliant ideas researchers at the Paint Research Laboratories have come up with, developed and passed on to industry for commercial exploitation, is a graphite-based paint which is electrically conductive. This means that, once a wall has been painted, the whole surface can literally be plugged in to a power source, switched on, and the wall itself becomes a central-heating element–no radiators, no blowers, no fires of any sort. Already it has been tried out successfully in test flats in Newcastle. The implications are enormous. The researchers are thinking of using it to fight the problem of icy roads. Road tests are going on with a $\frac{1}{4}$-inch-thick resinous layer to act as an anti-skid surface. They say that it could well be possible to put a layer of the paint under the resin, and when the icy weather comes–just switch on.

The paint has many other potential uses. It could be used, for instance, for painting the inside of concrete moulds to stop building work being disrupted by the concrete cracking in icy weather–and incidentally speed up the setting of the concrete. Or it could be put on hundreds of miles of oil pipelines in Alaska, to keep the oil warm enough to flow freely in the sub-zero conditions.

Space heating

ICI have developed another advanced heating system called 'Flexel' space heating. They found that silicone rubber filled with carbon particles was a good conductor of electricity. When current was passed through, the surface of the carbon-filled rubber attained an even heat, and the rubber itself did not deteriorate. 'Flexel' is designed not for walls, but for ceilings: it can be laid against the upper side of a plasterboard ceiling, or stapled to the underside of rafters during house construction before the ceiling boards are fixed in position. A light, 2-inch-thick blanket of insulating material is laid over the 'Flexel' so that all the heat is generated downwards into the room below.

Automated Hypnosis

Research into hypnosis by Dr John H. Clark at the Hypnosis Research Unit in Manchester University's Department of Psychology has produced an automatic hypnotising machine with its own built-in logic unit. Dr Clark hopes the machine will have many applications in medicine, particularly in midwifery, dental surgery and psychiatry. It also has a potential use in experimental psychology where it could be used to deliver a standard hypnotic procedure avoiding any possible difficulties produced by the relationship involved in employing a human experimenter. The advantage of the machine is that it takes a routine task off the shoulders of the human hypnotist, so freeing him for other duties and preventing fatigue. So the dental surgeon, for instance, will be free to prepare his instruments and the experimental psychologist to concentrate on his observations while the machine gets on with the spadework.

The instrument itself consists of a tape recorder with prerecorded tape, a logic unit for finding the right part of the tape at any given moment, and a loudspeaker. The subject settles comfortably in his chair with a control button in his hand. This is how a session is likely to go:

Tape:
Your eyes are very heavy . . .
They are so heavy you want to close them . . .
They are so heavy they are closing . . .
Closing by themselves . . .
They are closing . . .
If your eyes are closed, press the button.
They are closing . . .
They are so heavy they are closing . . .
Closing . . .
They are closing . . .
If your eyes are closed, press the button . . .
(And so on)

This monologue will go on until the subject presses the button. Once he does, the tape spins on to the next stage of the session. And if it has no effect at all and the subject never presses the button, it will eventually spin back to the beginning and the session ends. In fact only one in ten people are very susceptible to hypnosis. And although we know that hypnosis involves monotonous repetition, the real nature of the phenomenon is not yet fully understood.

The Holiday Computer

Bernard Shaw used to maintain that man had organised the working part of his life well enough, but didn't know what to do with his leisure. For anyone needing advice about holidays and leisure, an International Recreation Association has been formed with its headquarters in Geneva. In June 1970 it held its first exhibition of leisure activities. In addition to the exhibition, there was a conference on the recreation of today and tomorrow at which those concerned with making the non-working part of life more pleasurable exchanged ideas. There were discussions on the sociology of the leisure age, tourist directors exchanged experiences, and there was considerable discussion on environment, conservation and pollution.

There was also a specially programmed computer provided to show how organisations could help their workers or clients choose where to go for their holiday. The computer asked a series of questions, such as the kind of accommodation you needed, how much money you wanted to spend, do you prefer to fly, drive or sail; it showed pictures of different kinds of scenery for you to choose from–sunset on the waves, snow-capped mountains turning pink, a village belfry sinking into darkness.

Finally, at the push of a button, the computer presented a compilation of the holiday best suited to your personality.

Inflatables

A 26-year-old Australian, Jeffrey Shaw, who runs an organisation called the Eventstructure Research Group, has his own ideas about leisure and people's daily pattern of life. He is concerned with air art-inflatables.

At the 1970 Brighton Festival, the Group used twelve inflatable tetrahedrons, coloured green, red and gold, to walk on the water. Sealed into air envelopes 15 feet high, twelve Brighton men set off to walk across the sea between the West Pier and the Palace Pier–a grotesque and magical ballet as they somersaulted the inflatables to move over the water. The Group's aim in this case was to change people's preconceptions about water. Their overall goal is to make people think again about their surroundings.

An inflatable 'experience' in Holland

Jeffrey Shaw says: 'The existing environment is built on concepts that are geometric, over-permanent, over-stable, which don't respect the whim of individual people. It's got so the world is something fixed and you are just living in it; rather than that the world is your product and something that is in conversation directly with you. We want to show that it is transformable, easily changed.'

Participation is the key to most air art. And faced with inflatables, people's inhibitions seem to break down fast. At the Knokke Film Festival in 1968 ERG provided an air ground–a double cone 30 feet high and 36 feet wide–where people could fling themselves. They did–to extremes. They took off their clothes and romped over it until the Casino authorities cut off the electricity to the air pumps.

ERG have built a giant, 30-foot inflatable dragon and paraded it through the streets to 'bring art to the people'. They have set up huge tubes called pneutubes for people to walk, crawl, wriggle and push through on their way to work–to give them an invigorating start to the day.

Their intention is to upset the status quo–of museum art, of the art object, of the role of the spectator, of architecture's rigid conceptions, of technological alienation, even of the concept of property and its immutable effect on our lives.

Disposables

Just as plastics are having an increasing impact on our way of life, so will disposables of all kinds–many of them made of various kinds of plastic, as well as paper, metal foil and the like. There is already a huge market for disposable underwear for men and women, and a growing one for dresses and shirts. But they represent only a tiny part of the picture.

One of the largest users of disposables in Britain is the National Health Service. Currently they have disposable gloves, face masks, examination gowns, bags for used dressings and instruments, sheets, towels and wipes, as well as instruments such as syringes, scalpels and forceps. Their use leads to an enormous saving in time and labour, and therefore in costs.

Children's toys and clothing is another big disposable field. As well as nappies, there are tissues, towels, toothbrushes (with enough toothpaste to be used once and then thrown away). In the home, particularly in the nursery, items like lampshades, tables and chairs, made of cardboard, can have a very short life, as well as plates, cups, knives, forks and spoons.

Even housing is coming under the eye of the disposables experts. One idea is for an emergency housing unit made of air-inflated plastic ribs, which can be covered by an outer skin for greater insulation, with doors that zip into the end panels. It can be used for temporary housing in disaster areas, several units being joined together to house a family. It folds away into a lightweight pack.

Indisposable Disposables/Unfortunately, plastic disposables bring a considerable litter problem with them. A paper cup left out in a field or on the beach will gradually

Everything in the picture is disposable—even the Sophia Loren coat-hanger

disintegrate and disappear. A plastic cup will not. Its life-span, under normal conditions, whether it is left in the open or buried, is many years, and unless we find an effective way to dispose of plastic forms once they've been used, there will be a gigantic litter problem within the next ten years.

The basic problem is that plastics which are cheap enough to use today are not bio-degradeable in either air, soil or water–in other words they do not 'decompose'. But the practical possibilities of achieving this are now in sight, although research scientists still have a great deal more work to do to solve this problem.

Disposables bring responsibilities with them: the public needs to avoid litter and pollution; the designer needs to work with a full understanding of materials and functions; the manufacturer needs to make products not only cheap enough to throw away, but to ensure that his products can be easily disposed of; the paper and chemical industries need to research the short- and long-term effects of discarded materials; and local, national and world authorities need to devise effective methods for disposal of waste and to control pollution. Whether this is best done by stiffening penalties for the offender or by more public spending or by a mixture of both may well be one of the most argued social problems of the next ten or fifteen years.

Disposable Noise/Another environmental problem for tomorrow is aircraft noise, increasing as air traffic increases at 10 per cent a year. At an international conference in London, Britain, France and America agreed that an aircraft's noise ought to be a basic part of its performance and should therefore be examined, just as the engine's thrust or its fuel consumption is a crucial operating factor. The aircraft should have to pass a series of noise tests, just as it has to pass a whole battery of tests for safety and handling before it can gain a certificate of airworthiness.

The Americans have now taken the whole thing a lot further in spite of powerful lobbying from the aircraft industry. They have laid down detailed test procedures and maximum noise levels for all new American jet aircraft. The noise levels are lower than many of today's jets

could meet, but even so they intend eventually widening the regulations to take in all jets flying in America. So the onus is fairly and squarely on the shoulders of the aircraft designers.

Obviously there will be no sudden reduction in aircraft noise immediately, but assuming that Britain and France follow the lead set by America's tough legislation, there could be a general reduction down to a quite bearable level within the next fifteen to twenty years. The trouble is that reducing engine noise is a very expensive business –both in terms of engine design and aircraft operation. And of course, there's no profit in it for the airline operators themselves.

The signs are already there in Britain though. A ban on sonic booms from commercial aircraft was one of the recommendations of a White Paper, 'The Protection of the Environment', published in May 1970 by the Labour Government. It is little consolation for people living near the big airports, but at least it is a start–official recognition from a government that a problem does exist.

Tomorrow–like today

All in all, as disposables multiply, a day in your tomorrow is going to be a lot cleaner than it is today. It will be a lot less arduous, too, as labour-saving equipment becomes more widely used. It will be a good deal more colourful and pleasing to the eye, as we gradually do away with slums with which we have had to live for so long, unable to replace them until cheaper building methods became available. A day in your tomorrow will also very probably be a good deal more frantic. There doesn't seem to be an increase in the desire to gain rest in the time made available by more efficient systems, but to fill that time with more things to do. So you will travel a great deal more and in general spend a great deal more money (and you will eventually forget what cash is and regard debt as a very necessary means of keeping the national economy healthily productive). You will be physically healthier, but probably a good deal more on edge, as the tempo of living quickens. In fact, a day in your tomorrow will make parts of your life very much easier and other parts that you have never considered important very much more difficult.

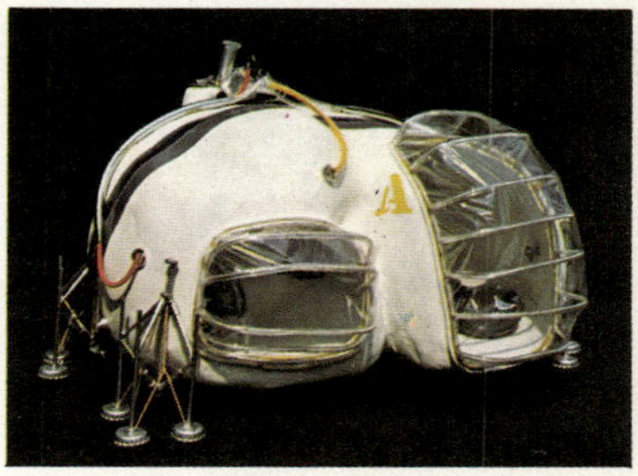

Top The Fun Palace at Girvan, Ayrshire *(see page 12)*

Above The Pod house *(see page 13)*

Thirty feet beneath the waters of Paradise Bay off Malta an aquanaut prepares to enter his underwater home. It is part of an experiment to discover the realities of underwater living. One day men may live in these conditions while they farm the sea-bed.

The Fourth Element

Mapping the Seas/Buoys/Breaking the Depth Barrier/Tektite II/Diving Gear/Underwater Jeep/Submersibles/Sea-bed Vehicle/Oil/On the Surface/Minerals from the Sea

From far out in space, as the crew of Apollo 8 came back from man's first voyage round the Moon, they described the Earth as no man had ever seen it before–a beautiful, shimmering oasis in the vast cold blackness of space, and dazzlingly blue. It is ironic that the blue of Earth's oceans, covering two-thirds of the planet, should have made the greatest impression on the astronauts coming in from a journey of exploration that had cost over £10,000 million, when only a fraction of that sum has been spent on exploring what may be man's last resource–the sea itself.

The oceans of the world are still virtually unexplored, but if we are to be assured of adequate resources for the future, man must learn to understand them and turn them to his advantage. We are beginning to reach the situation where the world needs more protein than it is capable of producing: the sea could be the greatest source of protein ever discovered. Our ability to create civil-engineering structures in the sea is still very primitive and our techniques have not really changed since the first London Bridge was built. Oil deposits under the sea, untouched sources of minerals and foods, the ability to bring ships of a million tons or more close into the shore, the increasing problem of pollution–all of these demand a much more sophisticated knowledge of undersea engineering and tools than we possess at the moment. It is a fairly simple matter to pressurise a human body to one atmosphere when it is in the weightless condition of outer space. To know how it reacts to the desperate pressures met underwater is something that we have yet to learn. We know less about the deepest parts of the ocean than we do about the floor of the Sea of Tranquillity. If the 1960s will be remembered as the years when man began to move into space, then our growing awareness of how important the sea is to us may mean that the 1970s and the 1980s will be remembered as the decades in which man first began to conquer the inner space of his own planet.

Deep Sea Search Vehicle: The $9\frac{1}{2}$'-diameter, 35'-long, 4-man sub at work 20,000 ft under the surface. Sonar and TV systems will locate crippled submarines or pieces of wreckage as little as 6 inches long. Telescopic 'feet' will steady the sub as the grab collects samples from the ocean bed. Battery power operates the single-screw propeller. Fine manoeuvring is by jet, 4-direction water-thrusters fore and aft.

Mapping the Seas

In the last few years technology has encouraged oceanographers and ecologists to develop new tools to aid in the investigation of the sea. So little is known of it that we are still finding ways to make undersea maps, let alone surveys of what the oceans and sea-beds can yield in the way of resources.

Acoustic Scanning/One of the latest techniques to be tested in undersea mapping is an advance even over closed-circuit television, until now the only way to get visual pictures of the sea-bed. The new system is called 'Acoustic Scanning', because it produces what is in effect a sound picture from under the sea. Unlike previous acoustic echo-sounders this equipment uses ultrasonic beams to produce signals that are then converted into TV-type pictures. The main advantage is that the beam does not rely on light, and can therefore 'see' in the dark and in muddy water. Ten thousand square yards of the sea-bed are scanned instantaneously.

The equipment is mounted on a tower on the ship's foredeck, directly over a shaft that runs through to the keel. The viewing unit looks much like a conventional ship's radar scanner, and includes a gyroscope stabilising head that counteracts pitch, roll and yaw to keep the unit in position in heavy seas. Before a scanning run begins, a lightweight plastic and metal dome is fitted over the bottom end of the shaft. Working like a ray dome, it protects the head from sea damage and keeps it free from water swirl. The head is winched down the shaft to operating position. Scanning involves two beams: one transmitting, one receiving. The transmitting beam sends out a continuous signal over a 30° sector. The receiving beam, one-third of a degree wide, scans back and forward over this 'illuminated' sector ten thousand times a second, picking up the echoes. The speed of the scan means that the tiniest movement of an object in the area is recorded.

A mid-water trawl net shown on the acoustic scanner screen. The cluster of white dots, *top right,* in each frame is a shoal of fish

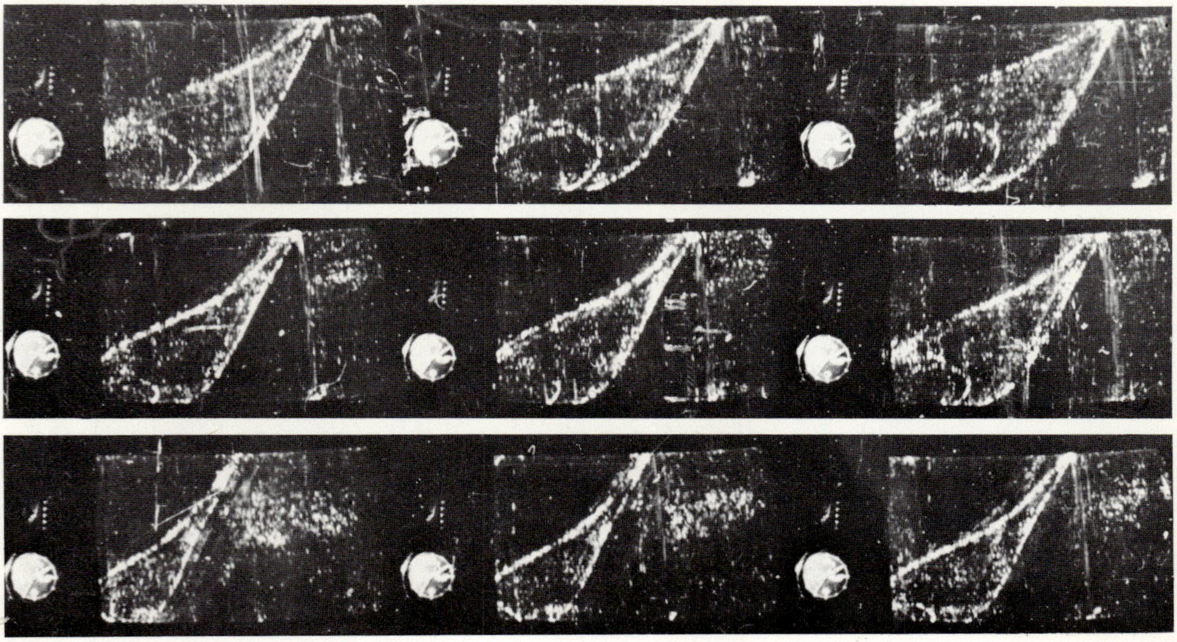

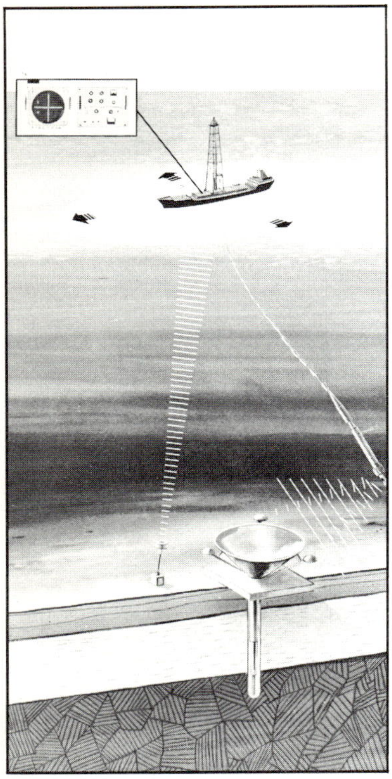

The *Challenger* drill-head being guided to 're-entry' by signals from the site and the well-head funnel

These echoes are then processed to produce detailed light-and-shade pictures of the area under examination, which can either be videotaped or filmed for later analysis.

For the first time pictures exist of a net moving through a shoal of fish. Analysis of fishes' behaviour and their ability to see and evade the net, the design of the net itself –all this is now possible. Shipping-lane surveys for the huge tankers of the future can be done to find wrecks and giant sea-bed sand waves that can ground a tanker and cost hundreds of thousands of pounds in delay. Geologists have already seen pictures detailed enough to chart the extent of potential mineral resources. For the first time man has a sight of where he is to go.

Four Miles Down/The first attempt to carry out a world-wide survey of the sea-bed was by HMS *Challenger,* which sailed from Portsmouth in 1872. It must be counted as one of the greatest scientific achievements of the nineteenth century. Fifty volumes of the scientific observations are still the basic reference work for oceanographers, and details from it are used by the astronauts navigating in outer space. But today marine scientists must know much more about the formation of the Earth's surface under the sea. The only way to achieve this is by drilling and examining cores of material brought to the surface, and studying the readings of delicate instruments inserted into the drilling-hole.

But even just getting to the sea-bed in some parts of the world means drilling 20,000 feet below the surface of the sea. We have little experience to go on but oil rigs, and they are always in comparatively shallow water. In 1968 a special deep-sea drilling ship called the *Glomar Challenger*–as a token of respect to the early one–began a series of experimental drilling runs off the coast of the United States.

It is operated by Global Marine Inc., which is a Los Angeles-based international off-shore drilling company, and it is scientifically controlled by the Scripps Institution of Oceanography of the University of California, San Diego. Using new drilling techniques, the most modern navigation facilities and highly sophisticated locating devices to keep the ship in exactly the right position, they have carried out a series of probes, drilling into the surface of the sea-bed in water nearly 4 miles deep.

A siting decision is made on available charts and the ship navigates to that position by satellite. The satellite emits a continuous radio signal. As it orbits the pitch of that signal gradually falls just as a train whistle does when it comes past. A navigation computer on board 'listens' to the signal and knows that there can be only two places on Earth, to either side of the satellite's path, where the pitch of the signal will be just at a given level. The navigator need only tell the computer which hemisphere it's in and his position on the Earth's surface can be calculated to within 200 feet.

The drill-head is lowered through the drilling-well on board the ship. Under the ship it is fed into a large funnel with a concrete base, and the whole lot is dropped to the sea-bed, where drilling begins. More pipe sections are

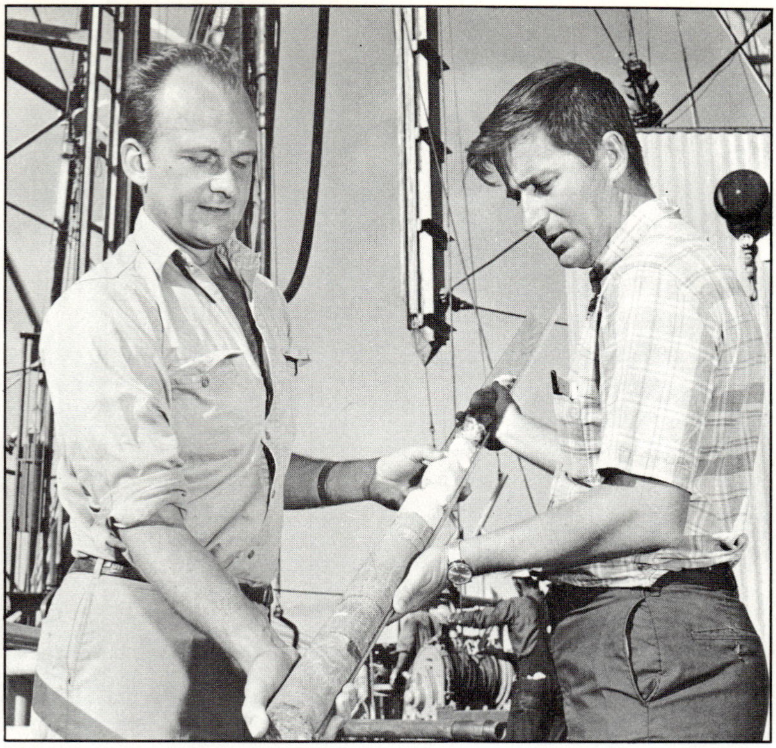

Scientists examine a drill core from 17,000 feet below the surface in the Western North Atlantic

added gradually, and when the drill-head encounters a harder or difficult rock or needs replacing, the whole thing is drawn up and the drill core samples are taken out for inspection.

The funnel and the sonar-transmitting devices attached to the drill are left at the top of the hole on the sea-bed. When the new drill is ready, it carries down in its head a tiny receiver, so that the funnel transmitters can guide the drill home. As the receiver 'hears' which way to go, it activates pumps in the drill-head itself that thrust water sideways at up to 1000 pounds' push. The drill-head moves from side to side as it goes down. Finding the hole again has been compared to a snake weaving and finally striking.

At the same time one of the sonar transmitters on the funnel is continuously telling the ship the position of the hole, and the ship's computer is activating thrusters on the side of the vessel, as well as the propellers, to hold the site.

For the foreseeable future the *Glomar Challenger* will provide the world's oceanographers and geologists with almost more material than they can handle. So far in two runs it has been established that at 200 million years the North Atlantic is the oldest ocean bed; that Europe and America are 60 feet farther apart than they were in 1900; that the ocean bed may contain large oil deposits covered by salt domes; that the Pacific sea-bed is in some places strewn with valuable minerals lying there in nugget form waiting to be picked up.

The refrigerated core sections coming back from *Glomar Challenger* for analysis ashore in the next few years may provide some of the most important discoveries about our planet and what it can do for us that man has made for centuries.

Flip moves to the final vertical position

Flip/The *Glomar Challenger* has bow and stern thrusters to keep her steady above the sea-bed. But sometimes scientists taking very delicate measurements at sea need to be far more steady even than that. So engineers from the Scripps Institution have come up with a novel idea–a ship than sinks itself and then turns through 90° and bobs up and down like a hydrometer. They call her Flip.

Flip is towed to position. She is approximately 750 tons weight, 355 feet long and almost cone-shaped (the base of the cone being the prow). At station Flip's ballast tanks (85 per cent of her length) are flooded with 1500 tons of salt water. As they flood, the stern slowly sinks and then finally the ship 'flips' vertical with the prow above water. 55 feet show, the other 300 feet go vertically down under water. The effect of this tremendous draught (the *Queen Elizabeth*'s is only 39.5 feet) is to use the still waters that lie just a few feet below the worst storm as a stabilising influence. In 35-foot waves in the Gulf of Alaska Flip swung no more than 3 feet from side to side.

Skamp/If you do want something that will drift at the mercy of the waves, the problem (if there is no one on board) is to move your drifter when you wish or to rendezvous with it, or simply to stop it drifting. These are simple problems for Skamp. It is a new disk-shaped automatic sailing craft with stiff foam-filled aerofoils instead of sails. The hull is just two circular metal plates with a thick layer of pliable foam sandwiched between them. The sails are designed to operate in winds up to hurricane strength. The ship steers itself by satellite. A small on-board computer will take it anywhere and keep it there to within 0.2 nautical miles. This would also make it ideal for a navigational buoy or meteorological station, as well as for an oceanography vessel. The prototype is 9 feet in diameter and 16 feet high–but its makers, RCA, say it can be built to any specification.

Gloria/Another new scientific device to map the sea-bed is Gloria (Geological Long Range Asdic). With this system acoustic mapping is possible along a wide swath of the sea-bed ten miles out from the track of the ship towing the device. Designed and operated by the British National Institute of Oceanography at Guildford, it will produce an accurate submarine map by bouncing sonar signals off the sea-bed in water nearly 4 miles deep.

Gloria is a 32-foot-long torpedo arrangement weighing nearly 7 tons, towed 600 feet under the surface of the sea behind the research ship, *Discovery*. Inside are 144 transducers which send out a series of sonic waves through the acoustically transparent fibre-glass body and then listen as the signals are reflected back by the sea bottom. The device has a depth range of 4 miles. Delicate electronic instruments built into the machine automatically steer it, keep it at the right depth and keep the sonic beam looking in the required direction. Steering is done by sending out a pulse that 'recognises' the strip of sea-bed irradiated by the last set of pulses. Since it takes 30 seconds for the beam to travel to the bottom and back, the maximum corrective action necessary to keep on course is only about half a degree. On board the research ship marine geographers can get an exact picture of the shape and texture of the bottom of the sea. And it can all be done while the ship is moving at 10 knots.

Gloria being lowered into operating position and *bottom* a diagram of Gloria showing the long centre panel of transducers

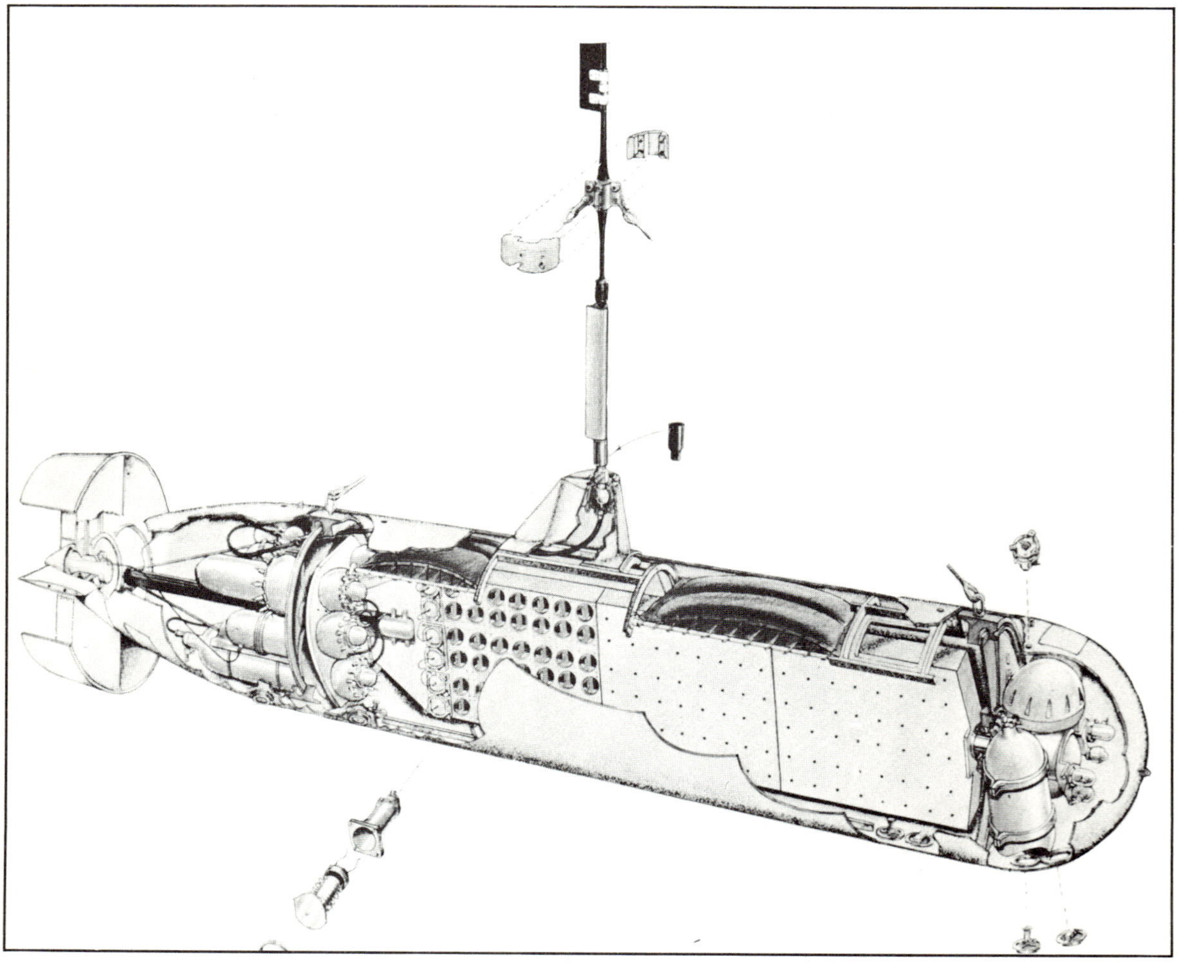

Free-fall Tide Capsules/With present methods, once you find an area of sea floor where the movement of the tide is of prime importance, you must either stay over the spot or return many times to observe the deep-sea behaviour in the area over a period of time. A new device, a free-fall tide capsule, has answered the problem. The first three have already been placed in Antarctic waters. Priority has been given to these waters because of their considerable influence on the tides in the entire southern hemisphere and their unique richness in sea life.

The capsule is a 'free' vehicle, i.e. it is not tethered to the ship from which it is lowered. It is made of aluminium spheres capable of taking 3-mile-deep pressures. Communication from ship to capsule is by acoustic signal: commands from the ship activate the capsule once it has reached the sea-bed. It is powered by a battery-pack attached to it which also acts as an anchor. Inside, instruments measure water temperature and pressure as well as currents, and the data is recorded on magnetic tape. When the mother ship returns, it sends a command to the capsule, which disengages itself from the battery-anchor and surfaces. The taped data is then either stored or run off into a computer. New battery-packs are added and the capsule returns to the sea-bed to record more data and wait for another surfacing command from the next mother ship to pass.

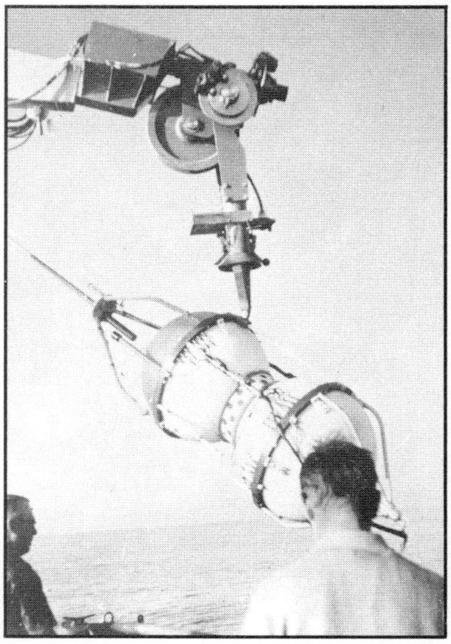

Above A free-fall tide capsule being lowered into the Antarctic from the research ship *Eltanin*

Left Free-fall tide capsule—the lowering rope is attached to the end of the three struts at upper left

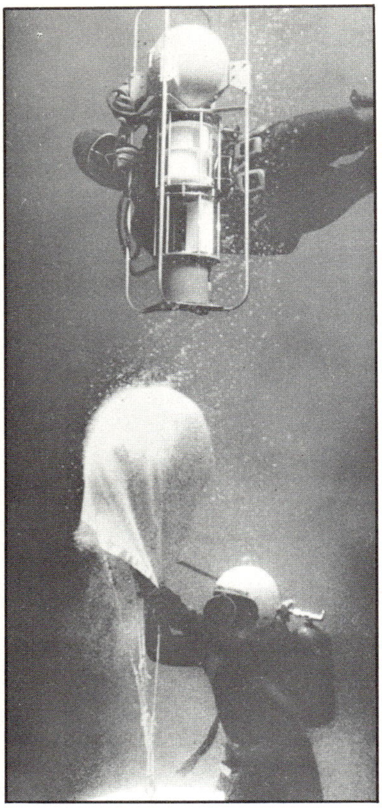
Divers collect a shallow-tide gauge and set its replacement

Buoys

Data-buoys/Not all the information we need from the sea will come from great depths. The shallower waters where man can go with ease to mine or to farm food are likely to be the areas that will give up their resources soonest. These shallow waters are now being observed and analysed with simple devices like data-buoys. They collect information and then automatically drift to the surface to be collected, or else they are brought up by a signal from a mother ship. Such tide meters are installed and replaced by divers.

Orb and Rum/The biggest buoy yet designed for off-shore research is a giant oceanographic research buoy, displacing nearly 200 tons, developed by the Marine Physical Laboratories of the Scripps Institution of Oceanography. The buoy is towed into position and then moored. It includes laboratory and living accommodation for twelve scientists and crew. It has a flat top which is used as a helicopter landing pad and a centre well which can be opened to allow instruments and equipment to be lowered through it. The well is 15 by 20 feet with hydraulically operating doors and can safely handle loads up to 12 tons. Orb has been specially designed to work together with a new underwater form of remote-controlled equipment called Rum, to which it is connected by power and control cable.

Rum–Remote Underwater Manipulator–has been developed by the US Office of Naval Research as an underwater tool for use on the sea-bed. Already they are thinking of operating it nearly a mile and a half under the sea. The vehicle is propelled by two independently controlled reversible electric motors–one on each track. It can be equipped with underwater television cameras, sonar devices, lights, depth-sounder, and can even be processed to follow a magnetic course. Various forms of manipulating arms will be used for collecting geological samples and carrying out simple civil-engineering tasks. At the moment the manipulator boom can work off either side or to the rear of Rum. At full arm's-length it can extend 50 lb. of force in any direction, and can pick up and move loads of up to 1000 lb. and swing them across the ocean floor through an arc of 300°. At the end of the boom the manipulator pivots on its mount at the shoulder and elbow, rotates at the wrist, and carries a grip with jaws 6 inches long, opening to grasp objects up to 7 inches across.

Autobuoy/Perhaps the most exciting development in buoys so far is now sitting in the mid-Pacific. Called Alpha, it is undergoing sea trials 3000 miles from land and in waters nearly six miles deep. Alpha is the first automatic device to measure the interaction of the ocean and the atmosphere–the vital part of our meteorological cycle that affects what our atmosphere is made of, and what our weather is like.

The buoy is 40 feet in diameter and weighs over 50 tons. Tests have shown that the disk shape is the most stable, although automatic pumps on board shift ballast in

Alpha in position in the Pacific. The aerial is 50 ft high

reaction to the sea state so that Alpha is always upright. Control machinery and electronics are stored inside the hull. Everything works either automatically or on radioed instructions from shore. Sealed generators, running off bottled gas, keep the main batteries charged–two generators, each containing twenty nickel cadmium cells. On board, a computer receives instructions and controls everything from a change in the schedule of scientific measurements to a blast on the fog-horn.

The buoy is kept in place by a mooring rope reaching 6 miles down to the ocean floor. Made of nylon and dacron it is only 2 inches thick, and in rough weather stretches just enough to prevent snapping.

There are over one hundred measuring instruments on board–up the mast, inside the hull and down the mooring line–taking precise readings from 45 feet above the surface to 1500 feet below it. Data is radioed to shore, and a duplicate stored on magnetic tape aboard. This means that the computer on shore can at any time receive information about what is happening at the moment of transmission or what was happening six months ago. The buoy also has a short-term memory bank into which new instructions can be placed, to be acted on up to 36 hours later.

If Alpha is successful, she could be the prototype for a network of buoys, seen only from time to time by passing sailors and deep-sea fishermen, stretching around the world and adding immensely to our knowledge of both the ocean and the atmosphere. Already one similar buoy has been installed east of Portland Bill to replace the conventional lightship.

Breaking the Depth Barrier

Diving is dangerous. If you make a serious mistake while you are under the water you will probably die. The future of all underwater exploration, as far as man is concerned, depends on a complete understanding of the effects of pressure on the human body as the swimmer penetrates deeper and deeper.

The human brain is kept alive by being fed with oxygen which is taken to it by the bloodstream. The blood extracts the oxygen through the fine tissue at the base of the lungs from the air we breathe in. Different lungs require different volumes of air to fill them and the amount remains constant. However deep a swimmer is, therefore, the cubic capacity of his lungs never alters. The brain also always requires the same molecular amount of oxygen. But as the swimmer dives deeper, the pressure compresses the air he is breathing into a smaller volume, and it becomes denser, so that at a given depth the body requires only half the amount of air that it needs on the surface because the air, including the oxygen in it, has been compressed to half its size. The lungs, however, still require the same cubic capacity to fill them during each breathing cycle; therefore, at a given depth, the body is getting twice the amount of oxygen it really needs. But oxygen is a toxic gas; too much of it will bring about oxygen poisoning which could lead to death. So the physical amount of oxygen a diver breathes is extremely important: it is directly related to the depth at which he is working and must be constantly regulated.

But as the oxygen is being compressed, so too is the nitrogen which makes up 79 per cent of the air you breathe. Below 200 feet this excess of nitrogen can become extremely dangerous, since it has an anaesthetising effect. The diver becomes drugged and may even pull off his mask. Known as the 'raptures of the deep', a secondary effect is that the nitrogen becomes so compressed and so dense that it becomes physically difficult to breathe in–like trying to breathe in sand.

Below depths of about 200 feet divers must use a less dense 'carrier gas' which is mixed with the oxygen. The normal one is helium, but it is expensive–it has to be processed and most of it comes from the States. Another is hydrogen, which has been used in some experiments and resulted in the death of a Swedish naval lieutenant several years ago; and a further gas is neon.

It is important that experiments are carried out to determine the effects of each of these gases, experiments which must be closely controlled and are usually conducted in hyperbaric test chambers on dry land. In March 1970 the Royal Naval Physiological Laboratories just outside Portsmouth carried out an experiment on two young scientists, John Bevan, 26, and Peter Sharphouse, who celebrated his 21st birthday during the experiment. Using helium the Royal Naval physiologists took the two down to an equivalent of 1500 feet.

The divers entered a special chamber, designed to simulate underwater conditions, and spent a total of thirteen days inside. During the experiment blood, urine, respiration, heart-beat and brain-activity tests were carried out, as well as checks on their ability to perform manual and mental problems. They spent a total of $3\frac{1}{2}$ days below 1200 feet. In this condition the pressure is nearly 60 times as much as at sea level.

The Navy doctors now believe that there is no reason why men should not be free-swimming 2000 feet under the water in the total darkness with pressures 60 times greater than they are on the surface. Soon the Royal Navy will go metric with dives of 500 metres and within a year they may well be carrying out tests at pressures equivalent to a depth of 2000 feet, when they will be using neon as the 'carrier gas'.

But getting down and working on the bottom is only one part of the problem–getting back to the surface is even more dangerous. There will be no dramatic 'splash-up' for the aquanauts coming back from deep inner space.

Under pressure the human bloodstream absorbs whatever gas is being used as the carrier. The deeper the body is taken the more gases the bloodstream absorbs, and if a diver surfaces too quickly these gases will bubble off in the bloodstream. Known as 'the bends' its agonising effects can kill a man. It's like opening a bottle of beer. The pressure changes violently and the beer froths over, and this is what happens inside the bloodstream of a diver if he surfaces too quickly. So the diver must surface slowly, staying for a certain amount of time at given depths to allow the bloodstream to clear itself of the gases naturally through the lungs. This is known as 'doing his stops'. Where and when he does stop depends on the

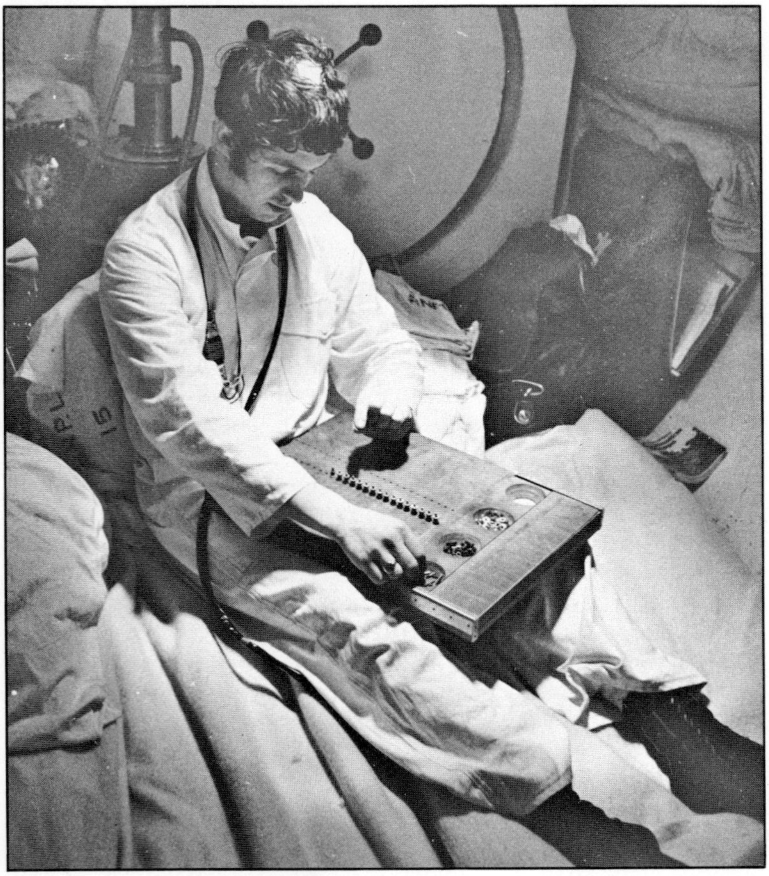

One of the two deep-dive test subjects carrying out co-ordination tests at 1500 ft in a pressurised helium atmosphere

decompression tables, and it is the compilation of these tables which is one of the fundamental reasons for carrying out this type of experiment. The varying depths and the times that the diver must remain at them depend on all sorts of factors–the depth he has been working at, the depth he started from, the time he has been on the bottom, the total time he has been underwater, and so on. Each factor is of critical importance.

Because of the factors discovered through the experiment on two young scientists, man's limit has been pushed farther. Before the test it was thought that the limit at which a human body could work on helium was 1200 feet, because a joint United States/French diving experiment in 1969 had abandoned the attempt after four minutes at 1190 feet when the diver's nervous condition began to deteriorate rapidly.

Tektite II

But if man is eventually to farm the sea-bed, to carry out experimental work, to create mining works, structures must be prepared from which he can venture out into the water, using his new-found ability to free-swim at great depth.

Already such a structure exists as a prototype for future underwater factories, and even cities for those whose working life may be below the surface of the sea. It is called Tektite, and the latest model, Tektite II, has been successfully used for six months in the sea off the Virgin Islands. It consists of two vertical structures, rather like oil tanks, 12.5 feet in diameter and 18.1 feet high, joined near the top by a tunnel 4.5 feet in diameter. The lower right compartment (wet laboratory) is a work area that is open to the sea at all times. Since the inhabitants are working in 'saturated conditions' (i.e. at the same pressure as the sea around them, breathing a special mixture of gases) the sea will not flood in as it would if the interior were at normal sea-level pressure. The upper right compartment is the food store and environmental control centre. Here the atmosphere, temperature, etc., are monitored constantly. The upper left compartment contains the main laboratory and communications centre. Below it are the living and resting quarters for the crew. Both cylinders sit on top of a box-like structure which is ballasted with scrap iron to keep Tektite II stable and submerged.

So far the habitat has been submerged to only 50 feet, with the crew breathing a mixture of nitrogen and oxygen. The conclusions are that for shallow-water work there is no reason why larger structures should not be built. Their greatest advantage is, of course, that since inhabitants live in them, day-long activity around the structure on the sea-bed and at the same depth will reduce time spent on depressurisation, as the crew will not return to the surface until perhaps the end of a working week.

Diving gear

Once you have your scientist, engineer or farmer on the sea-bed, the problem is to provide him with the most

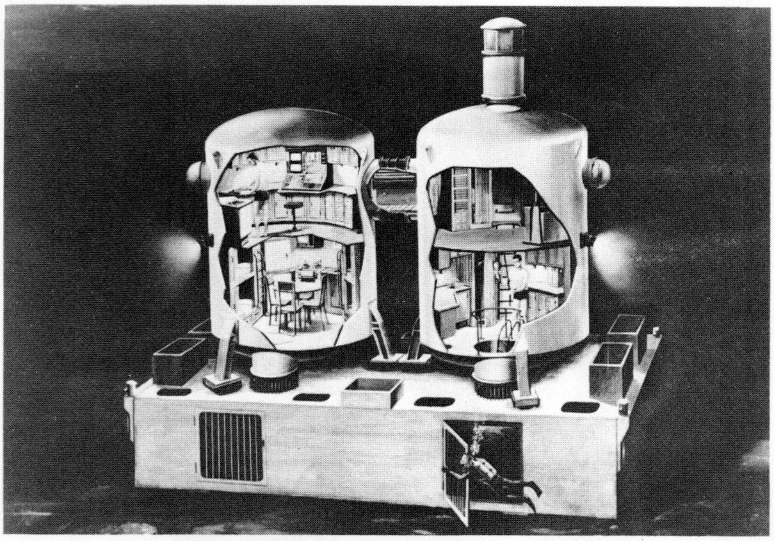

Left Tektite II on the sea-bed and *below* a diver's view. In the foreground a diver-scientist at work

Scuba Latest: Diving gear used in Tektite II includes a hot water-heated suit, a semi-closed breathing apparatus, and umbilical tube for replenishing water, oxygen and nitrogen

efficient possible form of personal breathing apparatus for work outside his underwater home.

Tektite Gear/Used during the Tektite II experiments, this system is designed by General Electric (USA). It is capable of working up to 6 hours and is lighter than the conventional scuba-diving gear. The most visible difference in operation is lack of bubbles. The exhaled oxygen is returned to a system that extracts the carbon-dioxide content. The unused oxygen left is then recirculated to the diver's oxygen supply, thus reducing the amount of oxygen needed in his tanks.

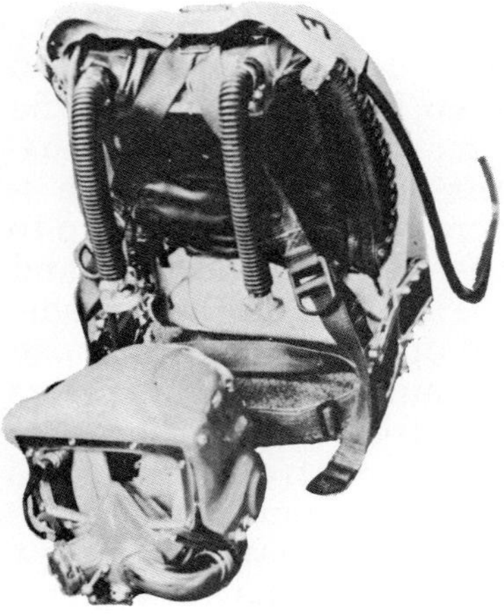

Abalone/This is a semi-closed-circuit system that is streamlined by putting all the equipment under a fibre-glass shell on the diver's back. Produced by Westinghouse, it should considerably reduce drag and therefore expenditure of energy on the part of the diver. The system is designed for helium-oxygen mixture used in shallow waters. The oxygen is 'scrubbed' of carbon dioxide as in the Tektite gear.

Top The Abalone fibre-glass backpack and mask. *Above* A diver prepares for work in the Windak gear

Windak/Once a diver begins operating on helium it becomes very expensive to allow the gas to vent in the normal open-circuit system. It has therefore been vital to produce a deep-sea helium breathing apparatus which operates on the semi- or closed-circuit principle, allowing the diver to rebreathe exhaled gas. One of the most effective has recently been produced by a team of textile manufacturers in Lancashire who previously had specialised in aircraft and space clothing.

The system is semi-closed-circuit. It has been designed for operating from an underwater habitat or submersible and receives a controlled supply by an umbilical cord. Emergency gas is available through a small cylinder carried on the diver's back. The equipment has had two months' operational trials on an off-shore drilling rig in the North Sea and it is tested to 660 feet. The apparatus is compact enough to allow the diver to squeeze through a hole 24 inches in diameter. The set is designed so that it cannot be assembled incorrectly and the soda-lime canister that removes the poisonous CO_2 breathed out by the diver can be changed without moving the set.

The breathing bags that operate as a diver's second pair of lungs have been positioned in such a way that the breathing effort is minimal. They extend from the back, over the shoulders, to the breast and are housed in, and protected by, a very tough outer skin. An outward-vent valve is situated on the diver's right shoulder in a position where the diver can see a constant small stream of bubbles giving him positive indication that his gas supply is coming through. This last feature is vital in any semi-closed-circuit system, because if a diver's supply fails he will get no warning. He will just continue to rebreathe the same gas mixture until there is insufficient oxygen to sustain life and in the perpetual darkness, 600 feet below the surface, it will be too late when he first feels the symptoms.

The helmet is a strong fibre-glass shell which gives good

protection to the head and fastens on to a standard type of drysuit. This drysuit is a light-weight, shaped, rubber diving-dress, contoured and tailor-made for each diver.

Underwater Jeep

At some point in the future, surveys will have to be carried out by personnel who may need to move from place to place at speed. An underwater vehicle has been developed by General Dynamics in San Diego, California. Called the Marex Diver Transport Vehicle, it carries four scuba-equipped divers down to depths of 300 feet. It has a 4-knot maximum speed and will cruise at 3 knots for up to 6 hours. Seventeen and a half feet long, 5 feet high and 7 feet wide, it is powered electrically. The three-blade, controllable-pitch propeller will operate from full ahead to full astern, and for low-speed manoeuvres bow thrusters move the nose up and down and from side to side. The small 'wings' operate like aeroplane control surfaces. Its whole structure is made of steel and aluminium and the forward transparent canopy opens upwards for crew entry and emergency jettison, as does the lower half. Once on board the divers plug in to the air supply in the vehicle.

Marex is designed mainly for use in off-shore oil operations–and there is even an underwater 'garage' for repair, maintenance and refuelling with air and batteries.

Submersibles

Deepstar 2000/For occasions when examination and survey is necessary without the divers being in actual physical contact with the sea-bed itself, the submersible Deepstar 2000 will provide an underwater transport capable of taking three men, in normal atmospheric condition, down to 2000 feet for a period of 12 hours, with a reserve for 36 more hours. Deepstar 2000 can either move independently or tethered to a mother ship

A diagram of Deepstar 2000

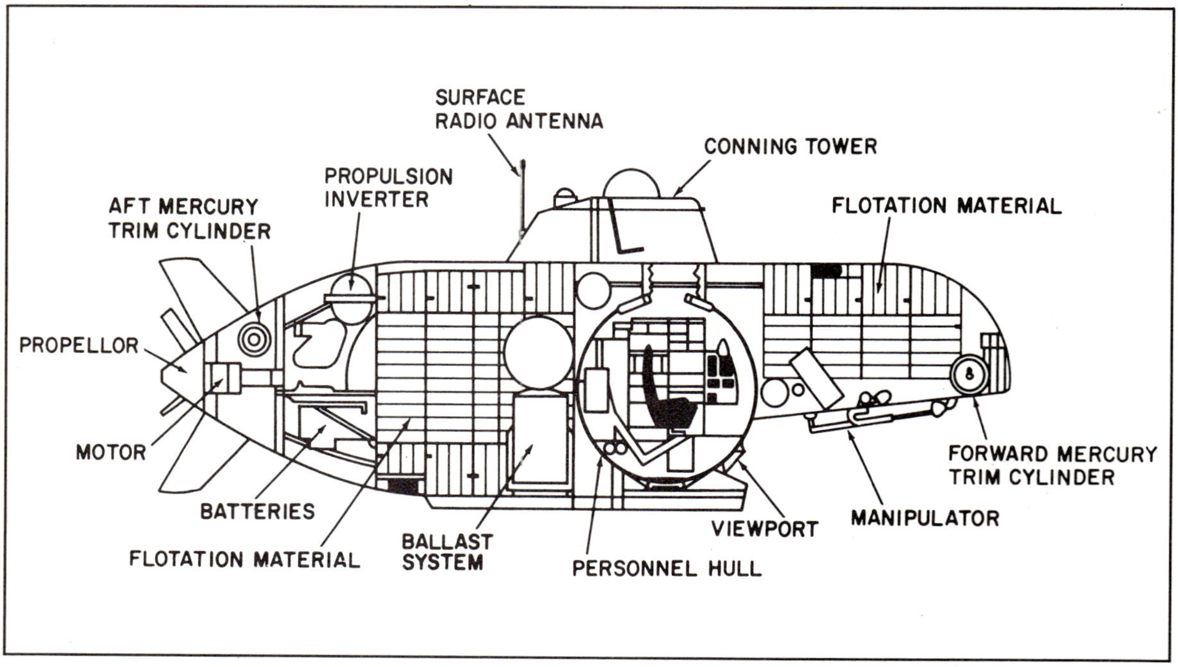

Deepstar 2000 being lowered into the Pacific during trials

on the surface. It is 20 feet long, $5\frac{1}{2}$ feet high (with a 3-foot sail), and $7\frac{1}{2}$ feet across. Power comes from three lead-acid batteries which provide up to 8 hours' continuous engine work. There are two main 20-inch shrouded propellers in the stern, as well as two vertical and two horizontal 9-inch ducted propellers in the bows. Just ahead of the bows is the 'brow' section which can hold up to 500 lb. of scientific equipment and is interchangeable with another 'brow' module. The conning tower (or 'sail') provides crew accommodation and has viewing windows. Crew transfer at sea can be made through the sail as in a conventional submarine.

The equipment on board includes closed-circuit television, movie and still cameras, radio, underwater telephone, acoustic trackers, sonar, gyrocompass, extendable booms for lights, and remote-controlled grabs for obtaining samples from the sea-bed.

DSSV/But the demand is already to go deeper. Lockheed have begun work on a Deep Submergence Search Vehicle, designed to go to depths of 20,000 feet. Due to go into operation in the late 1970s, the craft should carry a crew of four and would primarily act as locator for submarines in distress, as well as carrying out very low-level science survey runs. Several crews have already spent periods of up to 34 days in a DSSV simulator at Lockheed Laboratories in California.

Deepstar 20,000/At Westinghouse in California the hull of the deepest submersible yet planned is being completed now. The 3-man Deepstar should be operational in 1972. The pressure hull, made from high-strength steel, is 7 feet 5 inches in diameter and 2½ inches thick. The vehicle will eventually weigh 40 tons and will be used for deep-ocean research, carrying one ton of scientific equipment. With a maximum depth capability of 30,000 feet, Deepstar will be able to explore more than 98 per cent of the ocean bed. At such depths the pressures are so great that only a small section of the sub will be set aside for the crew, sitting in a super-strengthened cylindrical hull-within-a-hull. They will have one viewport, and a remote-controlled grab for use in sample gathering. At present the sub is of the simplest design–nobody knows what operating conditions are like at 30,000 feet down, or indeed what will be found at such depths, living or otherwise.

Pisces/Most of the submersibles designed so far are for underwater exploration. Few of them have been designed as underwater civil-engineering tools. Probably the most successful of these is Pisces II, owned and operated by the British Vickers company. Unlike most of the exotic American submersibles, Pisces is made for hard work and has already proved capable of operating for 60 per cent of the time in the North Sea–no mean achievement.

The sub weighs 11.2 tons, is 19 feet 4 inches long, 9 feet 10 inches wide and 10 feet 4 inches high. Carrying a crew of two, with room for a third, Pisces can operate down to 3000 feet for as long as 15 hours, and is capable of carrying out a wide range of tasks: inspection and maintenance of pipes and fittings; biological and zoological sampling; measurement of water temperature, composition and

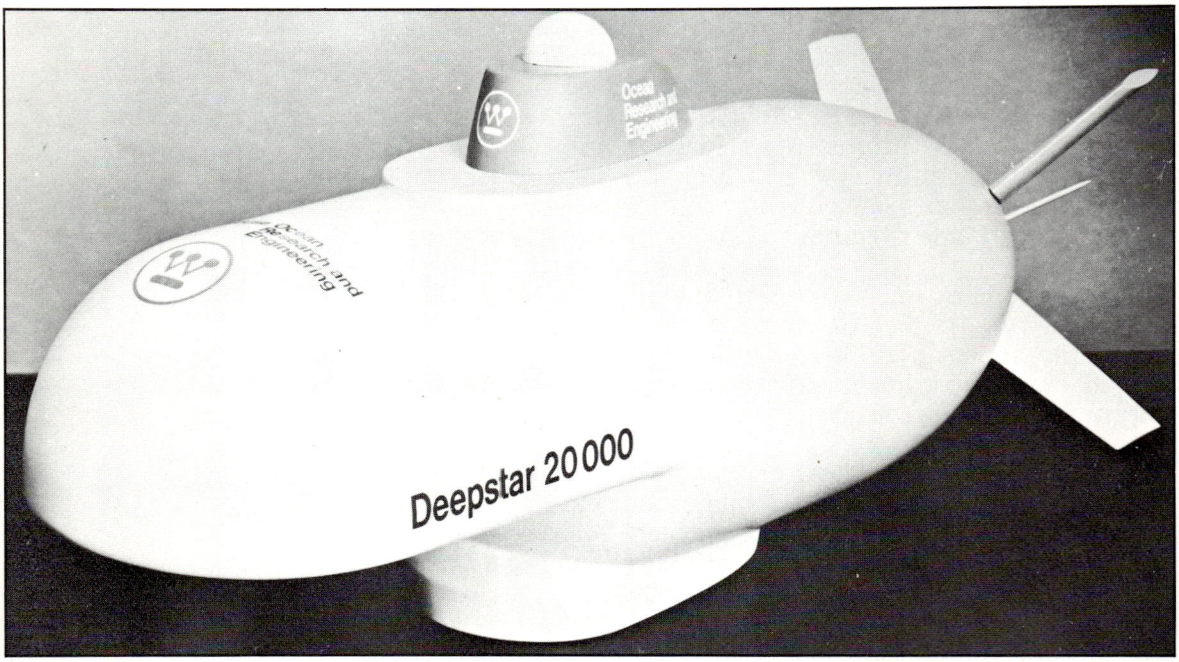

A model of the deepest-operating submersible yet built

DSRV in operation over a crippled submarine

current; underwater demolition; equipment transport; cutting and welding; as well as photographic and sonar work. There is a closed-circuit television on board as well as equipment for underwater and surface tracking and navigation. In addition there are two manipulators which can be used as mechanical hands and arms for picking up, locating, attaching, cutting, and even (when a weight is placed in one of the claws) for hammering. Pisces has completed extensive trials with government and private organisations in pollution investigation, observation of sea life, underwater geology and salvage–and the sub is now available for charter work.

DSRV/Every two or three years disaster seems to strike a submarine somewhere in the world. Now the US Navy is equipping itself with a very new device which could bring hope of life to the trapped submariner. It is called the Deep Submergence Rescue Vehicle and is the first rescue submarine ever built. DSRV is capable of operating down to 5000 feet, and rescuing up to 24 men from a submarine lying at anything up to a 45° tilt.

Its torpedo shape has an outer glass-reinforced plastic hull containing three interconnected, high-strength steel spheres, each three-quarters of an inch thick. The forward sphere holds two operators and an integrated control and display system using TV and sonar to provide onboard scanning. The second and third spheres hold a third crewman and up to 24 rescuees. DSRV is 49 feet 3 inches long and 8 feet in diameter. Propelled by lead-acid batteries, she can go for up to 10 hours at 3 knots, although there is enough oxygen/nitrogen for at least 24 hours. Sensors on board include TV cameras (2 zoom, 2 wide-angle, 2 normal lenses), 3 TV monitors in the control sphere, 5 viewports and sonars. Under the central sphere sits the transfer skirt which seals on to the crippled submarine escape hatch.

DSRV goes to rescue piggyback on a mother submarine (nuclear-powered, capable with DSRV on board of 15 knots). Near the disabled craft DSRV takes off from its mother ship and can make as many trips as necessary to take away survivors, recharging batteries and taking aboard air supplies if necessary between trips. The US Navy has announced that all other navies have been informed of the minor escape hatch modifications necessary to mate with DSRV.

Deep Voyager/Most submarines are made of high-stress metal, but there is one currently being built at the US Navy Research and Underwater Development Centre in San Diego which is made of glass–or at least the front of it is. It is a test vehicle which the designer–US Navy Engineer Will Forman–hopes will be the forerunner of a 70-foot glass and fibre-glass submersible, designed for ocean voyages 20,000 feet under the sea, with a three-man crew having a perfect vision of everything that is happening around them.

Glass has an extremely high strength to weight ratio: a steel hull of the same capability would be 50 per cent heavier. Its weakness lies at the edges and it is here, where the glass joins the metal hull, that Forman has researched the right shape of edge and the right type of rubber collar to act as the seal. The ultimate glass submarine–called *Deep Voyager*–will cost nearly two and a half million pounds, and it will have a nautical range of nearly 2500 sea miles. The vehicle will be 70 feet long and 7 feet in diameter, with a wing-span of 18 feet. The pressure hull will consist of three 82-inch-diameter glass spheres: one sphere will house the crew, and the other two will contain their life-support systems.

The submersible will employ a simple glide technique as its means of propulsion, movement being affected by a series of downward and upward swoops. Downward thrust is achieved by flooding the submersible with sea-water, creating a negative buoyancy and causing it to sink. When at the bottom, upward thrust is initially created through the use of hydrazine, a space-industry-developed fuel which combines with a catalyst to form a gas buoyancy. The glide angle is achieved through the use of wings located near the after sphere. The location and size of wing will determine the drag/lift capability of the vehicle. The optimum glide angle should be about 9 degrees. Manoeuvring ability on the sea floor will be by small propulsion units driving propellers. The fuel will also be hydrazine-generated gas.

This glide technique is simple and very economical. It was first thought of by torpedo designers over ten years ago. But it is the development of the space fuel that has made it possible.

Unmanned Sub/Most of the submersibles designed in the world are for use in the sheltered conditions of the American Pacific seaboard, with clear weather, still seas and good working conditions. The North Sea is a very different problem–stormy, fierce tides which constantly shift the sea-bed, and eroding trenches. Visibility is low, and it is cold and dangerous.

The British Ministry of Technology, working with the Marine Trial Support Unit at the UKAEA plant at Harwell, are carrying out research into a new way of observation work in these conditions. They have designed an unmanned submersible which will be equipped with sonar and television devices plus whatever telemetric system is needed. It will be connected to a controlling mother ship by an umbilical cord of perhaps two miles. Using remote control through video-pictures they will be able to guide the device to the exact spot they require.

Because there is no need to carry people or life support apparatus nor any form of power, the submersible will be small and can be operated in conditions when it would be too dangerous for human divers to work. In the event of bad weather conditions–which are always hazardous when recovering a manned submersible–the unmanned sub could be towed behind the mother ship back to port or, in extremely bad conditions, the cables could be slipped and attached to a marker buoy so that the mother ship could return at a later date to pick it up. It would have a working speed of up to 4 knots and could be used for research, observation and underwater searching prior to a salvage.

It should be possible to produce one for a third of the capital cost and to operate it, if necessary, for 24 hours a day. Design studies have already been submitted.

Deep Mobile Target/A new type of submersible has been built by Recording Designs of Camberley in Surrey for the Underwater Weapons Establishment of the Admiralty at Portland. It has been developed to play the part of a submarine during training of submarine-hunting techniques on the surface. Looking not unlike a torpedo it will react as a submarine reacts under attack. It is also packed with electronic recording devices to monitor and record all the moves that it makes. It is for use by the Royal Navy in training and has already been seen in action by representatives of many NATO countries.

But the real future for this craft lies in less warlike activities. Fitted with underwater television cameras and equipped with video-tape recording devices, along with its normal sonar and electronic telemetring, it could be preprogrammed to follow a set course along the sea-bed, automatically compensating for tide and drift, while carrying out, for instance, the survey of an underwater pipeline route.

This could be one of the most significant moves yet in new underwater tools for ocean industry. It could be fast, inexpensive and accurate. It could operate in all weather conditions and be completely free of the safety limitations imposed on man under the sea.

Sea-bed Vehicle

If man is going to carry out standard civil-engineering work on the sea-bed, he will need not only a mobile base

from which to operate, but also standard structural-plant equipment–drills, borers, trenching equipment, pile-drivers, concrete-mixing equipment, winches, and so on.

The most inventive project in this area yet conceived is a sea-bed vehicle being designed by Cammell Laird of Britain for work down to 600 feet. By spring 1971 it should be in a fully operational state. It is designed to have a crew of two who are kept at atmospheric pressure, and in a separate chamber a working crew of three divers with an operating period of up to 60 days. These two chambers form the internal main hull, with an airlock between them. At one end of the divers' chamber a vertical tunnel gives access to the water above and the sea-bed below. The vehicle is connected to a surface ship by an umbilical cord which carries air-supply, communications, power, and several other systems.

Propulsion is by wheels, each of which is driven by an independent hydraulic motor. There is an ingenious system which will always ensure that the submersible is level and even when crawling along the side of a slope, the downhill wheels float level. Negative buoyancy is obtained by a large weight held underneath, which can also be used as an anchor. When the vehicle needs to move it merely winches up the weight and moves forward on its balloon tyres, which are sufficiently large and buoyant to ensure that the vehicle does not sink into soft mud. Guidance for the pilots in the crew chamber is through highly sophisticated sonar vision and television cameras.

The vehicle is far more than just a mobile base for the divers. There is no limit to the amount of equipment and plant which can be carried on the outside–anything which can be operated by a tractor on dry land could be operated by the sea-bed vehicle under the water.

This is the first realistic vehicle that has ever been designed for sea-bottom use. It is the most significant advance yet in underwater technology.

Oil

Oil Sub/As the search for oil moves on out across the bed of the ocean, and the work parties drill wells thousands of feet below storm and ice, the oil reservoirs will grow like mushrooms along the hills and plains of the seas. With the transfer of the oil to the surface and a waiting super tanker comes the problem of connecting flow lines thousands of feet long and the inherent danger of breakage and pollution. General Dynamics in California has approached the problem in an entirely new way with their design for an oil-carrying submarine tanker.

Capable of carrying 170,000 tons of oil in its rectangular-shaped hull, the sub would be 900 feet long, 140 feet across and 85 feet high. Nuclear-powered, twin screws give her a speed of 18 knots through the calmness of the lower depths. She would carry a crew of 39 in a central cylindrical pressurised hull, divided into control and living space, reactor plant and engine room. Around the hull, running the length of the sub, are the oil tanks. Sensors would give precise, constantly monitored information of depth, direction, location, and the existence of possible obstacles such as icebergs, or ridges on the ocean bed. Surfacing

Opposite The hull of the Sea-bed Vehicle already mounted on its chassis in the Cammell Laird factory and as it might appear in operation

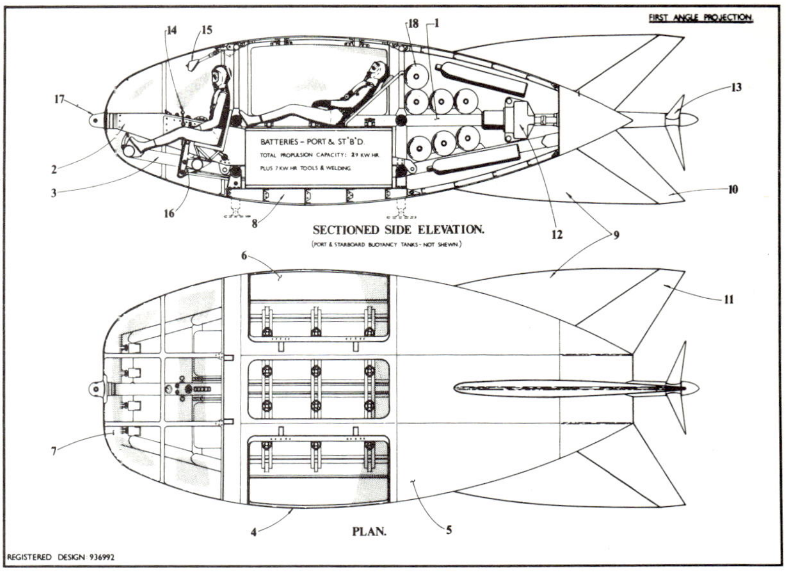

Four-Man Dive Transport Vehicle
1 *Primary Structure:* 4-in. o/d welded stainless steel. Entire structure is neutrally buoyant. Batteries, propulsion unit, fins and all other equipment installed on this structure.
2 *Forward Collision Beam:* Distributes collision shock loads to primary structure.
3 *Forward Structure:* Jettisonable, with seats in capsize emergency.
4 *Port-Starboard Collision Beams:* Internal not shown.
5 *Hull Panelling:* Double and single curvature marine light alloy. All panels have integral stiffening members; are backed with sintactic foam for neutral buoyancy and are quickly detachable.
6 *Side Access Doors:* Hinged for crew entry. Jettisonable in emergency.
7 *Forward Canopy:* Thick transparent perspex, upper half hinged for crew entry, jettisonable. Lower half jettisonable in emergency.
8 *Reinforced Steel Bottom Structure.*
9 *Fixed Fins:* Interchangeable, marine light alloy, thick maximum section, neutrally buoyant with foam filling, replaceable outer edges.
10 *Rudders, Upper & Lower:* Full power operation, 45 degrees port and starboard.
11 *Elevators:* Direct manual operation 10 degrees up and down.
12 *Gear Box:* High efficiency, single reduction, Mech. Ratio 10:1.
Motors (2): Simple series wound traction type. 20bhp max. 80° elevation at 2500 rpm at 6 hp.
13 *3-Blade Controllable Pitch Propeller:* 76° efficiency at 250 rpm.
14 *Propeller Pitch Control:* Infinitely variable from full ahead to full astern.
15 *Navigation Instruments:* On hinged arm.
16 *Elevator Controls:* Also moves mass trim for slow speed manoeuvres.
17 *Bow Thruster Unit* For slow-speed manoeuvres. Port-Starboard, Up and Down.
18 *Air Cylinders:* 4200 cu. ft free air.
The MAREX four-man Diver Transport Vehicle is a second generation design. It is now being constructed under contract to the oil industry by MAREX and COMEX. An outline of the specification is given below:-
Length (overall) 17 ft 5 in (5m 33)
Height (max.) 5 ft 0 in (1m 53)
Width (max.) 7 ft 0 in (2m 14)
Crew (max.) 4 divers
Life Support (excl. two emergency systems) 4200 ft^3 (120 litres)
Batteries Lead/Acid
Maximum Speed 8.7 knots
High-Speed Cruise 7.2 knots for 3 hours
Maximum range 50 nautical miles
Maximum depth 200 ft (61m)
An extensive series of model tests, with wind-tunnel and free-fall models and a radio-controlled sixth scale model, was carried out during the design stages.

takes place in the same way as in any submarine.

Although primarily designed to get the oil out of the Alaskan fields, closed by ice for many months of the year, the nuclear sub-tanker could fill up at any surface or underwater oil terminal.

Underwater Oil Production/The impetus for underwater heavy engineering is coming from two fairly familiar sources in the States–NASA and the oil companies.

The problems are not unlike some of those that have to be met in outer space–sophisticated materials, sophisticated electronics capable of withstanding severe physical trials, the physiological effects of isolation in a hostile environment. Space doctors from NASA are already running underwater experiments, and they are currently in charge of the Tektite II operation off the Virgin Islands (see p.40). But industry, too, is becoming involved.

The big American companies that have been responsible for getting the United States to the Moon are now practically all turning their immense research and development potential to inner space–even if only to diversify their commercial interests because of possible cut-backs in the space programme. North American Rockwell, Grumman, General Electric, Lockheed, Westinghouse–all are turning their attention to the sea-bed. But the main impetus at the moment is coming from the oil industry which desperately needs to develop more effective methods of extracting oil from undersea deposits.

The present system of building a drilling platform, drilling a wildcat pole, making a strike and then converting that drilling platform into a surface base for either storing the oil locally in great underwater tanks or piping it to the mainland is proving extremely uneconomical and almost impractical in certain parts of the world, because the oil-rigs cannot stand up to local sea conditions. Within the next decade the above-water oil-rig may disappear altogether, with the entire production taking place on the bottom, hundreds of feet below the surface.

Already Lockheed Missile and Space Company have

The 900-ft, nuclear-powered oil sub-tanker will carry 170,000 tons of oil under the Arctic icepack

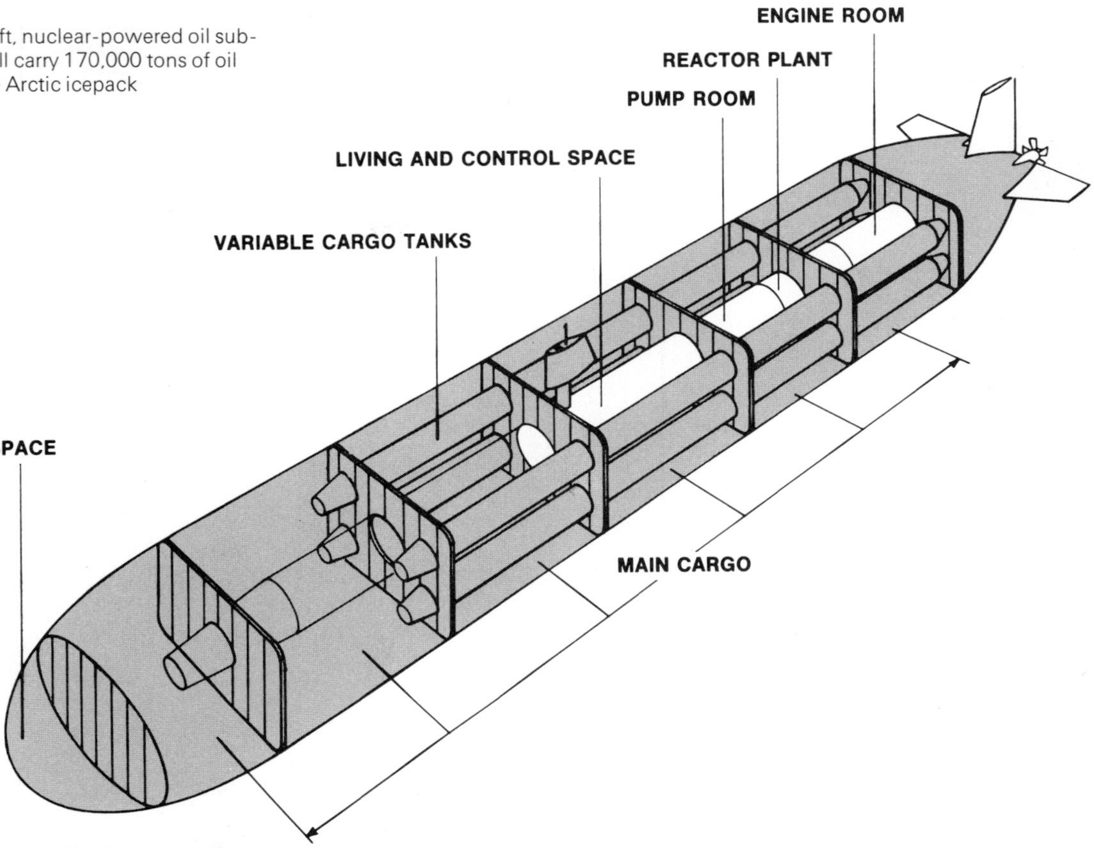

The underwater well-head cellar
Far right oilmen at the capsule controls arriving above the cellar. *Right* the connected capsule/cellar working area above the well-head

Below A number of underwater rigs feeding oil to the manifold centre where a nuclear, oil sub-tanker could load up

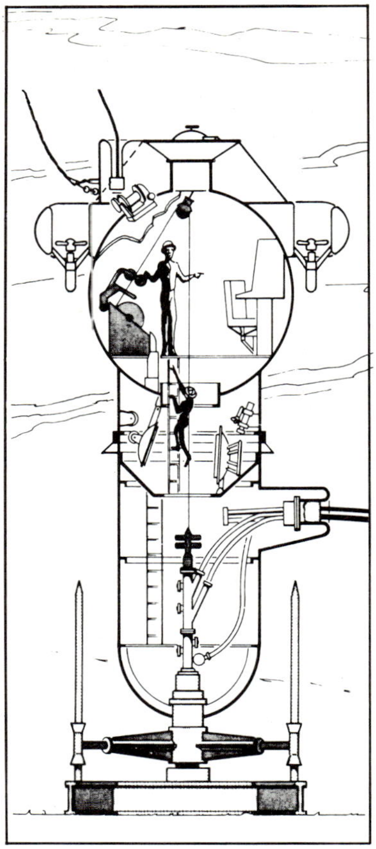

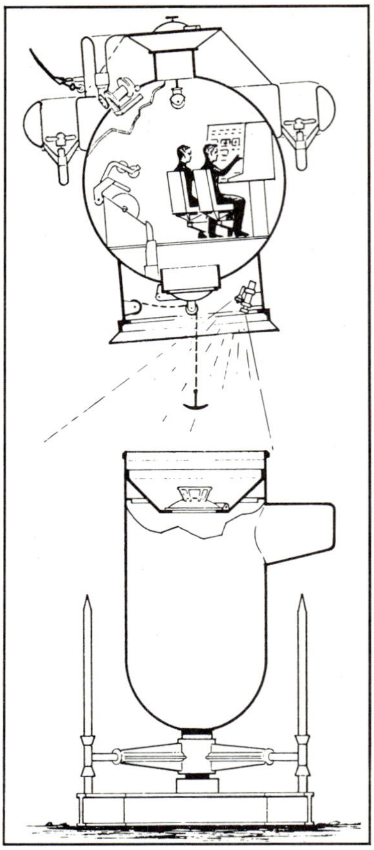

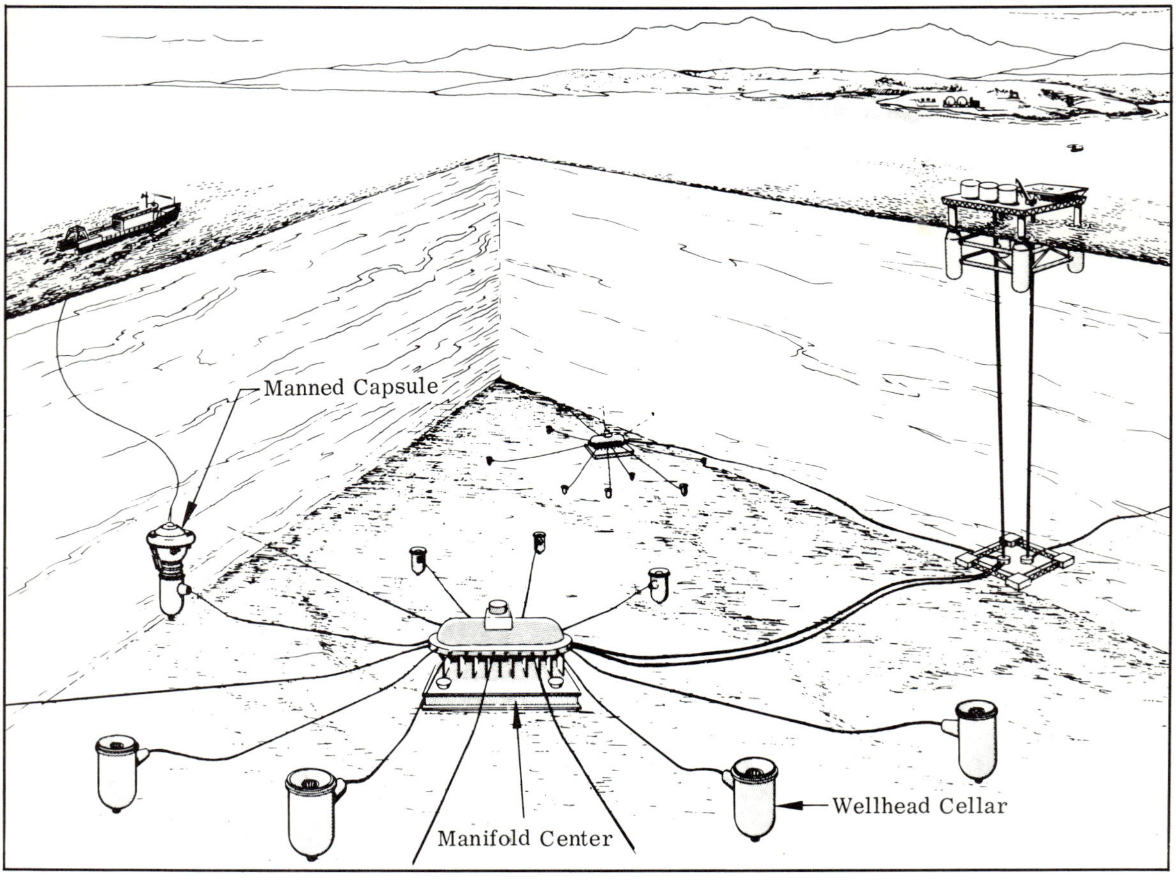

carried out a design study on this sort of oil production that has resulted in a set of underwater structures which are at the moment being tried out off the Pacific coast of Canada, close to Vancouver Island in British Columbia. The key component is a 21-foot steel 'cellar' that will be fitted permanently over each well-head in a sea-bed oilfield. It has been developed by Lockheed and will enable oilmen to work in and around the wells.

The 'cellar' would fit onto the well-head. A crew would drop down from the mother ship in a 'utility capsule' which would mate with the top of the cellar. Once the hatchway was open, the two structures would become one working space about 38 feet high and 10 feet wide. The well-head 'Christmas tree' fittings would be placed on the well-head and lines and tubes would be fitted and carry the oil and gas away to a manifold centre. This centre, taking supplies from a number of well-head cellars, would deal with control and intermingling of the crude products from the wells and would be a habitable, normal-air, permanent steel structure sitting on the bottom of the ocean. It too would have its utility capsules–in an emergency the capsule could be cut free of cables and lines and float rapidly to the surface.

The system could give the oil companies an economical means of extracting oil far offshore in water 1200 feet deep.

On the Surface

Hoverjet/On the surface, one of the newest techniques in propulsion is the hoverjet, developed in England. In combining the advantages of the hydrofoil and the hovercraft, it may turn out to be their successor and, since it does not have some of their disadvantages (extra weight and fuel for the power to 'lift' the craft; control-problems in flight), the hoverjet may also turn out to be commercially more successful.

A hoverjet development

Its basic principle is to 'surface-skim' on two wedge-shaped planing surfaces fitted underneath the hull. Towards the forepart of each wedge a channel is introduced which 'gulps' air as the craft begins to move. The air, travelling the length of the wedges, lifts the hull above the water, while a jet above the back of the wedge mixes the air with its exhaust and the water and jets the whole thing out backwards, giving forward thrust to the hull. The hoverjet has so far given noticeably better results than the conventional hydrofoil/hovercraft.

The Silent Hovercraft/The hovercraft appears to have been the white elephant of the sixties: ten years ago hopes were very high, but in the intervening decade many of the companies who have pinned their faith on this revolutionary new form of transport have gone out of business.

But a hovercraft has now been produced that is different. To start with the VT-1 is almost silent, since water propellers have been chosen as the means of propulsion and they are set back at the end of two skegs or keels which extend down below the air cushion into the water. The gas turbine engines driving the propellers also provide the air cushion. The whole craft, weighing 83 tons, is made of aluminium alloy. The standard version will carry both passengers and cars at a cruising speed of about 40 mph.

The skegs give good grip on the water and greatly improve lateral control, lessening the tendency of conventional hovercraft to skitter sideways. A third and smaller skeg is fitted forward of the other two, on the centre line, as an aid for 'sitting down' the hovercraft on a hard or landing strip. VT-1 rides $5\frac{1}{2}$ feet above the water and, since supplies are crossfed, failure of one of the two engines does not mean that one side of the hovercraft tilts. Evaluation trials carried out recently are being followed by simulated passenger routes to assess reliability and operating costs – and the hope is that VT-1 will be in service by the summer of 1971.

The silent hovercraft during speed trials and one of the skegs with rudder and propeller mounted aft

The Mobile Base-Ship taking a smaller vessel for repairs and refuelling. At speed the dock area is pumped dry of water

Mobile Base-Ship/One of the most successful small warships ever produced is the gas-turbine fast patrol boat built by Vospers of Portsmouth. She travels at nearly 60 mph–faster than anything in the British or American Navies; she has a low capital cost and can be fitted out with almost any modern weapon system. She can be used for coastal patrolling, anti-smuggling, anti-slavery patrol, mobile defence and on occasion it can even double as a presidential barge.

But it is one thing to have a fast, modern warship; it is quite another problem to look after her. Most of the developing nations have poor harbour facilities, and those that do exist are old-fashioned and vulnerable to air attack or commando-type raiding parties.

A completely new type of ship is being designed which could change all that. It is a mobile shipyard that looks like a warship, can move at nearly 15 knots, and has a cruising range of over 3000 nautical miles. It is a complete mobile base with dry or wet dock, workshops, cranes, training facilities, emergency power, fire-fighting equipment, laundry and even a small hospital. The ship displaces 2200 tons and is 324 feet long. The rear section is cut into to provide the docking facility for ships up to 120 feet long (e.g. patrol and torpedo boats). On board there is accommodation for over 100 people, and up to 1000 tons of fuel, water, stores and ammunition can be carried. A distillation plant makes 20 tons of fresh water a day, twice the ship's own needs.

Planesail/If you feel like moving really fast on holiday, then the latest in high-speed yachting may satisfy your desires. It is a trimaran (3-keel) which incorporates three moulded wings cantilevered above the main hull section as a means of propulsion. The lift, generated by the wind blowing over the wing sections, drives the boat forward and direction is changed simply by twisting the aerofoil sections to face the wind.

P2, as this latest version is called, will be 52 feet 3 inches long, carry six people, and be handled with ease by somebody who can drive a car. The side floats are made extra buoyant and fold neatly into the main hull for entering port. One small point–P2 costs £17,000.

Gyrofin/The aerofoil section is also being used elsewhere on board ship–but on a much bigger scale. Gyrofin is a stabiliser. As tourist traffic increases, more and more people prefer to go on a ship with stabilising fins. These are like small aeroplane wings jutting out of the side of the ship. Controlled gyroscopically, they react to keep the ship steady. As the gyroscope swings, indicating a roll to the right, the fins change their angle and deflect the water passing over them downwards on the right, upwards on the left, and so stop the ship from rolling to the right.

Gyrofin does that too–but very much faster. In fact, it does it before you know you've even begun to roll. Both fins, weighing 26 tons, are produced by Sperry Rand in Britain and are worked by computer. Strain-gauges on each fin measure the lift and drag on each fin and sense roll almost before it happens. The signals from the gauges are sent to a computer that compares them with signals from a stable platform controlled by a complex series of gyros and accelerometers. This platform in effect tells the computer which way is vertical, and the computer then moves the fins to achieve the necessary roll correction. The effect, rather like the autopilot on an aircraft, is to blanket out the movement continuously. All that's needed now for the queasy passenger is some way of stopping the pitching motion of the ship!

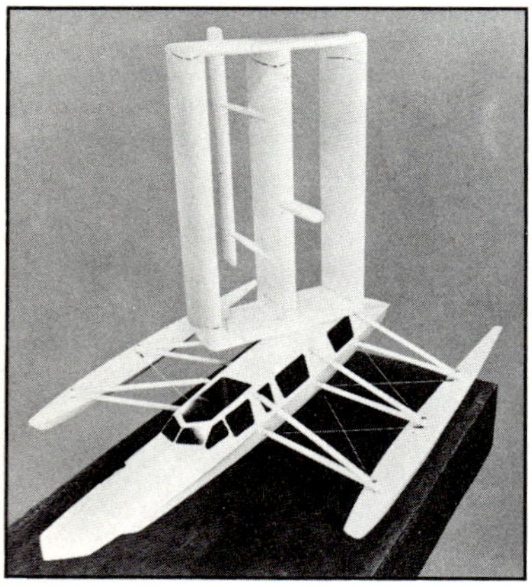

Above The Planesail P2

Right A gyrofin section ready for mounting

Minerals from the Sea

When all is said and done it may not be the techniques of mining the ocean bed and even growing food on it, or the ways of bringing whatever we find down there back to land, that will prove to be the key for providing enough resources for the needs of our growing world population. Perhaps the greatest discovery of this century will be the means of obtaining all the food and minerals we need from the waters of the seas themselves. There is no doubt that millions of tons of those materials and the greatest store of protein we shall ever find lie locked in the salt-water solution that covers two-thirds of our planet. All we need is a way to process the water to remove what we need from it, and we shall be provided with more than we would be likely to need for the foreseeable future. And it is an inexhaustible source; almost as fast as we take from it, it takes from the land, the air and the ocean floor in its own vast processing system–the cycle of natural developments that we only dimly understand. But it is a cycle that we must understand one day if we are to have enough resources to use. We are only just beginning–with our desalination plants and our mineral extraction systems. Perhaps ultimately the most exciting and rewarding successes of this century will not come from what we can do *in* the fourth element, but from what we can do *with* it.

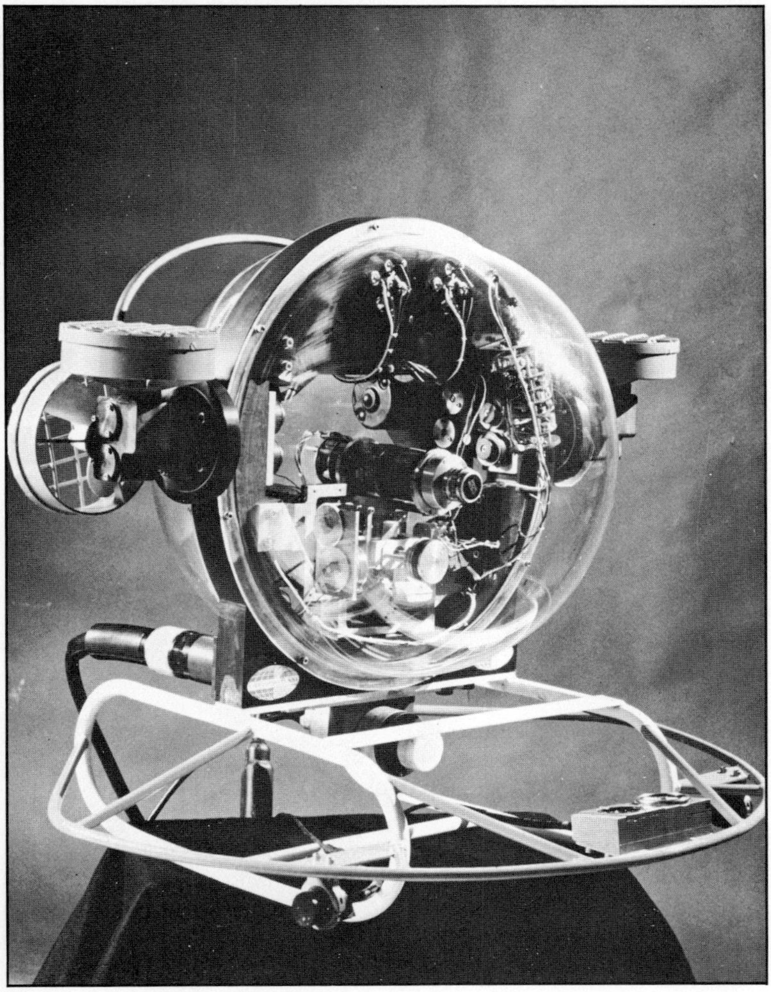

Above The Missing Middle: 185 ft long, bow and stern sections of an ore carrier. Split down the white dotted line, the separately built mid-section will be added later. The completed ship will be 1000 ft long and carry 52,400 tons of cargo on America's Great Lakes

Left Anthro: A head-controlled search sub using TV cameras for visual sensors and ear-like side sections for acoustic scanning. The sphere and camera mimic the head movements of an operator on board ship at the surface. Two reversible propellers on each side manoeuvre Anthro in any direction

Park this in your garage—and stay *ahead* of the Joneses. The Aerocar, the latest line in dual-purpose vehicles, does a comfortable 40 mph on the road and a luxurious 150 mph in the air. Its designers at Wagner Helicopters in West Germany believe that it is ideal for farmers, policemen, or anyone else who needs to fly one moment and drive the next. On the road the rotor blades fold away neatly along the length of the car's body. Today's price: a mere £15,000. Future models may include a version powered by a small gas turbine engine designed in Britain

The Shrinking World

Moving Pavements/Rapid Transport Systems/Magnetic Mobility/Riding on Air/Back to the Wheel/Power/APT/Vertical Flight/Supersonic Travel/Call Confravision/Faster than Thought

Speed is relative. Kneeling at the controls of a tiny racing hydroplane, 30 mph on the water with an eye level 3 feet above the surface feels like at least 130 mph on land. The passenger in the jet airliner has no sensation of speed unless the aircraft is climbing or descending through cloud. Only when the plane is near the runway, and paradoxically flying as slowly as safety permits, does the average airline-passenger find the sensation of speed in the least degree alarming. Even the astronaut travelling at a steady 17,000 mph or more has no physical realisation of velocity. It is only when he is still accelerating after blast-off, or slowing down during re-entry, that there is any uncomfortable experience associated with the incredible speed of space flight.

There are some, but not many, who love speed for its own sake. This is a basic requirement for the successful racing driver. Many people think they love speed, only to discover subsequently that they are in love with the emotional illusion of speed. All too often, they kill themselves in the pursuit of this misplaced romantic ideal.

Be that as it may, since the invention of the wheel, man has spent a very large proportion of his technological effort in the pursuit of the ability to travel from A to B as quickly as possible. And there is no doubt that one of the most dramatic manifestations of 20th-century technology has been the sensational scaling-down of the problems of time and distance. The practical limits of a day's journey during the last century are plotted across the British countryside by the siting of the old coaching inns. Railways exploded that map. Today we see how the great air staging posts of the post-war boom years are already becoming obsolete. Places like Prestwick, Shannon and Gander no longer occupy the key positions which they enjoyed as late as the mid-1950s.

But this revolution in travelling times is by no means confined to the upper speed range. Concorde, and the other supersonic air transport projects, may command the limelight, but are not necessarily the most significant developments on the horizon.

Transport and communications are often grouped under the same heading, but their relationship can be compared to that between medicine and surgery. The better the medicine gets, the less surgery may be needed, and at some point in the future the need for surgery may disappear totally except in cases of accident and physical defects. The same is virtually true for transport. The better communications get, the less the need for passenger-travel. But the point at which passenger-transport research becomes superfluous is not even visible in the day after tomorrow's world.

On the other hand the transport of goods, essentially a distribution problem, presents a different set of values. Containerisation, bulk-carrying, pipelines and conveyors are beginning to point the trends of the future.

Meanwhile, we struggle to get to, or at, each other faster, cheaper, and more comfortably. Briefly, developments in passenger-transport are conditioned by those priorities and in that order, be it on land, sea, or in the air. At the same time the main features which have a negative influence on the rapid advancement of transport are costs and sociological considerations, including safety and the extent of both chemical and noise pollution.

The fixed installations of transport–railways, docks, airports and so on–are expensive in capital investment, and as often as not require more expensive structural facilities to bring a marginal increase in transport speeds. In modern society, the question of costs is being assessed more closely in relation to immediate and long-term benefits, because increasingly it is the tax-payer who foots the bill, and in the light of this cost assessment, certain forms of transport are being totally abandoned rather than being modernised. The widening of the British canal system, necessary to make it a viable commercial proposition today, is a case in point. So are the branch lines of British Rail, about which there has been such outcry.

The sociological responsibilities, too, become increasingly more complex. Increasing transport-speed brings with it new phenomena in sound, and this is not confined to the sonic booms of Concorde. New and more aircraft will increase the ambient noise-level not only around airports but all over the country, and create logistic hazards of major proportion as they have already done in the New York–Washington air corridor in the United States, for instance.

Hover trains, and hovercraft in general, by their nature require large quantities of air to be blown through small gaps. This inevitably produces a large amount of what is known as 'white' sound. Vehicles moving at or above the speed of sound produce a wash commensurate with their size; but current research appears to suggest that the problems of the resultant sonic boom may have been grossly overexaggerated. In any case, whether the anti-Concorde lobby likes it or not, supersonic travel is with us now, and is here to stay.

New motorways dissecting the countryside are already causing major social and other natural problems, quite apart from that of noise and the hazard of swallowing the already small quantity of agricultural land available. Traffic engineers and technologists are therefore looking with increasing interest, and indeed alarm, at the sociological responsibilities which their own developments have placed upon their shoulders.

By no means the least of these is safety. As speeds increase, the problems of safety increase in direct proportion. Again, the problem is one of relativity; the relative speed between the static countryside and the rapidly moving vehicle. As this relative speed increases, the chance of the total destruction of the vehicle and its occupants in the case of mishap becomes an absolute certainty, and the available period of avoiding action disappears. This inescapably brings us back to the question of costs, because if money is to be spent to increase speed, the cost of providing commensurate safety increases even more alarmingly. For these reasons alone, it would appear worthwhile to concentrate at least a substantial research effort into the problems of comparatively low-speed, run-of-the-mill transport.

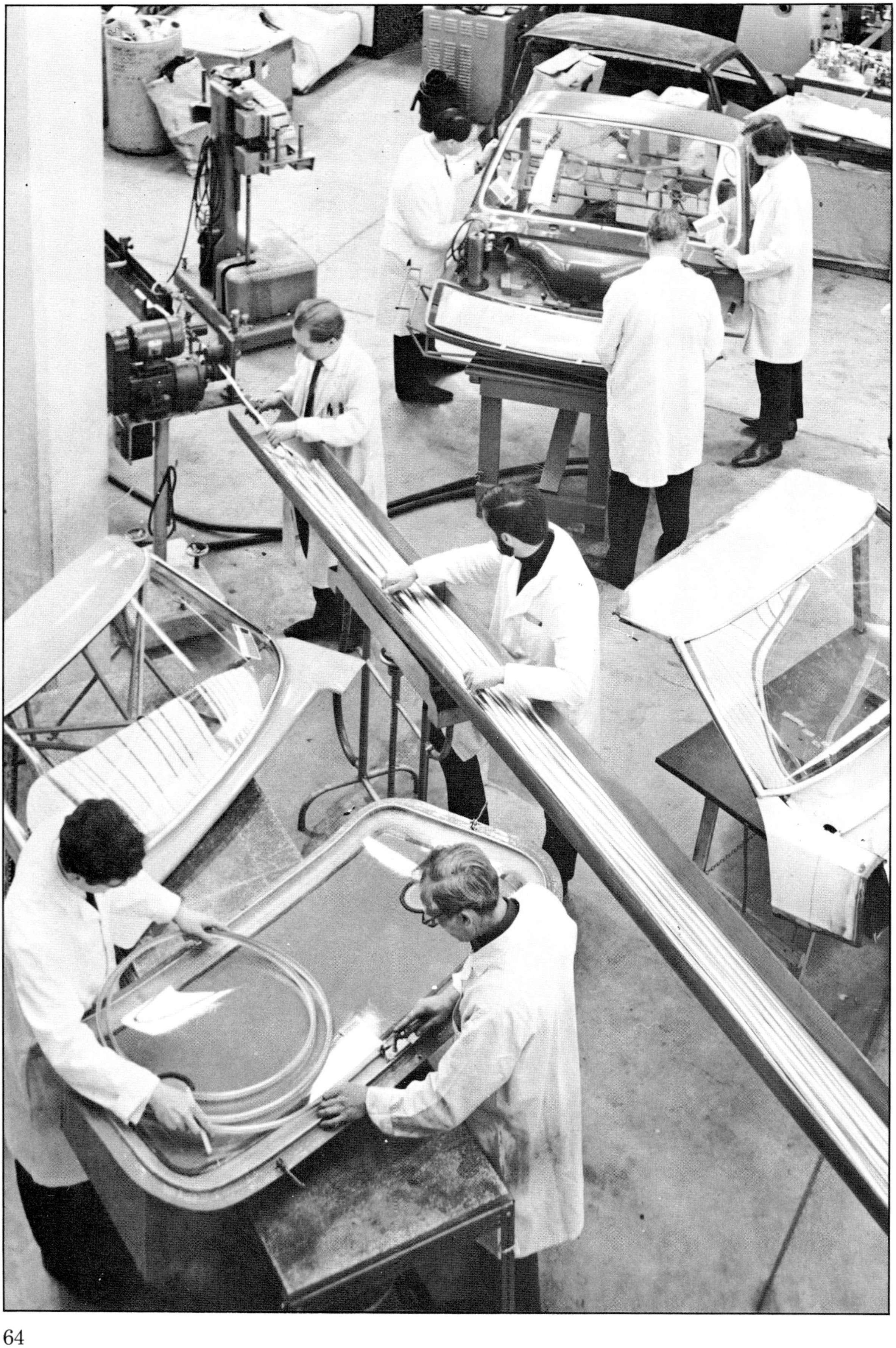

The manufacture of a new car windscreen that will not fall out in a severe crash. The system involves sealing and trimming the windscreen into the frame of a vehicle with adhesive and an external plastic trim which is fixed in position by butyl tape. It is not only crash-proof, it is weatherproof, too. The inventors claim that it is better looking because it allows the side posts of the windscreen to be slimmer, since the windscreen area no longer needs an edging of thick rubber

Moving Pavements

One of the most promising developments in this field is the concept of the passenger conveyor. The average speed of buses, and even taxis and private cars, at peak periods in towns and cities is today approximately 3 mph. Clearly any system which improves on this will be a major breakthrough. The change from moving capsules to moving roads is not a new idea and has been tried out for many years. The conventional escalator is a case in point. However, if speeds in excess of walking pace are to be considered, modern technology is necessary to produce accelerating areas, as well as conveyor belts which can withstand the load and wear of an intended speedway of 10 mph. In tomorrow's world, there is no doubt that this type of walk-way will exist, and be in daily mass use in every urban area. One of its major advantages is that it can be incorporated in any existing town without causing major traffic or structural upheavals.

Passengers must, of course, be able to use the conveyor, either walking or standing still, without physical effort or discomfort, and for this reason the troughed conveyor commonly used to transport bulk material is impracticable. The conventional moving staircase, consisting of a series of rigid plates connected by swivelling joints, also has disadvantages for long-distance, high-speed passenger conveyors. Experiments were conducted using a flexible belt supported on a low friction steel pad, but in terms of speed and distance the friction involved proved to be unacceptable.

Passenger Belts/As is so often the case, the requirement called for a new material. A British company found the solution in a new type of conveyor-belting. Supported by rollers only along its outside edges, the central load-carrying area (which is normally 90 per cent of the belt width) is unsupported. This ensures that the area on which passengers stand is completely free from hard and uncomfortable supports and gives a smooth ride. The belt has been designed so that it has a very high lateral rigidity across its width, supporting the load between the rollers, like a bridge. It also has a good degree of flexibility along its length, so that it can wrap round the terminal winding-drums.

This 'one-way rigidity' has been obtained by designing a belt on the stressed-beam principle. Essentially it consists of a composite rubber-textile construction sandwiched between two outer layers of transverse steel cords, which are in turn enclosed in a rubber casing. The textile reinforcement provides longitudinal strength to take the drive tension, and the combined action of the upper and lower transverse steel cords creates lateral rigidity across the belt to limit deflection. A 42-inch-wide belt of this type, supported at each side by 2-inch-wide rollers and loaded at the centre with a 150-lb. weight, will deflect no more than $\frac{3}{8}$ inch at the centre.

By closely aligning the individual transverse steel cords, sufficient longitudinal flexibility is maintained to let the belt run over conventional driving-drums. Belting of this type is already carrying passengers all over the world. The Dorva Airport at Montreal, developed to cope with the increased demand of Expo '67, was one of the pioneer installations. There are similar passenger-carrying belts at Heathrow Airport, London, at the British Transport Commission Docks in Hull, in Seattle, Los Angeles, San Diego, Melbourne and several other places in Britain and the United States.

Researchers are now looking at ways of bringing passenger belts into the central areas of busy cities in the form of moving pavements. Most solutions to the problem of moving people so far proposed have been restrictive in their nature–the elimination of the car from city centres, road pricing, parking limits, and so on. It is refreshing, therefore, to discover this positive approach directed to the provision of a new facility, which will separate pedestrian from vehicle while still providing the pedestrian with all the freedom needed in his or her movements in congested shopping-areas.

Battelle Integrator / The new system has resulted from joint development work carried out between the Dunlop Company and the Battelle Memorial Institute in Geneva, and a feasibility study was completed as long ago as 1968. Given a material capable of withstanding the loads involved, the basic problem is to design a system by which the passenger can gradually be accelerated from, say, 2 mph up to 10 mph, at which point he can easily board a belt moving at that higher speed. A similar system must also of course provide for easy slowing-down at exit points. The joint development work already carried out has shown that a means for providing the required acceleration and deceleration, so that passengers are synchronised with the speed of the main belt, is a practical proposition. It is called the Battelle Integrator.

Integrators will be provided at the terminal points of the system, and at intermediate points along the way to provide convenient entry and exit to the moving roadway. The passenger will be brought to these integrators by any conventional means, such as an escalator or a low-speed pedestrian belt. The integrator will be in the form of an angled passageway, the floor of which consists of equal-length ribbed platforms coming slowly through an entrance comb, just as they do on an ordinary escalator. Stepping on to this platform the passenger can either stand still or walk forward at any speed he chooses along the flat integrator corridor, guided by handrails at either side. As the platform moves forward it starts to slide progressively to the left, following the one in front. The platform movement is therefore in two directions, forwards and sideways, and a passenger standing on it moves along a parabolic path, the net result of which is a constant acceleration along a continuous track. The sliding movement between adjacent platforms is relatively slow, at 20 inches per second, and the resulting acceleration along the parabolic path towards the main passenger belt is also relatively low, at as little as .05 g.

The passenger will eventually find himself travelling parallel to the main conveyor-belt at exactly the same speed, so that all he has to do is to step sideways on to the main belt. Disembarkation is naturally the same process in reverse, and the passenger eventually leaves the system at the same low speed at which he entered, having travelled for any distance up to 3 miles at 10 mph over the ground, even if he stands on the same point of the main transporter for the entire distance of his journey. Add the pace of a brisk walk to the 10 mph average speed of the conveyor and a point-to-point average of 14 mph can be achieved by the pedestrian with a minimum of personal effort and inconvenience.

Considerable thought and care has been put into the overall appearance of the system, as well as its practicability and safety. The passenger-conveyor can be built above street-level and fly over existing traffic routes, at normal pavement level, or under the ground, depending on its needs. It can be made to follow a 900-foot horizontal radius where necessary, so that curves and corners present no problem, and it can also be elevated or lowered through vertical curves to change its level.

This is probably the most significant potential solution currently available to one of the most pressing problems of our time–the strangulation by congestion of city centres all over the world. Not only does it enjoy the advantage of offering total freedom for the individual, but old bugbears such as pollution are eradicated by the use of electric or other fumeless power to drive the system.

As to costs, studies published in March 1968 by the Stanford Research Institute for the US Department of Housing and Urban Development showed that a transport

Computer-aided design. The designer uses a light pen to modify the shape of a car on the screen. With this system designers use the computer's memory and its ability to calculate. The result is faster checking of blueprints

system of this type had a potential value in 45 urban areas of America. In a typical central business-district one mile square, with 40 per cent of all travellers using the system and paying a 7-cent fare (say 3 new pence), the revenue would be sufficient to cover capital outlay in less than a year. No comparable study for the United Kingdom has yet been completed, but its cost would be very competitive with that of the underground.

It also seems likely that the basic speed of 10 mph could be substantially increased–perhaps even doubled. In the United States a safety code has been issued for the design, construction, installation, operation and testing of 'moving walks'. Competition between belting manufacturers for business in the new field is already keen and there can be little doubt that new materials, making higher loads and higher speeds possible, will be developed in the immediate future.

Rapid Transport Systems

While the passenger-conveyor concept is forging ahead in the United States, at certain major American airports, it has also been decided to use specially built moving vehicles on their own permanent way to handle the vast anticipated increase in passenger demand. It is calculated that by 1975 there will be as many as 150 million passenger trips a year over United States domestic routes alone. This means that up to 10,000 passengers per hour will have to be accommodated at the main airports.

STS/The way ahead is indicated by a $5.3 million contract for 'the most sophisticated automated transport system ever ordered for an airport', which has been signed by the Port of Seattle Commission and the Westinghouse Electric Corporation. The Satellite Transit System, as it is called, will soon be in service. It will consist of nine computer-supervised, 106-passenger transit vehicles which will serve six intra-airport stations. It is virtually a 'no delay' system with the longest interval between vehicles predicted as 100 seconds in rush hours. The Seattle–Tacoma system will consist of two loops and a shuttle. It will be capable of transporting 1200 passengers one way on each of the two loops every 5 minutes. During slack periods the system will work 'on call' like an automatic lift.

The nine light-weight, rubber-tyred, electrically-powered vehicles will operate on their own concrete permanent way. They will follow an 'I'-shaped guide beam located in the middle of the running surfaces, to which they will be positively locked through their guidance systems. There will be a fully computerised, automatic train-control and communications system, backed up by a television monitoring circuit.

At Tampa, Florida, a similar overhead system has been installed at the airport. Eight air-conditioned units will shuttle automatically between the main terminal and four individual loading terminals about 1000 feet away. Each electrically-propelled vehicle will be able to move up to 100 standing passengers to their destinations in about 40 seconds.

With these two advanced systems the Westinghouse Corporation is gaining invaluable experience in the problems of high-efficiency, automated vehicle transport. Engineers in the United States have not been slow to realise the vital importance of bold and revolutionary approaches to the problems of moving passengers in accordance with the particular needs of their environment. These Rapid Transit Systems can not only influence the operational efficiency of an airport but dictate the development of entire urban areas.

Post Oak is an urban centre within the city of Houston, Texas. When it is completed it will include office-towers, a hotel, department stores and a shopping mall complete with entertainment, restaurant, financial and other facilities. Included in the original development plans is a Rapid Transit System which will serve the entire area in the horizontal plane in much the same way that a lift serves a high rise tower. The first phase will consist of about a mile of double elevated guideway, separated from pedestrian and vehicle traffic and passing directly through some of the buildings it serves. The vehicles themselves will be comparatively small, capable of carrying 16 passengers seated, with a further 16 standing. They will be rubber-tyred and electrically powered, and needless to say, there will be a totally automatic train-control system with television supervision. The cars may be operated as single- or multiple-car trains according to demand, and passengers will never have to wait more than 75 seconds for a vehicle to arrive.

Magnetic Mobility

The ideal propulsion for all three systems would appear to be the magnetic induction motor or linear accelerator. It has the tremendous advantage of being able to produce very rapid acceleration without having any moving parts in its components. The principle is that an electromagnetic field is generated along a static guide rail by the motor installed in the vehicle, which is then literally drawn forward by magnetic force alone. Braking is achieved by reversing the direction of the field.

Despite the exciting potential of such an arrangement, it has yet to be incorporated in any large-scale transportation system, but advanced research projects are being conducted in Britain, France, the United States and Japan.

The Japanese interest in the very high-speed, frictionless drive of the linear induction motor is no doubt encouraged by the phenomenal success of their New Tokaido line. This has proved conclusively that trains cruising at 130 mph, and offering all the comforts normally associated with airline travel, can not only be run at a profit, but also have a profound influence upon the development of areas which they serve. Along the 340 miles linking Tokyo and Osaka, Japan's two biggest cities, the journey time has been reduced to 3 hours. The trains travel along a crowded corridor which houses 40 per cent of Japan's population and 70 per cent of her industry. Despite the extremely high average speed, some of the

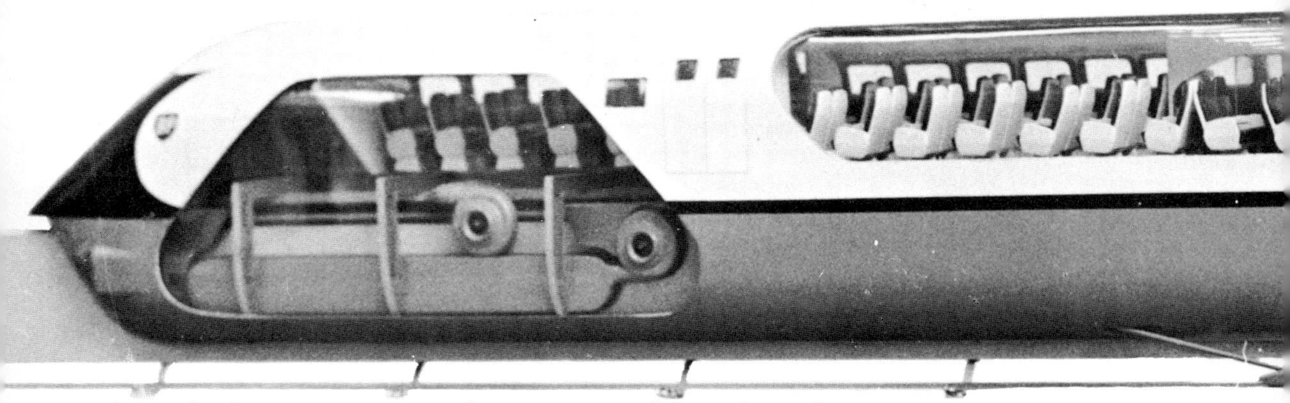

The model of a Tracked Hovercraft that one day might link central London with the third London Airport. Support and guidance would be by air cushion, propulsion would be by linear motor. The all-electric vehicle is designed to carry 100 passengers at cruising speeds of 250 mph. Present estimates suggest that it could be operated at a profit with fares no higher than current British Rail first class

trains take in as many as a dozen stops, although each one is limited to a maximum of 2 minutes. As a result, along the new track of the super-train, local industries have sprung up, and a new phenomenon in ribbon development now follows the Tokaido line.

Tubed Travel/While a 700-mile extension will halve the time between Northern and Southern Japan, the railway technologists there are by no means resting on their laurels. Professor Kiunogo Ozawa believes that tubes rather than rails are the answer to the problems of the super-train of tomorrow, and he has his eyes on target speeds in excess of 1000 mph. His idea is comparable to the old pneumatic system which was used in department stores to convey bills and money from counter to cash desk and back. A container is propelled along a tube by compressed air.

At least one British expert, Coleman O'Flaherty, Professor of Transport at Leeds University, considers such a system feasible for passenger travel. He argues that solids such as coal, as well as gas and oil, are being moved by pipeline in industry. 'It is not too far from this to see the idea of people moving in suspension,' he says. 'By this I don't mean feeding people into a pipe and having them go scurrying helter-skelter on their way. But rather having them in some form of a capsule, such as a train, and having it move within an enclosed tubing.'

Professor Ozawa has gone further. He holds a world speed record–of nearly 1600 mph–by firing a rocket down a pipe. He claims that already, with only a few 'minor details' to be ironed out, people could be sent by tubed flight from London to Glasgow in 12 minutes.

Professor Ozawa's tube is as yet only one mile long. His passengers so far have only been cockroaches, each wearing its own anti-G safety harness, but his work at Meijo University has attracted the interest of the American Government. It is financing a project which hopes, more modestly, to send a hover-train in a tube down the centre of motorways.

Riding on Air

Urba Train/In France, too, there has been rapid progress in the field of hover-trains. Like the linear induction motor, the hover-train enjoys the advantages of frictionless transport. Coupling the hover principle to a guide-

way system also overcomes the lack of inherent directional stability to which almost all forms of free hovercraft are subject.

The Urba train has already amassed a very considerable test mileage, and has carried over 1500 passengers on its test-track near Lyon. It is a monorail, underneath which the passenger-carrying cars are suspended. Each car is supported on 'air-bogeys'; these work on precisely the same principle as any type of hovercraft, with lift-fans which pump air on to the lower flanges of the monorail, causing the bogey to float. Friction is thereby reduced to a minimum, noise is virtually eliminated, and an exceptionally smooth ride is achieved.

Since the Urba train is also propelled by an induction motor, it enjoys to the full the technical advantages offered by both these British inventions. The maximum speed so far achieved is a modest 45 mph, but quite clearly far higher speeds present no major practical problems. Trains of up to eight 30-seater cars are envisaged, and cost analyses suggest that the Urba train would be 50 per cent cheaper than the conventional underground railway, and 10–20 per cent cheaper than conventional elevated railways. The construction costs of a double-track system work out at £300,000 per mile.

These figures are, however, disputed as a result of work currently being undertaken by a British company, Tracked Hovercraft Ltd, in East Anglia. Civil engineering contracts worth some £680,000 for a 17-mile-long test-track had been concluded by November 1969, and another contract has been placed with the British Aircraft Corporation's Space and Instrumentation group at Bristol for an advanced data-handling system.

Studies so far completed by this group suggest a maximum overall construction cost of £750,000 per mile of twin track, including stations and all control and communication facilities, which, they claim, is not much more than the investment required for a motorway of equivalent capacity. After careful assessment of the propulsion possibilities, ranging from conventional propellers and ducted fans to jets and rockets, the group has chosen the linear induction motor as its energy source. They have so far produced a design for a single-car passenger-container riding above the surface of an elevated permanent way.

Of the scheme as a whole Professor Colin Buchanan says, 'In all essentials tracked hovercraft are a new form of transport. They are new in the manner in which they are supported, guided and propelled. The likelihood is that their impact would be considerably more than a mere extrapolation of current trends would suggest. We are dealing with something which could influence considerably the way of life and the social habits and organisations of a significant proportion of the population of the country.'

The system envisaged for the late 1970s would be essentially the same as the one which will soon be tested – a vehicle, supported and guided by an air cushion, and propelled by a linear motor, with fully automatic control-systems, running astride a concrete beam-way at speeds of up to 250 mph.

Professor Buchanan considers the most promising application to be its introduction as a form of high-speed transport between main regional centres of population. This could be either in competition with existing modes of inter-city transport, or as a replacement for one of them. 'There are trips where the return journey cannot be completed, and the business transacted, on the same day.

Below Transport of the future by Hovercraft Development Ltd

Right Professor Eric Laithwaite, Professor of Heavy Electrical Engineering at Imperial College in London (left) checks a quarter-scale, single-sided linear induction motor rig on a test track at Cambridge. This motor achieved its designed speed of 45 mph and measurements of its thrust and overall performance indicate that the full-scale passenger vehicle will operate at speeds of 250 mph or more

By tracked hovercraft, however, it would be just possible to make even the farthest return journey in this country (say, Glasgow to Southampton) in a day and still leave a reasonable proportion of the time for doing business.'

But as Sir Christopher Cockerell's brainchild, the Hovercraft, enters its third decade, one must ask whether or not it has justified the major excitement which greeted its announcement. The concept of hovercraft running on guide-tracks is, after all, a far cry from the total mobility and amphibious characteristics promised by the original work.

Although large load-carrying, high-speed craft have been developed for cross-Channel operation, it is significant that the rest of the world has been cautious in its response to the British sales drive with vessels of this type. In Russia, Italy and the United States the commercial applications of the hydrofoil have been exploited in favour of hovercraft. In practice it has proved difficult to generate sufficient lift to create hover-height capable of overcoming the passenger discomfort associated with even moderate seas. The problem is that, in order to obtain increased hover-height, the weight of the neces-

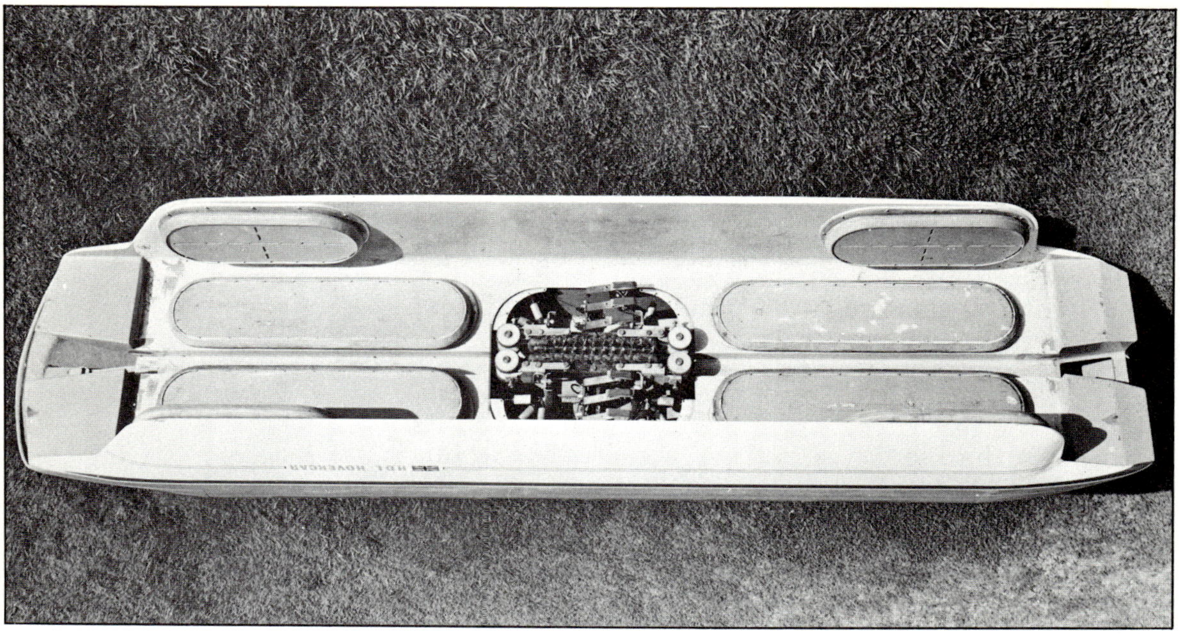

The Works. The belly of an HDL Hovercar. Note the guidance pads

sary machinery increases, and the designer finds himself in a vicious circle. Large vessels capable of lifting their main hulls well clear of the surface, including wave-tops, have been designed to ride successfully at speeds in excess of 50 mph on a submerged hydrofoil. This acts in exactly the same way as the wing of an aircraft–its curved surfaces generate lift as a result of its forward propulsion through the water. The battle between hovercraft and hydrofoil was joined in earnest some 10 years ago, and its outcome is by no means yet resolved.

On the other hand, it seems that the true commercial reward of the hovercraft principle may be found in its more off-beat applications. Lightweight single- or two-seater hovercraft, designed and marketed essentially as fun vehicles, are in mass production in at least one factory in this country, and hover-dodgem cars are earning their keep at a fun-fair in the Channel Islands.

Magnetic Repulsion/The hover principle adapted to an air-bogey offers an exciting alternative to the wheel. Even more exciting, perhaps, is the concept of the air-bearing. Modern precision engineering has made it possible to 'float' a shaft of considerable diameter within a bearing, using compressed air as the supporting medium in place of roller or ball bearings. The almost total abolition of friction which results will, of course, increase the mechanical efficiency of virtually every type of motor and machine, either static or locomotive.

The reduction, or indeed abolition, of friction is a sort of Holy Grail to the mechanical engineer. An alternative to air as the supporting medium is the magnetic field. At least one British inventor has been engaged in experimental work in this field for several years. It is now possible to make bricks of a ferrous substance so that the main faces of each brick present opposing magnetic poles. Since like poles repel, if a line of such bricks could be laid to present a pavement or guideway with a succession of

South poles uppermost, a flat plate of metal, with its South pole downwards, could be made to hover above, and indeed skate along such a line of bricks using the magnetic forces as the supporting medium.

A man-carrying prototype using this principle has been built in Buckinghamshire and development work proceeds. It seems at least highly probable that the adoption of this system will present yet another opportunity to 'float' conveyors of one sort or another with a remarkable gain in mechanical efficiency.

It is also possible to use a similar arrangement to provide a friction-free magnetic bearing, offering all the advantages of the air-bearing but using electricity instead of compressed air to generate the necessary 'lift'. We have already demonstrated conveyors using both principles in the *Tomorrow's World* studio.

Hover Conveyor/The 'hover conveyor' uses a system of valves which control air vents placed in the centre of the conveyor channel. As the load approaches, these sensing valves release compressed air, which supports the load for the length of its travel over each individual valve. When magnetic forces are used to generate the 'lift', it is of course necessary to mount the load on a suitable, ferrous carrying-plate.

Another potential of the hover principle has been demonstrated in the form of a hover truck, or carrying platform. In this, compressed air is fed to a flat load-carrier so that it generates enough lift beneath it merely to raise it sufficiently to overcome the contours of a reasonably smooth factory floor or other working area. As a result, it is quite possible for a man to push a load of many hundredweights with a minimum of physical effort. Control is neither difficult nor hazardous: shutting off the supply of compressed air will bring any load to an immediate stop. This principle is applicable to almost all types of containerised loads in the loading and off-loading problems now being experienced at factories, transport depots, airports, docks and stations throughout the world.

The potential of this type of application of the hover principle has been dramatically demonstrated at Wythenshawe near Manchester. It was the result of a combined operation involving the Central Electricity Generating Board, Vickers Ltd at their South Marston works, and the British Hovercraft Corporation. As a result, a particularly tricky problem in heavy transportation was overcome. The multi-wheeled, low-loader, articulated, heavy transporter, towed by a powerful tractor–or sometimes a combination of two–is becoming an increasingly familiar sight on our already overcrowded roads. In this case, the central load-carrying section of the vehicle was designed as a self-supporting hovercraft, fed with compressed air from an attendant vehicle. For convenience it retained its conventional multi-wheeled bogeys at either end. It succeeded in carrying a one-piece load–a 159-ton transporter–at a steady walking pace, making no more noise than a London bus. On its journey it had to cross a bridge which would have had to be rebuilt at a cost of £30,000 in order to carry this enormous load. A mere 1.5 lb. per square inch air pressure supporting the hovercraft load-carrier gave a weight reduction at each corner of the transporter equivalent to 10 tons, even though no perceptible lifting occurred. In the near future such loads are likely to reach 300 tons apiece, and by means of air-cushioned vehicles, in conjunction with roll on/roll off, shallow-draught ships, the products of British heavy engineering could be exported to almost anywhere in the world.

Back to the Wheel

Despite all these fascinating developments, it would be absurd to assume that the days of the wheel are numbered. For example, the dramatic increase in the performance of contemporary racing cars is largely attributable to the research and development work which has gone into the Grand Prix racing tyre.

This tyre technology has also done much for the safety, comfort and performance of the everyday family car. Microwaves are now being used as a heat-source in vulcanisation processes. The problems of aqua-planing–the reduction of friction between tyre and road or runway surface which is produced when the tyre tends to run up on to the top of a layer of water–have been largely overcome by research into tread pattern and the rubber mix from which tyres are manufactured. F. J. Kovac of the Goodyear Company has offered us the following predictions:

1976 A new material used for tyres. (The quest for a more efficient replacement for rubber which retains the necessary qualities of wear and elasticity is being pursued with such energy that such a breakthrough must be considered at least probable.)

1980 Elimination of spare tyre; 150-mph guarantee on tyres.

1981 Basic new tyre shape. (The flat, roller-like profile of the contemporary racing tyre has already demolished many of the long-cherished theories.)

1982 Tyre-pressure maintenance system–either a built-in air reservoir or some sort of tyre-pressure warning device with a display on the dashboard.

1983 Fail-safe tyre for all vehicles.

1984 Totally puncture-resistant tyres. (These have already been developed for heavy-duty military vehicles.)

1985 200-mph guarantee on tyres.

1990 100,000-miles guarantee.

Power

Kovac has confined his prophecies to the realm of vehicle tyres. Recently some less authoritative prophets have been chancing their arms in the field of power units, particularly in respect of the electric car. Our own view is that the future of the electric car is at this time severely inhibited by the absence of a breakthrough in battery technology. The contemporary electric-storage battery carries too high a penalty in terms of power/weight ratio to make performance acceptable to present-day standards.

Without compulsory legislation on grounds of air pollution, it is difficult to see how even an electrically-powered city runabout could be made a saleable proposition at this time.

However, thanks to the enormous impetus provided by space research, there can be little doubt that a new type of battery and/or fuel cell suitable for family cars will make its appearance within the next decade. Again, thanks to space research, the potential of solar energy to power the everyday family runabout cannot be overlooked. Meanwhile, electric power continues to be used in specialist applications.

An interesting example is a battery-powered ambulance from the Westinghouse Corporation, designed to get to the scene of an accident complete with first aid, rescue equipment and a stretcher-deck to transport the injured to where a conventional ambulance can make a pick-up. The seating arrangement allows a passenger or nurse to ride next to the stretcher so that aid can be administered continuously. The Model 254R rescue unit is intended for situations in which it is difficult to get a patient out to an ordinary ambulance–factories, airports, stores, hotels and similar large enclosed areas. The rescue unit is available either with its standard equipment, which includes oxygen cylinders and resuscitators, rescue kit, splint kit, stretcher unit and first-aid kit, or it can be fitted out to particular requirements. Nine feet 7 inches long, 3 feet 10 inches wide, and 3 feet 8 inches high, it makes a practical proposition even in restricted space. Its maximum speed of 12 mph is considerably faster than anything within the compass of even the most fit and agile stretcher-bearer.

While the Wankel rotary engine makes its way steadily

An electric ambulance designed to get accident victims out of difficult places. It carries first aid and rescue equipment and there is a stretcher-deck to transport the injured to a point where a conventional ambulance can make a pick-up. Such a vehicle can operate easily in airports, shopping centres, hotels, and holiday camps

Opposite Single-seater racing cars for three-year-olds and upwards. The driving controls can be adjusted for anyone from 3′ 8″ to 6′ 3″. Top speed is nearly 60 mph but for learners the engine can be governed to 15 mph. The car comes with a remote-control attachment for teaching toddlers to drive. There is a version with a sports car body

Opposite below Under the body a solid chassis and a 4-stroke 205 cc engine. The chassis is made of a single sheet of steel shaped in a press and reinforced with steel cross-members. It will support a 13-stone man

into an increasing number of ideas for both land and water transport, many distinguished automotive engineers including Sir Alec Issigonis consider that the days of the carburettor and the reciprocating engine are by no means past. Meanwhile, all over the world, engineers continue to file patents for designs of varying complexity which attempt to obtain the advantages of a gas turbine or a Wankel engine as well as the undoubted reliability and simplicity of the more conventional approach. Very often the principal stumbling-block is that of tooling costs. Major manufacturers are naturally reluctant to scrap their investment of millions of pounds and many years of experience based on the well-tried reciprocating engine.

But the pollution problem can be largely overcome by the use of cryogenic natural gas as an alternative fuel to petrol. Already a conventional car engine can be converted for this purpose at a cost of $120 for a kit of 'off-the-shelf parts' now on sale in the United States. Given the dramatic increase in the availability of natural gas and a chain of filling stations capable of providing the super-cooled liquid as easily as petrol, it is difficult to see how this solution will not gain ground rapidly. While it is easy to get carried away by the exciting possibilities of completely new techniques, immediate short-term benefits in almost every field of transportation are more likely to be secured from the adaptation and improvement of existing facilities and experience.

APT

A significant case in point is APT–Advanced Passenger Train–British Rail's approach to the Super Train, which has been hailed as 'probably the most important and significant development in railway history'. It enables existing track to be used at speeds 50 per cent higher than at present without reinvestment in permanent-way structures. In plain figures this would allow a train to operate on a good permanent way at over 150 mph. Gas-turbine engines will be used for propulsion, but the concept of APT involves several totally new ideas including the use of a 'worn-wheel profile' and a carriage with a new suspension which banks when cornering and therefore produces only small 'g' forces on the passengers even when travelling at these very high speeds.

Conventional railway wheels are designed so that their inside rims, or flanges, do not touch the rail when they are running on a straight piece of track. When cornering at high speed, however, the retaining rim of the inside of the wheel contacts the rail edge and this naturally puts a maximum on the safe cornering speed. Research established that by giving the profile a new shape, contact between flange and rail could be prevented. Curiously enough, the new shape is virtually the shape that a conventional wheel ends up with after running for a long time–hence its name, the 'worn-wheel profile'.

Thus enabled to corner on existing track at far higher speed as far as the wheels were concerned, the engineers were still left with the problem of the lateral 'g' forces which would not only subject passengers to considerable

BARNARD
6
6
BARNARD
4
6
4
BARNARD
RACING
3
BARNARD
6
BARNARD
RACING
2
THOMAS B
GTX

discomfort, but also roll the carriages outwards off the permanent way. When the pilot of an airliner executes a properly co-ordinated turn, the passengers experience no sensation of lateral 'g' because this is compensated by the angle of bank assumed by the aircraft. Precisely the same solution has been adopted by the research engineers of British Rail at Derby for APT. They have developed a system of roll control which enables each carriage to bank to a maximum of 9 degrees: a system of automatically-operated hydraulic rams bank the carriages into each successive turn in exactly the same way as a pilot would bank an airliner. Together with the banking on any curve that already exists, this enables the speed to be increased by about 50 per cent without exceeding the passenger comfort limitation. A completely new system of suspension in the bogeys of APT has also been designed.

Because of the low rolling resistance of steel wheels on steel rails, the aerodynamic drag is an important factor even at speeds well below 100 mph. The design of APT is therefore as clean as can be practically achieved–the high drag gaps between coaches are filled, and there is a complete fairing under the body. A normal Advanced Passenger Train would have a 1500-hp, 42-foot power coach at each end of the train, with between 6 and 10 passenger cars, each 63 feet long. Wide passenger access doors are placed in the 5-foot joining module between coaches. Passenger cars weigh 17 tons and 22 tons fully laden, and the power car weight is only 33 tons. So with 6 passenger cars the total weight of a train would be 198 tons, with a power to weight ratio of 15 hp per ton. Present main-line diesel trains have locomotives developing 2700 hp and with the same number of vehicles as APT are approximately twice the weight with a power to weight ratio of 5–6 hp per ton.

In each of APT's two power cars 4 Leyland 400-hp gas turbines will be installed. This engine, incidentally, is also under development for heavy road vehicles and appears to offer considerable advantages in terms of low weight, good fuel consumption and low initial and maintenance costs. The loss of any one of the eight engines should not affect performance, and the design enables an engine change to be made overnight.

For electrified lines, it is proposed to develop an electric version of APT with a similar performance.

With cruising speeds of 155 mph in mind, it is essential to have sealed windows and full air-conditioning. Power for these will be provided by alternators worked by separate gas-turbine engines in each power car.

In order to use the existing track without expensive changes in the signalling system, high-speed trains must need no greater stopping distance than present trains. APT will be able to stop in 6000 feet from a speed of 150 mph. A hydro-kinetic system will be used and slowing down will be achieved by driving a vaned turbine in hydraulic fluid, rather like reversing the process of the automatic transmission in a car. Disk brakes will take over when the speed has been sufficiently reduced.

The average speed of present British inter-city trains is normally between 60 and 75 mph; detailed timings for

APT over a number of routes, using a maximum speed of 155 mph, give an average speed of 90–105 mph.

The Board of British Rail and the Ministry of Transport have financed a joint five-year development programme at the cost of £5 million. The work which has been done is successfully following the predicted programme, so the first experimental train will run in 1971 and prototype trains will be available for passenger service in 1974. To carry out development work, the disused line from Nottingham to Melton Mowbray is being converted to a high-speed test-track.

All this makes a great deal of economic sense, since the major costs of a railway system are usually the fixed installations: track, signalling, station and terminal facilities. Because the costs of the fixed installations are substantially independent of the amount of traffic, this means that an increase in the volume of revenue traffic will create a disproportionately large increase in the profitability of the line.

Interest has already been aroused among a number of overseas administrations, and an American consortium recently reached agreement with British Rail for licences to exploit APT technology in North America.

As to the future, APT is capable of further development to operate at speeds of over 200 mph, although considerable improvement in the track and signalling would be required to maintain a high standard of safety and comfort.

The train-control system now being developed in the British Rail Research Laboratories at Derby is computerised and, with little additional development, can be used for very high-speed trains. It will incorporate telephonic communication between the train and a central control or power signalling box. To maximise the advantages of high-speed passenger travel from city centre to city centre a fully computerised seat-reservation system will also have to be provided, as with airlines. Such systems are already in service in Sweden, Japan, Spain and Italy.

Britain is leading the world in this particular field, and the APT may easily bridge the next stage of linear passenger transport. It is thought that its life will be another 50 years, by which time railways as we know them will be superseded.

Vertical Flight

VTOL/Clearly competition between airlines and railways for the bulk of passenger traffic from city centre to city centre will intensify in the immediate future. Hence the mounting interest of aircraft manufacturers and operators in vertical take-off and landing (VTOL).

The British aircraft industry secured a world lead in this field by the introduction of the Harrier, the first VTOL aircraft in the world to go into squadron service. The Harrier is, however, essentially a military aeroplane, and the appearance at the 1969 Le Bourget air show of the Dornier DO 31, designed as a load-carrier, was not without significance. Whereas the Harrier uses a vectored thrust engine for both its lift-off and forward flight, the

Dornier has two propulsion engines and eight separate lift-engines, four in each of two pods slung beneath the wings. The Dornier consortium chose specially-developed Rolls-Royce lift-engines for the DO 31, and the same partnership is to be continued for the next step in the programme, the Dornier DO 231, which will have two forward and twelve lift-engines (four in a pod on each wing, two in the nose and two in the tail).

For this and similar advanced projects Rolls-Royce have developed the RB 202, an advanced lift fan engine. It is designed essentially as a lift-engine only, and any VTOL using RB 202 will have a forward-flying engine system as well. The RB 202 is still in the early development stage, but is expected to enter commercial service in VTOLs in the late 1970s.

In order to change direction from lift to forward flight, it is necessary to deflect the angle of the lift-engine thrust gradually, and thus allow the cruise engine system to be brought into operation. This is achieved by tilting the lift engine up to a maximum of 20 degrees.

The arguments for and against separate engines for developing lift or forward thrust are by no means confined

Left The RB 202 fan lift unit which will provide vertical lift-off for the Hawker Siddeley HS 141 VSTOL airliner

Right The Rolls-Royce RB 211. The most advanced gas turbine aero engine in Britain for use in conventional short-haul airliners. It powers the new Lockheed Tristar and is planned for the BAC 311

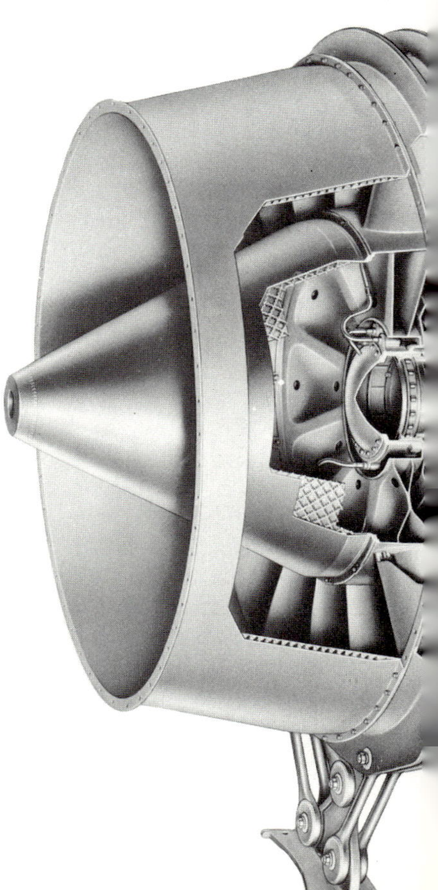

purely to technological considerations. As ever, cost is critical. R. G. West of Rolls-Royce, in a paper on fan lift in VTOL design, states: 'The choice of the optimum lift fan design requires complicated study, in which the noise level and the engine cost are of the same importance as engine weight and fuel consumption. If the target noise levels are not achieved, a fan lift VTOL aircraft will not be permitted to operate from the proposed urban "vertiports" *(sic)* for which it is being designed. Likewise if the lift fan unit cost is not a fully commercial proposition, bearing in mind the comparatively large number of engines in each air frame, no operator will be able to afford the first costs required to launch a fan lift VTOL airline.'

Aircraft and aero-engine designers are obviously taking into consideration the problems of noise at the earliest possible design stage. For some years it has been evident that a VTOL medium-range airliner could make a major contribution to the growth of civil airline transport. But in order to gain the maximum time-saving which would be available with a VTOL system, aircraft will have to be operated from specially constructed airfields within easy

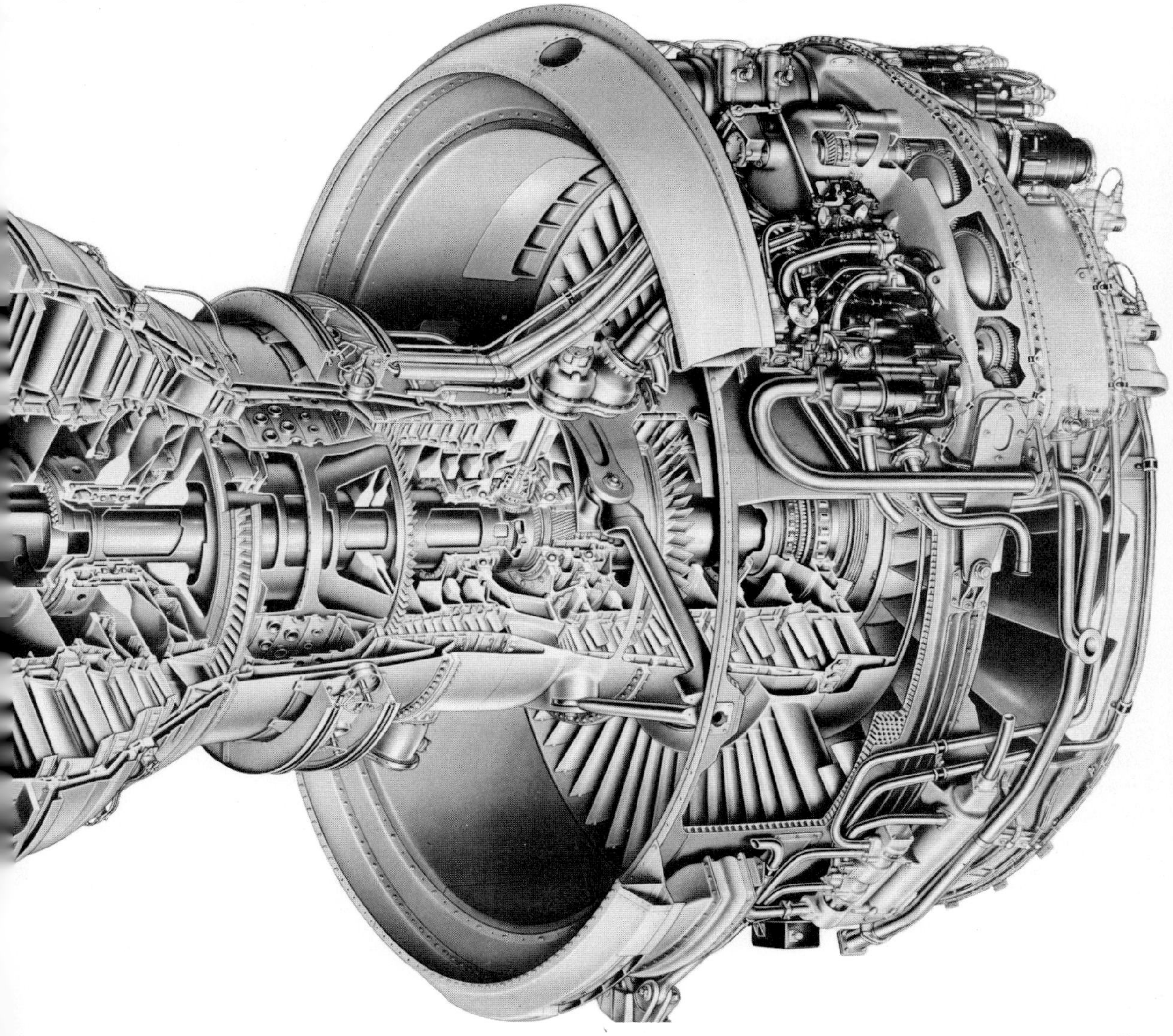

reach of city centres. Recent work at Rolls-Royce and elsewhere has shown that it should be possible to keep to the required noise-levels in such environments.

But a decrease in passenger-time from city centre to city centre is not the only advantage offered by the VTOL airliner. Airport congestion, both on the ground and in the air, is now a very serious problem all over the world. The efficiency of many major airports in the United States is already severely impaired because of the large numbers of landings and take-offs on each runway and safety requirements which impose separation limits in the approach and departure corridors.

The production of a competitive VTOL airliner is one answer to the overcrowding in the airport control-zones. But the commercial operators demand at least three basic operational criteria: a VTOL airliner must be able to fly slowly and safely to and from small pad-areas, making its approach and departure along any desired direction; in flight it must be as fast and as comfortable as present turbojet transport aircraft; and to be acceptable commercially it must have similar operating costs.

Rotor craft–helicopters of varying types–which have been projected to meet the future civil VTOL requirements will have an optimum range of about 200 miles. The lift fan VTOL aircraft can be designed for medium ranges from 400 to 1000 miles. There does not seem to be any practical limit to the size, and designs have been considered for 50–100 passengers, with a cruising speed of about 85 per cent of the speed of sound and full VTOL capability at both the origin and the destination of any individual flight. When operated from urban vertiports this type of aircraft would allow a door-to-door time between London and Paris of about one hour in almost all weather conditions.

One of the many complexities of VTOL airliner design is the necessity to provide for the emergency situation in which one or more of the lift-engines loses power during take-off or landing. The Harrier's single vectored thrust engine, providing both lift and forward thrust after take-off, is clearly an unacceptable arrangement for civilian requirements. The provision of a large number of individual lift-engines under each wing, and indeed in the nose and tail of the aircraft as well, would provide the pilot with full control even in the event of the loss of one or two engines. However, each lift-engine is a useless additional weight once the transition to conventional forward flight has been achieved, and so the search for 'added lightness' and new materials has been intensified.

Important research has also been completed in the attempt to confine noise-levels, and at Rolls-Royce and Farnborough it has been proved that noise nuisance can be reduced by designing for high-frequency noise generation only. This high-frequency effect is obtained by designing special turbo blades on the lift fan engine which are literally 'tuned' to give the best compromise between structural efficiency and high-frequency noise.

VTOL/STOL/ Clearly there will be very keen competition during the next ten years between VTOL and STOL (short take-off and landing). For a variety of reasons, it seems likely that in the United States the strongly established helicopter industry will concentrate on rotor aircraft of varying types, including the 'compound helicopter' which employs rotors for its vertical take-off but also has an engine plant to provide its forward-flight thrust, and wings for forward-flight lift. In Japan, too, active development work is in progress on STOL potential, and throughout the world at least ten different configurations are still actively under consideration as possible STOL civil transports.

The European aircraft industry, which is becoming increasingly international as a result of its consortia involving British, French, German, Dutch and Italian companies, appears, on the other hand, to be already committed to the lift fan configuration.

Using their experience with the Harrier and its predecessors, as well as collaborative work with the Dornier group, Hawker Siddeley have produced their design for the 141 VSTOL airliner, shown in model form to the aviation public for the first time in April of this year at the Hanover Show. The HS 141 will be able to carry 100–120 passengers, cruising at over 600 mph. Operating from a vertical take-off to a vertical landing it has a range of up to 500 miles, and can therefore operate over most of the high-density routes of Europe.

But the keynote of the design philosophy is flexibility in operation and application. The HS 141 will be able to operate with a commercial load from small restricted sites, from existing heliports or from peripheral airfields, using its short take-off as an alternative to its vertical take-off. For example, from a 1000-foot field a full payload can be carried over distances up to 1350 miles, enabling the aircraft to cover an extensive network of European operations from the Baltic to the Mediterranean. It will be Rolls-Royce-powered with RB 202 fan lift units and a Rolls-Royce fan jet for forward propulsion.

According to D. H. Jagger, of the Hawker Siddeley Future Projects Office, the years 1973–5 will see the appearance in the USA of low-speed turboprop STOL, and higher-speed jet rotor operations, primarily as a response to airport congestion. In Europe, on the other hand, where the immediate pressures are not so great, there will be a less enthusiastic adoption of STOL in available low-speed forms such as developments of the Breguet and D.H. Canada. Since European VTOL technology is in advance of that in America, the European airlines may seriously consider a direct approach to high-speed VTOL, missing out the 'intermediate' stage. By the mid-1970s he predicts the introduction in various parts of the world of significant compound helicopter operations, particularly on short, over-water routes. The parallel development of a chain of heliports, guidance equipment and advanced air traffic control techniques will naturally follow.

But perhaps the most interesting of his predictions is an increasing commercial and social pressure throughout the 1970s which will impel the full adoption of VTOL technology in high-speed, low-noise form by the end of the decade. This technology will by then be founded on

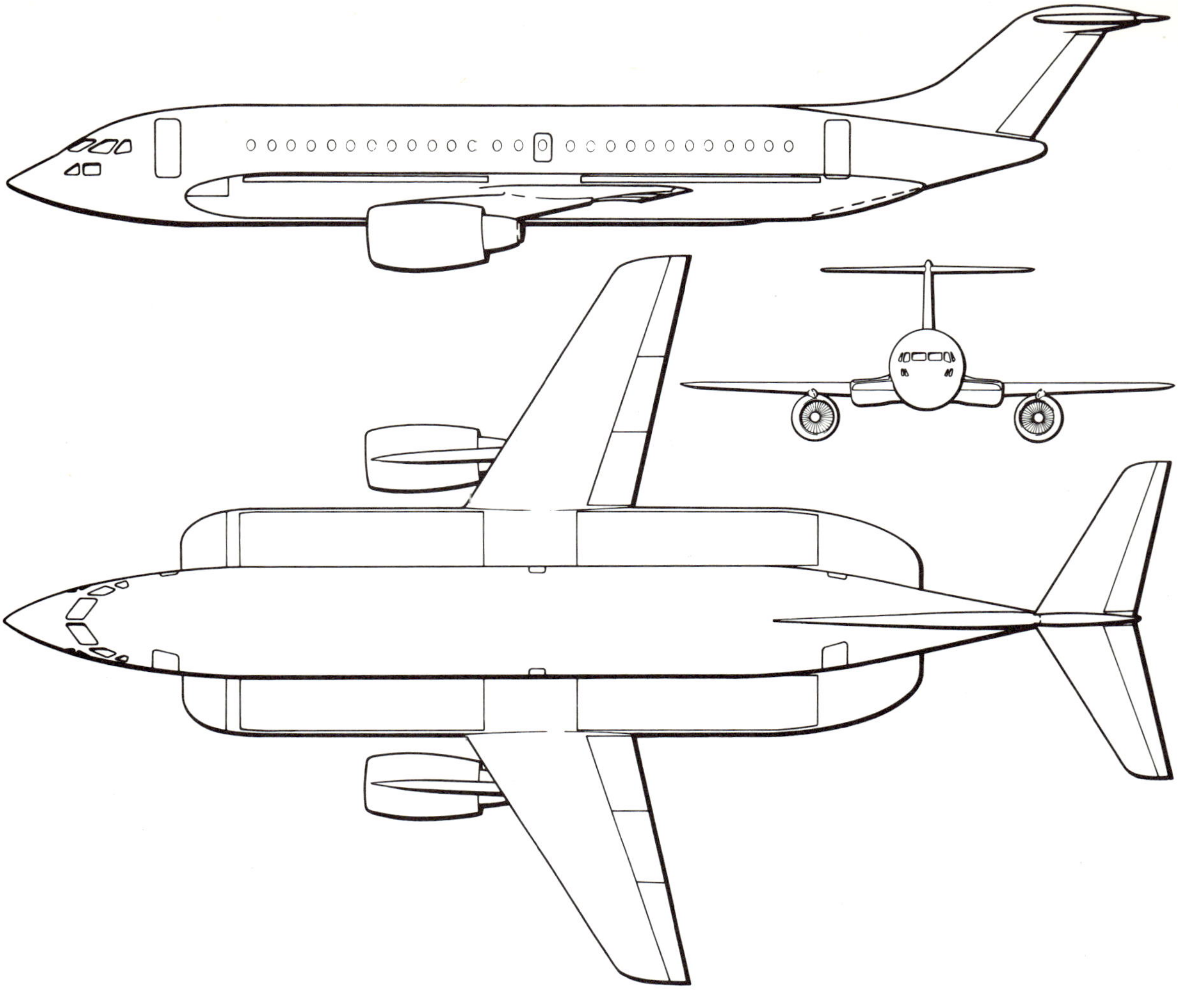

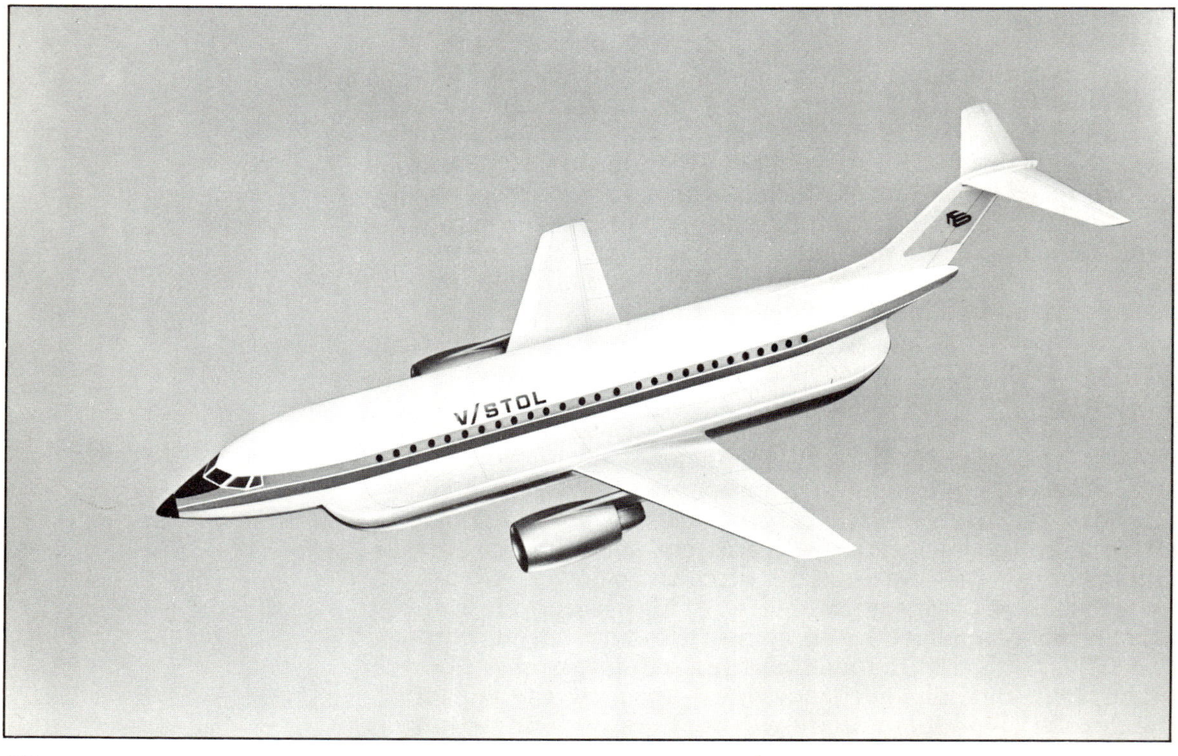
V/STOL

25 years of aircraft and lift-engine experience and will reach its practical application through development of the lift fan engine designed in this country in the late 1960s. He suggests that airlines using advanced VTOL aircraft will begin by operating from conventional airports and from the heliport sites established for the compound helicopters.

These prophecies, coming as they do from a leading world authority, are clearly relevant to those who are suggesting at this time that the need for a third London airport, for instance, has been miscalculated.

However, while no expert in the world of aviation is prepared to deny either the desirability or indeed the inevitability of the move toward vertical take-off and landing in civil aviation, the difference of opinion in how best this is to be achieved is not confined to international argument alone. Within the British aircraft industry three conflicting recommendations were submitted to the Ministry of Technology in 1970.

The Hawker Siddeley HS 141 VSTOL airliner as envisaged by its designers. It would seat 100 passengers and cruise at over 600 mph. Aircraft like this would not need long runways. Powerful fan lift units would push it vertically into the air, whereupon its conventional jets would take over for horizontal flight

STOL/BAC's interest in the VSTOL airliner is a logical extension of the company's specialisation in short-haul transport aircraft which can be traced back through the Viking, Viscount, Vanguard and the One-Eleven family to their projected Three-Eleven. The Market Development Division at Weybridge recognise the need for a new short-haul airliner for production in the 1980s in order to continue this progression.

BAC has, of course, been in close contact with every US domestic airline concerning their Three-Eleven project and as a result are particularly conscious of the impact upon operators which has resulted from air traffic control congestion in the United States. Their conclusion is that a STOL aeroplane is needed rather than a VSTOL or VTOL. Because an aircraft with 'V' capability must have higher engine-generated thrust, the argument is that it will be inherently noisier than a STOL machine, and that since it will have higher fuel consumption, it will also tend to make more smoke. They also believe that since a STOL looks more like a conventional aeroplane, it will be more attractive to travellers.

BAC also argues against the inescapable technical complexity of an aircraft capable of vertical take-off and landing, because of its requirement for a zero-flight-speed control system as well as conventional controls. They claim that a STOL can do the job as effectively as a VTOL but sooner, more easily, and more cheaply.

Based on their study of 23 different aircraft types, each suitable for operation in the north-eastern corridor of the United States, BAC concludes that a VTOL using rotors would be very expensive, a fan lift VTOL would be more expensive than a conventional aircraft, but that a STOL could be cheaper than existing conventional aircraft over ranges of 350–500 miles. Significantly, the costs do not come down with increase in size, as would be the case with conventional aircraft.

Studying the needs of the consumer for door-to-door transportation, BAC found that by the mid-1980s the London area alone might generate as many as 33 million short-haul passengers per year, while the US studies showed 37 million passengers a year likely to use 14 STOL airports in the north-eastern corridor by 1975. These figures of course assume the provision of such suitably located airports as well as an effective area-navigation system. Given similar facilities in the London and Paris area, the forecast predicts 5 million passengers a year on twenty different routes between five points in London and four in Paris.

To meet this demand, BAC believes it reasonable to assume that 1000 STOL airliners will be required, over half of them in the United States. They will be used first on routes where existing airline operators have scored sufficient profitability to ensure a reasonable return on investment in the new aircraft. On this basis, the United States north-eastern corridor and the United Kingdom domestic routes appear to be good starters for early introduction of STOL aircraft. There would be no need for international agreement, inevitably a time-consuming process, since both areas are within their own national jurisdiction.

BAC's specification for a suitable aircraft is a design capable of operating from 2000-foot runways in cross-winds of up to 25 knots, carrying full passenger payloads over a range of 600 miles at a minimum cruising speed of 460 mph. Even with the 'high degree of passenger comfort' which is considered a prerequisite of success, this is by no standards an ambitious specification. Numerous configurations have been studied, but at this time no detailed design has been submitted. The choice, however, seems likely to lie between an aircraft relying solely on the aerodynamic lift of its wings to achieve STOL performance, and one with some direct-lift engines to assist take-off and landing. To further research in this last idea, BAC plans to install the Rolls-Royce RB 202 lift fan in wing pods on a One-Eleven flying test-bed.

Rotors/Not surprisingly, in view of their long involvement in VTOL aircraft design and operation in the form of helicopters, Westland has long been aware of the opportunities offered for inter-city operation. Studies have been conducted into various types of convertible rotor aircraft, aimed at meeting both military and civil requirements, but with special attention to the inter-city operation.

Again Westland take account of low noise-levels and lack of smoke as well as the other economic and operational parameters. They argue that the *only* viable line of development for a compound VTOL design is to make use of large-diameter, lightly loaded rotors to achieve the 'V' component. These must be combined with a fixed-wing layout to obtain acceptable cruising speed and economy, and Westland have concluded that the most acceptable compound design would use tilting rotors and tilting wings. Further design studies, however, suggested that tilting rotors, while acceptable for comparatively small aircraft, were not suitable for larger machines. A full-scale mock-up of the WE 01 tilting rotor version was wind-tunnel tested.

Against the background of this work, and the theoretical arguments attendant upon it, the design office of Westland at Yeovil has evolved a tilt-wing aircraft which will meet the requirements of a 100-seater airliner. The WG 22 has its rotors mounted in the same way as a conventional, propellor-driven aircraft on the leading edge of the wing. For vertical take-off and descent the entire wing is rotated through a right angle, thereby bringing the rotors into the horizontal plane in exactly the same way as they are mounted on a conventional helicopter.

The WG 22 is a fair-sized aeroplane with a 101-foot wingspan, and two 48.5-foot-diameter rotors. The wing hinges about the main spar to a maximum angle of 100 degrees, and incorporates a special aerofoil design which is currently being tested by the Royal Aircraft Establishment at Farnborough. During the tricky transition stage from vertical to forward flight, the entire leading edge of this wing is tilted down by about 30 degrees in order to preserve the efficiency of the air flow over the wing as it is rotated.

Two engines are carried side by side in each of the wing-mounted nacelles, and each engine supplies power through a reduction-speed gearbox mounted at the front of each nacelle adjacent to the rotor hub. In the event of an engine failure, remaining power is equalised between the rotors by a super-critical cross-shaft in the wing. Each of the huge rotors consists of six blades and Westland proposes to use carbon-reinforced plastics for the main blade construction.

The WG 22 would carry 100 passengers a distance of 450 miles with sufficient fuel for another 100 miles. The cruising speed is planned at 460 mph.

Among the points in favour of a convertible-rotor aircraft, claim Westland, is its operational flexibility resulting from the low power used in hover; low-speed flight, permitting stand-off delays with little economic penalty; the reliability of using four engines at a steady rate throughout every flight; low fuel-consumption for minimum air pollution; and low external noise through low power-setting.

Supersonic Travel

Concorde/The most glamorous aspect of passenger travel of the future–the supersonic air transport–is already with us in the form of Concorde. Russia's Concordski, the TU 144, is not a serious commercial threat –an attitude in no way triggered by political attitudes. The plain fact is that the Russian aircraft industry is not designed to meet the requirements of capitalist operation in the rest of the world, nor has it so far shown any particular desire to compete with the West in the battle for orders from the major airline operators in other than Communist-controlled countries.

The American intention to build a supersonic airliner, having suffered a major set-back in the cancellation on the drawing-board of the 1700-mph Boeing 2707 swing-wing, means that Concorde has an effective lead, thanks largely to the tenacity and conviction of Sir George Edwards and his team at the British Aircraft Corporation, and their French partners of Sud.

Below and right The shape of things to come. Passenger and transport aircraft that could make the length of present-day airport runways unnecessary

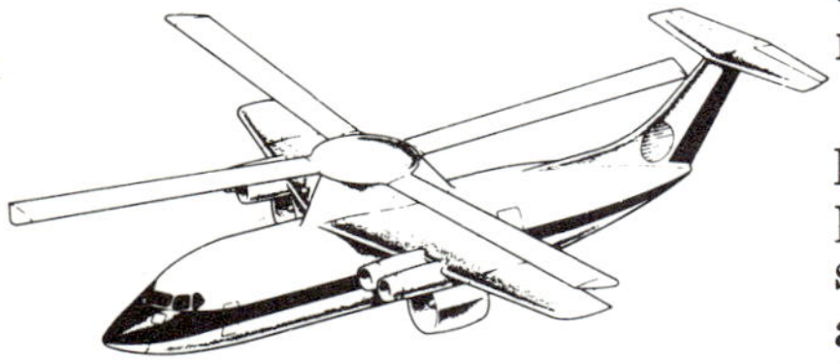

Compound Helicopter

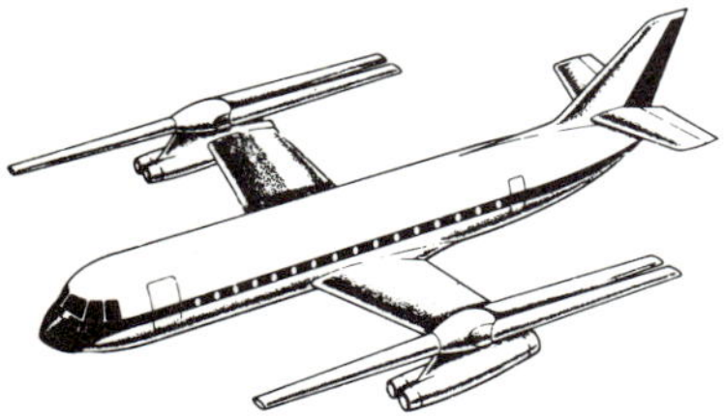

Controlled Circulation Rotor

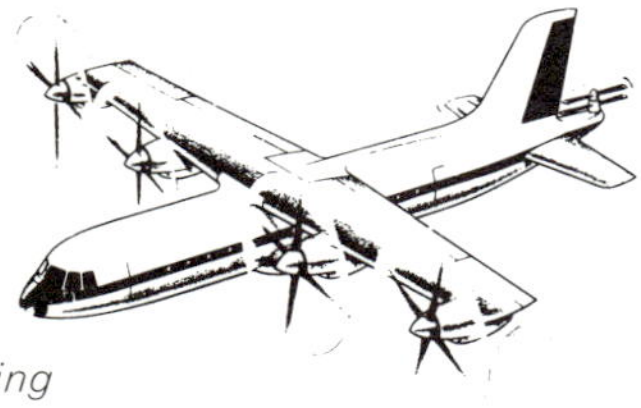

Tilt Wing

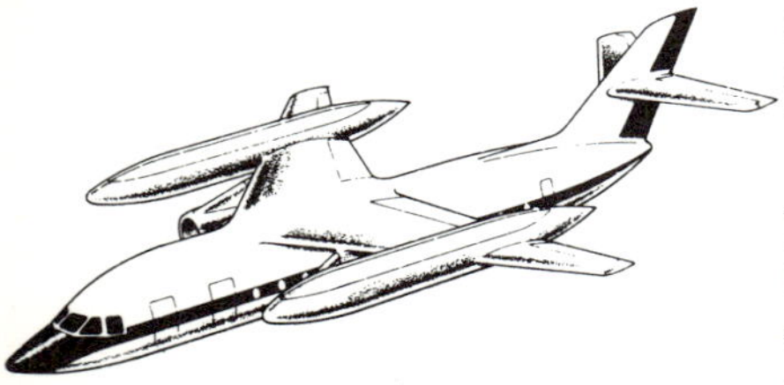

Fan Lift

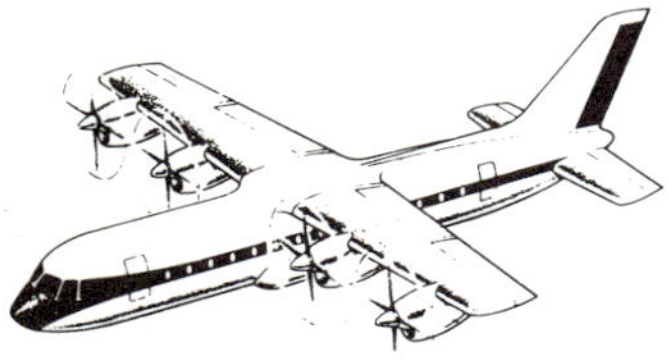

Deflected Slipstream STOL

Convertible Rotor

Yet the fight to get Concorde into airline service has probably been the most difficult political battle which has ever bedevilled the birth and development of a civil aircraft, and the battle is not yet won. Yet many of the doubts, fears, and downright condemnation which have built up around the Concorde project seem to be founded on misunderstandings of varying degree.

Even the basic design philosophy, though clearly expressed from the outset, has been misinterpreted. It is a philosophy of evolution rather than revolution. Supersonic flight poses new technical and operational problems, and the Concorde structure and systems have had to be designed to meet an unprecedentedly wide range of purely technical challenges–for instance, the temperatures in which it will operate. It has a number of interesting new features, including the 'droop snoot', a fuel-transfer system, which will maintain the fore and aft trim of the aircraft by pumping fuel to appropriate tanks, the swivel engine air intake and exhaust nozzles, designed to improve take-off and landing characteristics, and the inertial navigation system. Basically, however, the aircraft is of conventional construction and design. For instance, the choice of a speed 'limit' of just over twice the speed of sound made it possible to build the airframe almost entirely in aluminium alloy, a familiar and well-proven material, rather than the more expensive and comparatively unknown metals which would have been demanded by higher intended performance.

The size of the aircraft has also been subjected to unfavourable comment. The Concorde's comparatively modest capacity of up to 144 passengers should, however, be considered in relation to its supersonic cruising speed. The two combined give the Concorde a seat/mile productivity which is equal to that of a 250–300-seat subsonic airliner with, of course, the additional advantage of great operational flexibility, enabling it to fly profitably on a wide range of long-haul routes.

There has also been the question of Concorde's sonic boom. The fact is that the whole operational plan for Concorde is based on the assumption of a total ban on sonic booms over land masses throughout the world. Out of London and New York there is no problem because of their proximity to the sea, over which Concorde will climb and accelerate, or slow down and descend. After take-off from Paris Concorde will make her sonic boom off the Channel coast west of Brittany, and a similar track down the Channel will be followed by aircraft from Germany; the Scandinavian operators will presumably go north after take-off and therefore lose their booms over the North Sea. But it is over the Pacific that Concorde's flight time potential really assumes extraordinary advantages.

This of course leaves the problem of flying across the United States, and the American authorities have not yet stated their position on this issue. The lobby of the American aircraft industry is powerful enough, and sufficiently unscrupulous, to use this as a weapon against the operation of non-American aircraft.

Another fact about the Concorde, which is little known

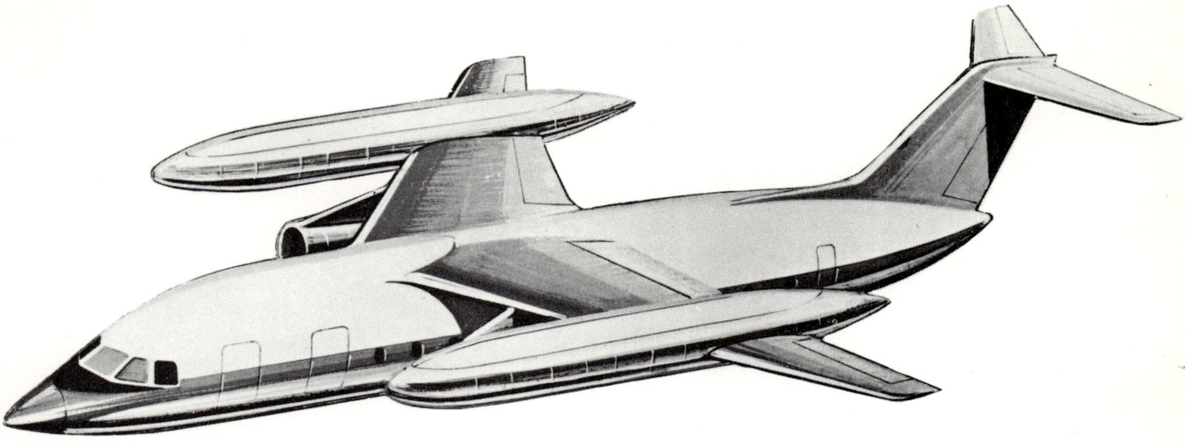

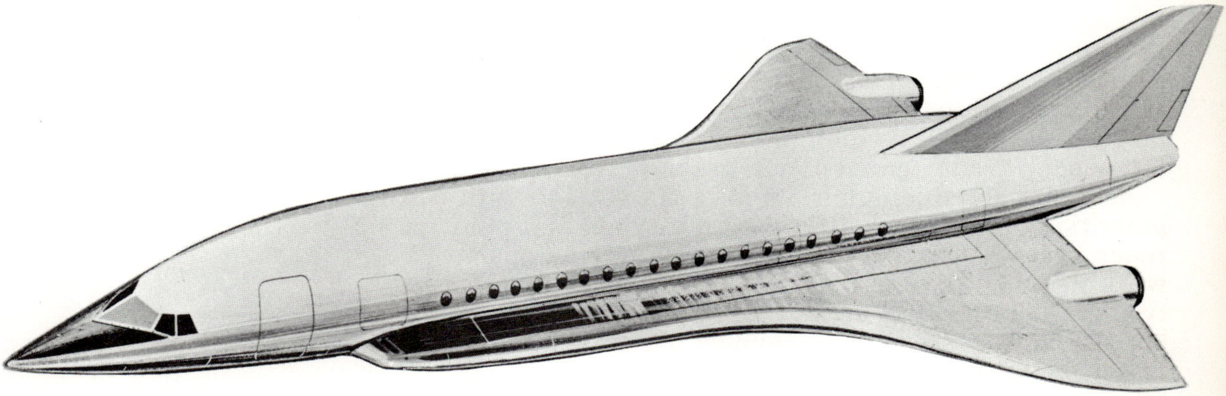

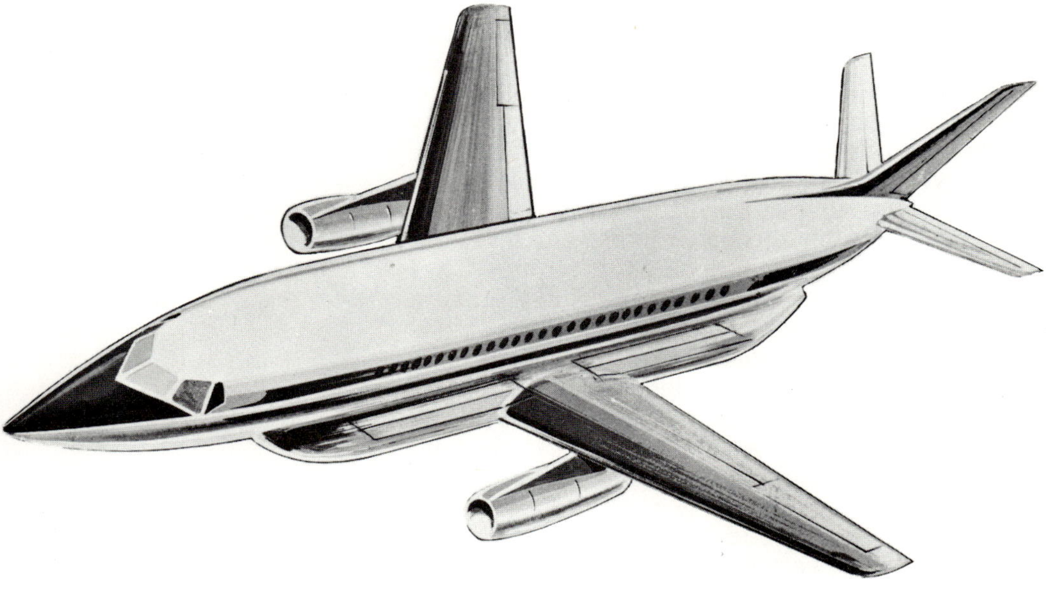

to the potential fare-paying public, is what is known as the 'uni-class system'. This marries Concorde to the Jumbo Jet in a manner which would appear to be extremely attractive from the operators' point of view, to say nothing of the fare-paying passenger. Under this system the Jumbo would offer subsonic passage across the Atlantic in economy-class conditions and at economy-class rates. For an excess, approximately equivalent to the present first-class fare, the passenger who needs to get to his destination in a hurry will travel by Concorde, in all the luxury currently associated with a first-class airline ticket. The commercial logic of such an arrangement would appear self-evident, and the commercial success of a uni-class system has been demonstrated by several of the world's leading shipping companies.

Meanwhile, the actual performance figures being acquired on the two prototype Concordes, 001 and 002, suggest that the theoretical figures evolved by the design team and their computers may have erred on the conservative side. In the past, whenever the airlines have been able to offer increased speed, air travellers have been eager to buy. The last major step in cruise speed came with the introduction into service in the late 1950s of the larger subsonic jets which helped to spark a substantial increase in traffic, particularly over the North Atlantic.

As Professor Buchanan noted in connection with the potential of the tracked hovercraft, when the point is reached at which a journey comes into the time-space of a working day, there is invariably a dramatic upsurge in traffic. Concorde will breach this 'time barrier' not only across the Atlantic but on a number of Pacific routes, and this area, with its long sector-distances and seaboard or island airfields, is tailor-made for Concorde operation. With a cruising speed of approximately 1400 mph, and a range of more than 4000 miles, including all present-day operation reserves, and with its capacity to operate into and out of any airport in the world that can accept present large-capacity subsonic jetliners, the potential impact of Concorde upon the travel habits of airline passengers all over the world is hard to overestimate.

There is good reason to believe that Concordes may still be in the service of the world's airlines right up to the end of the century. By then, no doubt, they will have been joined by larger, faster rivals who will skim the cream off the classic routes. In the past great aircraft have had long and profitable lives. There is no reason yet to suppose that Concorde is not in this category. BAC and Sud Aviation estimate the sale of 250 Concordes by 1975. On that basis alone, Britain's balance of payments would benefit over the aircraft's operating life by at least £2000 million. Calculations based not upon wishful thinking, but the hard realities of commercial aviation, suggest that by 1980 the number of Concordes sold, the world over, could be as many as 400.

For the European aerospace industry, struggling as it does in the face of the intense competition of the American giants, and governed largely by the swings of political fortune in the countries of its international partnerships, the future of Concorde is the most important thing that has happened since the Wright Brothers went to Kittihawk.

Call Confravision

The difference between a $3\frac{1}{2}$-hour crossing of the Atlantic and one taking 7 hours is of little significance to the tourist, who will almost certainly opt for the cheaper Jumbo Jet. But what makes the businessman have to cross the Atlantic in a hurry? Generally speaking, it is because he wants to talk to someone, or a group of people, in a face-to-face situation. A long-distance telephone call can achieve a great deal, but it has its limitations in terms of the exchange of information and points of view normally associated with a business conference. It is to provide all the advantages of face-to-face conversation that the British Post Office research engineers have developed their 'Confravision'.

Although television is accepted today as a universal medium of entertainment, its application in business and industry is still relatively restricted. Capital costs, technical know-how, only occasional need, rather than consistent requirement, have all limited the applications of closed-circuit television. As will be seen elsewhere in this book, where television has been applied in teaching, in medical work, undersea exploration, various security operations, traffic control and industrial processing, its function has been informative rather than communicative, that is to say it has in general been used as a one-way system to impart information, rather than establish a two-way link for the exchange of information.

It has been said that meetings are the curse of contemporary commerce. Therefore any saving in the time and travelling costs expended when senior executives spend hours in travelling to attend conferences with customers or colleagues or technical specialists or other business must appeal to any board of directors. A major purpose of contemporary business conferences is the launching of new products. Any sales director must be interested in a facility which would enable him to launch a new product nationwide without having to travel miles to do so. The product could be demonstrated to each regional sales force in turn, questions could be answered on the spot, sales charts and graphs could be shown and explained, and in short a major sales campaign could be mounted in a way never possible before.

All these advantages are the target areas of the feasibility study mounted under the name 'Confravision', to explore the potential co-operation of industry and commerce towards a service which could bring the convenience of the telephone call-box into the far more subtle and satisfactory realms of audio-visual communication.

Within the Post Office organisation two special television studios have been set up. They are not very far apart – one at Gresham Street in the City of London, the other at the Post Office research station near Wembley. But a fully developed network of such studios could link all the principal centres in the United Kingdom – centre talking to centre in a two-way exchange in sound and vision.

An important aspect of the design is that it has been

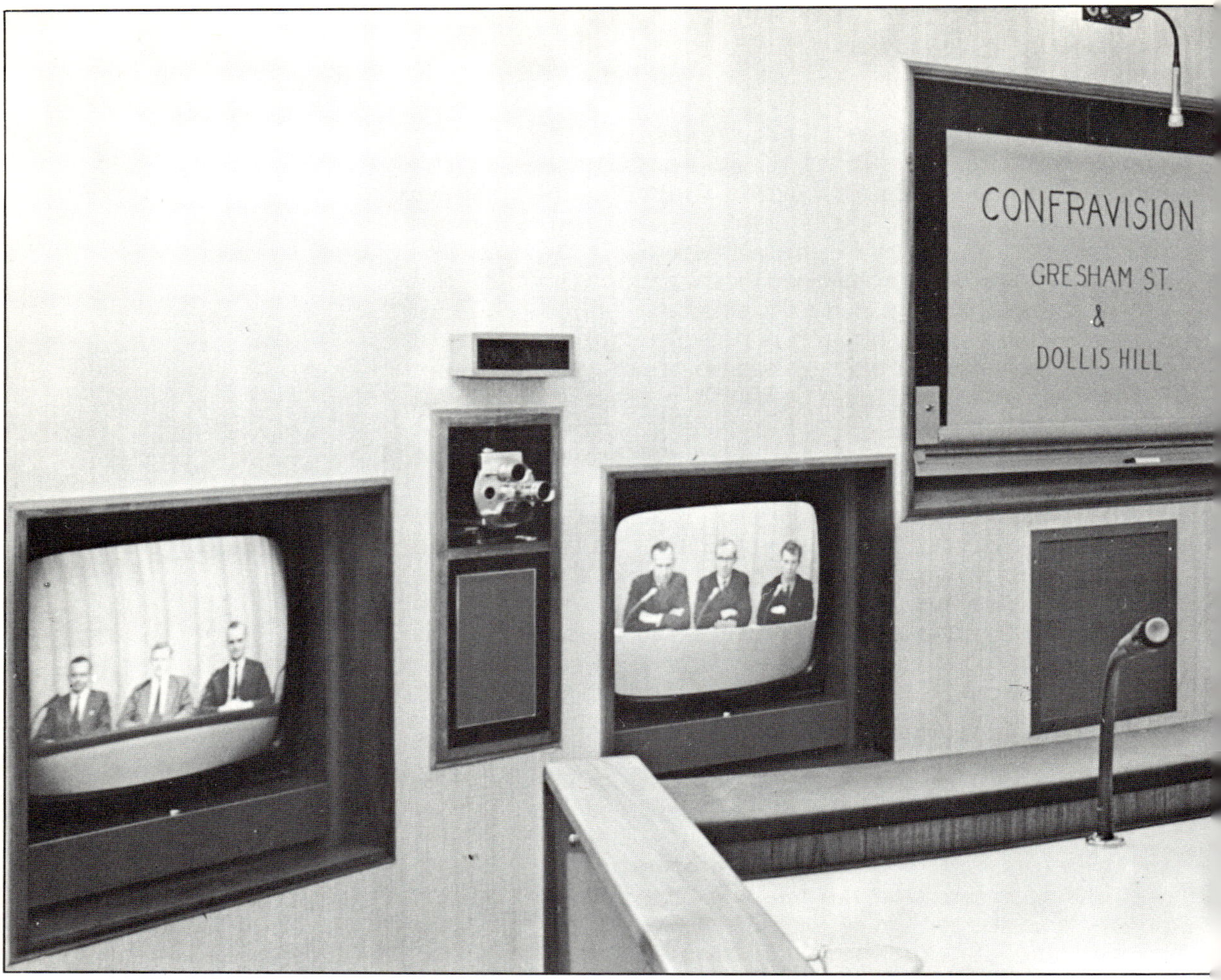

developed on 'do it yourself' principles. The businessman hiring the facility of a Confravision hook-up will not be faced with the additional complication of securing the services of technicians to operate the cameras, microphones, and all the attendant control panels associated with normal television transmission. Each studio has a conference camera which can be set for one, three or five participants. When set for one, for instance, the picture is suitable for demonstrating quite small objects. As yet, the cameras do not have a zoom ability, but this would not be difficult to develop. As it is, the high quality of picture ensures that gestures or changes of facial expression are easily observable at the receiving end.

In the *Tomorrow's World* studio we have demonstrated various techniques which enable simple graphics to be shown to the camera. A plain piece of perspex, if lit from the outside edges, can produce extremely vivid characters when written upon with an ordinary wax pencil. But that is perhaps the simplest of an increasingly expanding range of visual aids which could be brought into service by the Confravision user.

A tape recorder is connected into the system, and any part of both sides of a conference can be recorded, or prerecorded material can be played into the circuits for transmission to both groups. The equipment installed

Top Business conference by closed-circuit television at the British Post Office's Gresham Street centre in London. The picture on the right-hand television screen is coming from a similar room some miles away at Dollis Hill. Each speaker has a microphone in front of him and the television camera observing this half of the meeting is set into the wall between the two television monitors

Left Behind the scenes, the mini television camera (centre) which is the eyes of the other half of the meeting which could be on the other side of the world also provides for the transmission of documents which can be photocopied for distribution immediately upon receipt at either end.

A further advantage of this type of communication, where it is entirely probable that material of an extremely confidential nature may be exchanged, is that television communication is inherently secure. With Confravision scramblers can be incorporated into the sound circuit which will ensure absolute confidence in all proceedings between the studios.

Of course, there is nothing new about closed-circuit television. What is new about Confravision is the flexibility of this proposed service, which could be available when needed simply by booking time through the Post Office.

A very considerable amount of thought has been put into the more subtle problems involved, in order to make Confravision a truly attractive proposition to the businessman. The cameras, for instance, have been made as unobtrusive as possible. They are placed in the equipment room adjacent to the studio with only their lenses on view through wall-apertures. On either side of the lens of the main camera, two flush television screens are installed. One acts as a monitor for the studio output and shows that group. The other receives vision from the distant studio.

The Chairman uses a simple switch-box to control sound, vision and the tape recorder. Alternatively, the switch-box can be plugged into a special socket on the secretary's desk–a fully equipped secretarial console within the studio but 'off-screen', where minutes can be taken and the facsimile transmission and photocopying processes can be controlled. Within minutes–the time taken for an electronic scanner in one studio to 'read' and transmit the largest document–copies can be in the hands of the recipients.

An interesting psychological theory is that the use of equipment of this kind will impart a sense of urgency to the business in hand. It is thought that Confravision might streamline discussion and reduce irrelevancies.

Confravision is believed to be the first public service of its kind in the world. Tariffs have not yet been fully worked out, but it is clearly recognised that, if the service is to be financially successful, it must be commercially viable for the hirer. Eventual charges will obviously depend on the extent to which the service is employed, but a tentative costing suggests charges at a level of £120 per hour between London and Birmingham, rising to about £200 for somewhere as far as Manchester.

There is no technical reason whatsoever which would prevent the Confravision facility being used internationally, employing satellite communication links as might be necessary. The financial advantages would clearly increase with the distances spanned, with the savings in terms of travelling and hotel bills as well as the indirect costs of top management production time. But there is a newer and much more localised problem in business communication which results from the present-day tendency towards diversification and dispersal. Regions divorced from the accepted centres of industry are being developed. Confravision might well be a key bridge in communication between the old-established sectors and those springing up in new towns and even comparatively remote rural areas in which a particular company might choose to set up its own Confravision studio in its country-house headquarters.

Faster than Thought

Already, of course, the exchange of information by purely electronic processes constitutes one of the major technological innovations of our time. The introduction of the computer into the routine processes of banking is a classic case in point. Lloyds Bank, for instance, has 1454 branches and 781 sub-offices. All will be linked directly to one of their two computer centres by February 1971. Each branch has ten terminals, feeding information and requests to a 'concentrator'. The concentrator accepts requests and information from the terminals, checks that they are in the correct computer language, and condenses them by discarding superfluous comment or information. The concentrated material is then sent by telephone line to the main computer centre, where the enquiry is dealt with and any reply is sent back to the concentrator. Here it is expanded before being printed out at the appropriate terminal. In terms of commercial reality, it is likely that similar streamlined communications via computer, within businesses of all types, will be a major feature of commerce in the next decade.

Communication Hardware/In order to meet this potential requirement, the computer-manufacturers already have on the market an extremely impressive range of communication hardware. There is, for instance, the display console facility which visually displays graphs, tables, letters, figures and the lines and curves of drawings directly and automatically from the computer's store of information to the person requesting its display. The maximum character-display capacity is 52 lines with 74 characters per line, giving a total display of up to 3848 characters. In other words, the precision and accuracy of the display available is sufficient to meet the requirements of the most exacting scientific or technical standards.

Alongside the display console, a communications terminal will be installed. This works in precisely the same way as a normal electric typewriter, except that the characters typed are fed instantly and directly to the computer. The machine also reverses the process so that printed information can be received alongside and in synchronisation with graphic information shown on the display console. So versatile is the modern communications terminal that it can be used as an ordinary typewriter by the office secretary when she wants to transmit information on behalf of her boss in the old-fashioned way.

These two computer tools–the electric typewriter and the display console–can be harnessed in a single unit known as a 'visual display station'. This can be used in either direction; as an input medium, the information

A fibre optic plate used in night vision equipment is inspected for flaws. It is made of thousands of cladded glass fibres sealed together into a solid block. The multiple fibres collect maximum light and direct it along what are, in essence, parellel pipes, so ensuring maximum image reception, which is of critical importance in night vision equipment. Ordinary glass faceplates allow light to pass through, but they cannot control its direction

composed on the keyboard appears on the screen as it is being typed, and can be verified and edited before being entered into the system. In the reverse, information can be displayed on the screen in response to an enquiry, and the data can be added to, altered or edited before being returned to storage.

The use of a visual display station obviates the need to produce pages of printed output, and avoids the difficulties of supply, storage, handling and disposal of such paper records. Nevertheless, if a permanent visual record is required, a recording device can be incorporated in the system.

An immensely valuable tool available in connection with the visual presentation aspects of computer-communication is the light-pen. If the cathode ray tube or television screen is the blackboard of the computer, the light-pen is the chalk. When an optical image unit is incorporated in the communications system, touching a light-sensitive probe to the segment of the screen where pertinent information appears transmits that information directly back to the computer. This single step takes the place of many operations normally associated with computer-data entry. Because of its simplicity, all the operator needs is the ability to recognise illustrations or words and phrases or sentences of everyday language, and therefore no specialist training is required. Conventional filmstrip can even be used as the input source of information to be fed to the computer.

There is also on the market a computer plotter. It converts the statistical information stored by a computer into appropriate graphs as required by the enquirer.

With an audio-response unit, you can dial a number on an internal telephone and be put in touch with your computer in order to receive the spoken reply to your enquiry. The response unit sends the enquiry character by character to the central processing unit, the computer processes the data, composes a coded response and returns the message to the unit, which then interprets the coded reply, selecting the proper words from its stored vocabulary. What the response unit can 'say' is determined by your requirements. You can choose a vocabulary tailored to your business, in sets of 32, 48, 64, 80, 96, 112 or 128 words. And what a computer can say in 128 words has to be experienced to be believed.

But a very serious problem associated with the development of all these dreamlike communication facilities is that they all depend on the establishment of either a telephone or radio link between the ends of the communication chain. The resultant increase in demand for telephone lines alone is already causing serious overloading. The installation of a computer, if it is used to its optimum capability, can entirely disorganise hitherto adequate telephone communications in a rural area.

Fibre Optics/As a result, communication engineers are paying particular attention to the laser, amongst other alternatives. Laser light, like radio waves, can be used to transmit information at incredible speed. The laser beam

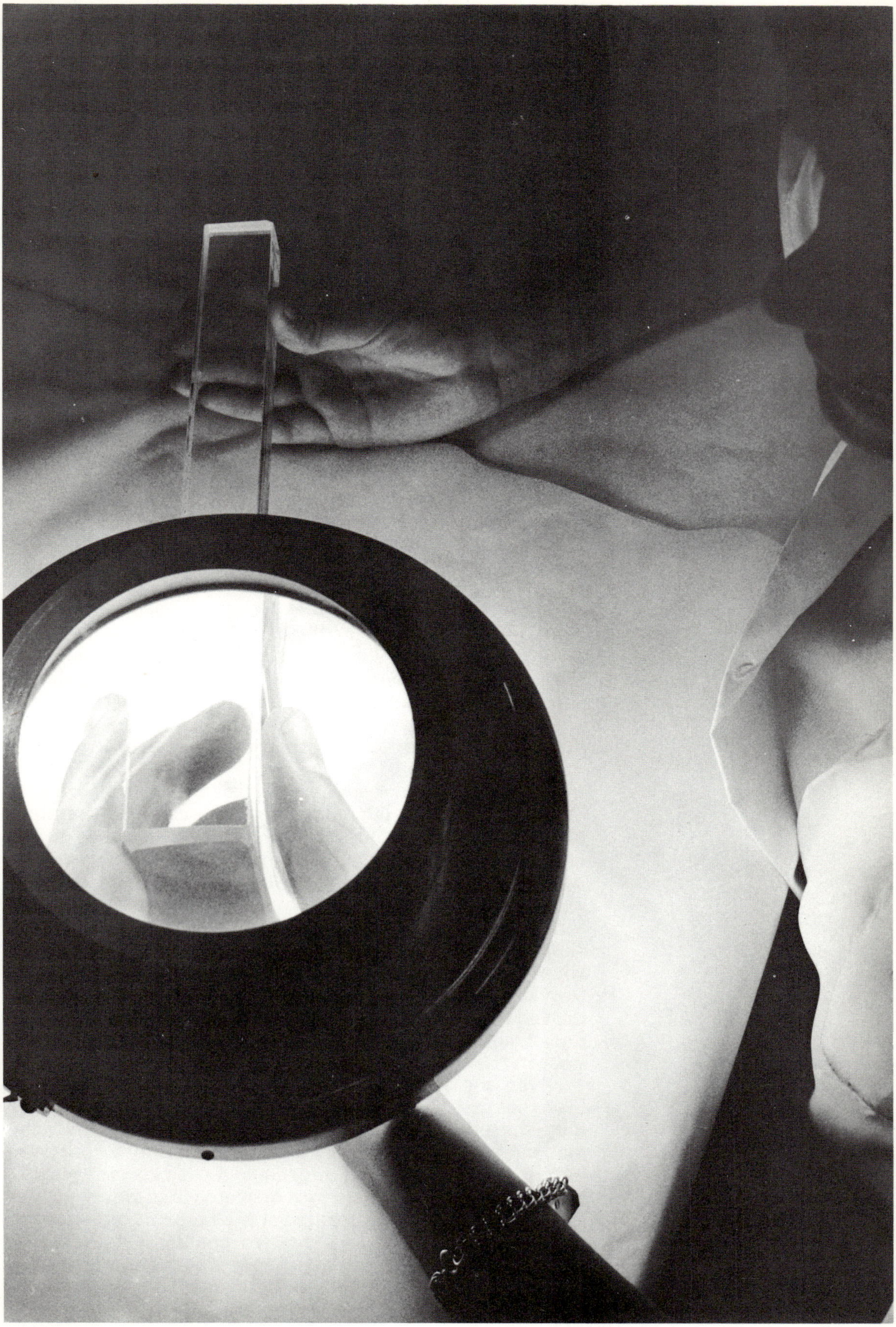

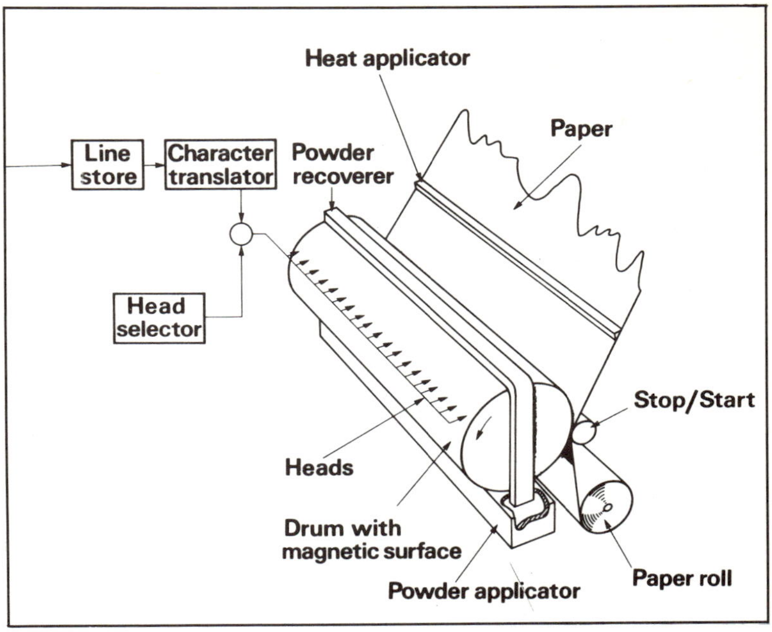

This apparently simple system is capable of printing at the incredible speed of more than 60,000 letters a second. A single line of recording heads (160 per inch) is applied to the drum which is plated with high remanence magnetic alloy. As it revolves, magnetic impulses create the characters using a matrix of dots. The surface of the drum then passes through a bed of minute resin-coated ferrite particles which stick to the magnetised areas. It then comes into contact with a moving roll of paper, the particles adhere, and moments later are permanently fixed when the resin is melted with radiant heat

can be aimed or conducted by means of confocal or gas lenses, by using a reflective pipe, or by di-electric wave guides, which all make it possible for a beam of light to be bent around corners.

Our studio demonstration of the technique of 'fibre optics' appeared to create a lot of interest in this new development. In essence, the technique consists of making a rope of glass fibres in such a way that each fibre occupies the same position in relation to its neighbours throughout the length of the structure. Each fibre then acts as a wave guide and the result is that an image induced into the rope at one end will be faithfully reproduced at the other.

We were able to demonstrate this remarkable quality by connecting the lens of a television camera to a fibre-optical rope, the other end of which was attached to the lens aperture. Effectively we put the camera's eye out on to a flexible stalk. In order to demonstrate that the 'organised light' transmitted into the rope by the lens was faithfully conducted to the camera, regardless of what shape the rope itself was bent into, a knot was tied in the rope. The image, which had been excellent throughout the experiment, remained undisturbed. Not only had we proved that fibre optics exploded the old myth about light only travelling in straight lines, we had literally tied a knot in the theory.

Given the ability to transmit the organised light of a laser beam, its possibilities as a communications carrier immediately become evident. To illustrate what the modern communications engineer means by 'ability to handle large amounts of information with great rapidity', a device called Ferrodot provides a case in point. It was developed to meet the requirement of a print-out speed of at least 60,000 letters or figures per second. Ferrodot surpasses that figure. It uses an electro-magnetic system and a resin-coated, nickel-zinc solution of ferrite particles as its ink. The particles are magnetically attracted to the dot pattern, and transferred to a moving roll of paper by light pressure contact. Radiant heat completes the bond. Unlimited numbers of copies can be made without re-recording, and any grade can be used. The same machine can plot graphs and statistical records in any desired form, and can also be used merely as an ultra-high-speed copying machine. It can also be made to display, rather than print out, its received messages. Unlike the type of display which uses a television screen, Ferrodot does not flicker and can therefore be viewed in any lighting conditions. Any number of displays can be run simultaneously, and their dimensions can be measured in feet rather than inches.

As you will have realised, machines such as these communicate or receive information at a far higher speed than the human brain. In fact, they are primarily designed to establish communication from one machine to another. However, attached to the 'face-to-face' conference capability of Confravision and similar systems there is the prospect of conferences being held between participating individuals, separated by enormous distances, and assisted by visual aids, relevant statistics and other records, on a scale hitherto undreamed of. The situation could be compared to a conversation between two experts, one seated in the library of the British Museum with all the information it contains instantly available, and another genius similarly placed in the Smithsonian Institution in Washington. The effective proximity during their conversation would be as if they were sitting opposite each other at a single table, with instant access to something getting very close to the sum total of human knowledge.

During the Second World War, one of the many slogans which were publicised to the extent of becoming national jokes was the simple query 'Is your journey really necessary?' The indications are that in 'Tomorrow's World' travel will be an experience undertaken for its own sake–for pleasure, for the enjoyment of new and wider horizons, for the truly face-to-face situation between individuals which no amount of electronic wizardry will ever replace. The tedium of the business letter, the sales conference, the technical consultation between experts will be replaced by conversations between machines, or by plugging into the information banks which are already being assembled all over the world. The traditional problems of time and distance will be virtually eradicated, and with a new totality of speed at his command, is it too much to hope that man may at last have time on his side?

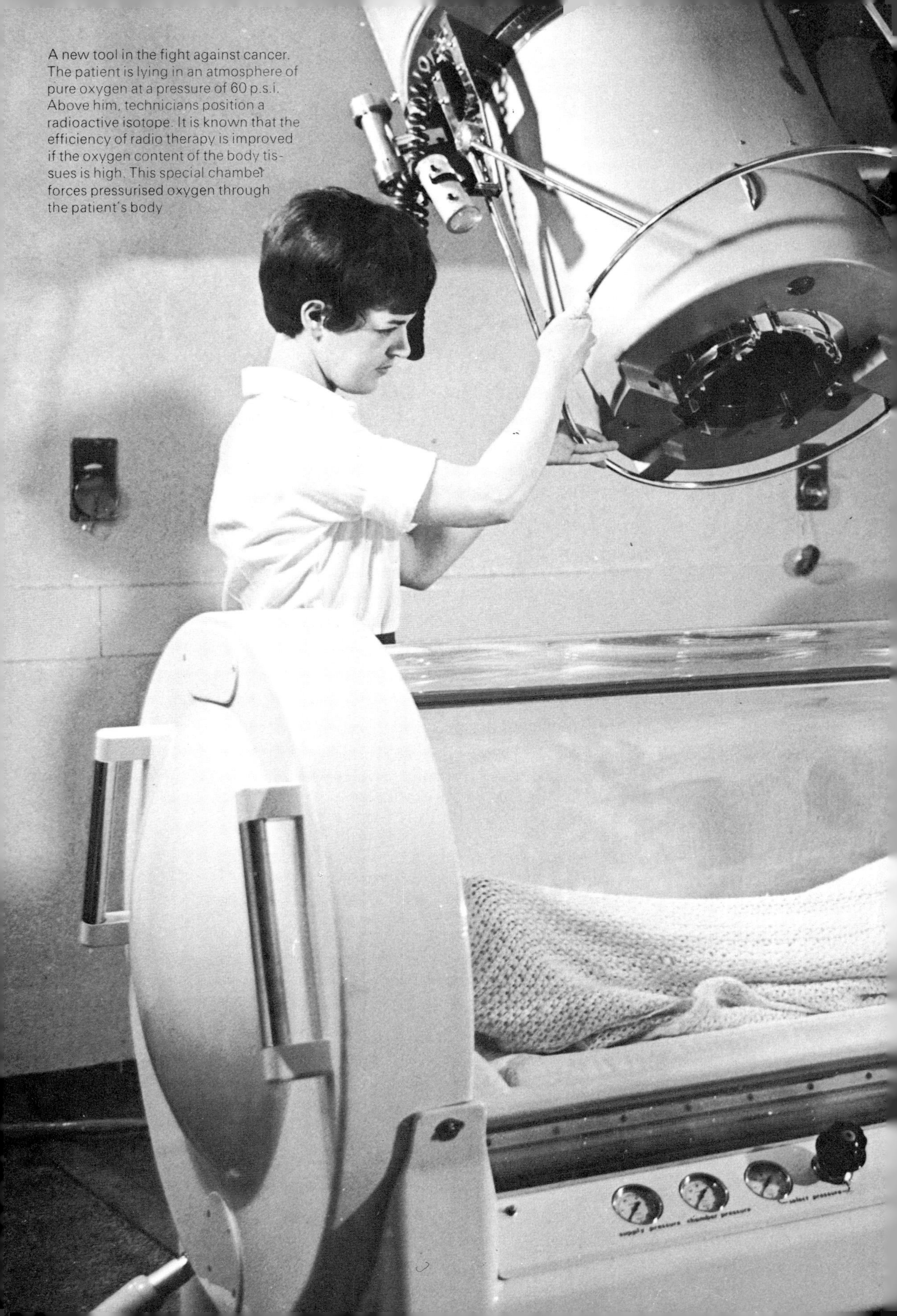

A new tool in the fight against cancer. The patient is lying in an atmosphere of pure oxygen at a pressure of 60 p.s.i. Above him, technicians position a radioactive isotope. It is known that the efficiency of radio therapy is improved if the oxygen content of the body tissues is high. This special chamber forces pressurised oxygen through the patient's body

Dr Tomorrow

The Cell Stripper / Multiphasic Health-screening / Autochemist / Mass Heart-screening / The Telltale Pill / The Bright Red Kiss / The Jet Injector / TV Doctor / Light-Knife Surgery / Soapy Blood / Oxygen Pup Tent / The Burn Camera and Skin Bank / The Swinging Patient / The Runabout Wheelchair / The Artificial Brain / Artificial Mouth / Sound Pictures / Artificial Hand / Artificial Arms / The Toe Thumb / Portable Paunch / Electric Babies / Baby Warmer / Germ-free Babies / Medical Crystal Ball

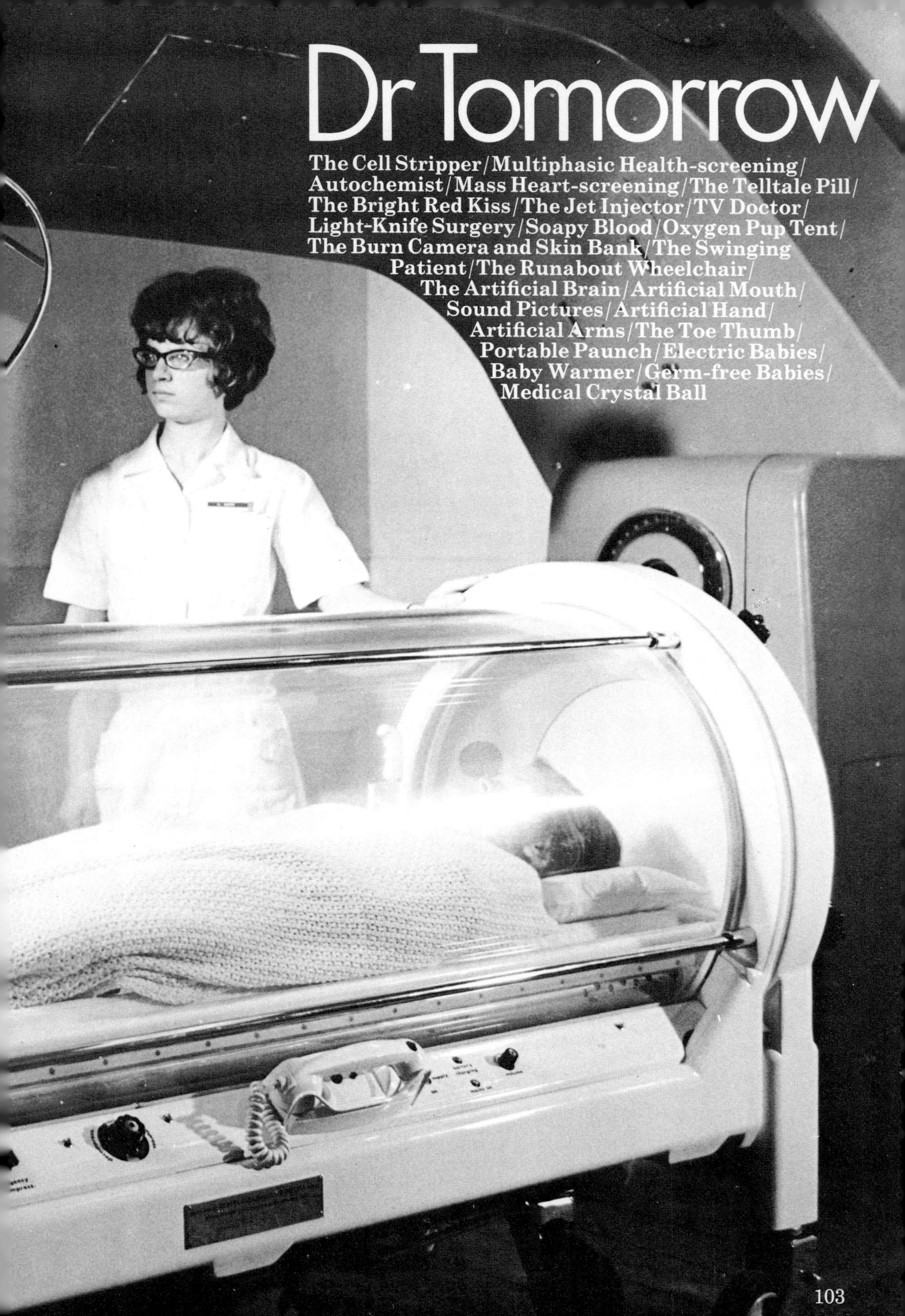

It is hardly likely that the science-fiction world where disease is wiped out will ever be more than a sci-fi dream. Even apart from those who suffer injury and need treatment for ailments that they bring upon themselves, the more we learn about curing the human body, the more diseases we are going to find to cure.

As scientific research reveals more and more about how the body works, technology produces, almost daily, new applications to comfort, to bolster, and even to replace bits of a human being that are attacked by disease or wear out through the process of ageing.

These 'mechanical aids' are only a half-way measure; and indications are that it will be a very long time before we can replace them with treatment that will alter the natural body processes themselves, and thus prevent breakdowns at basic cellular level before they begin to develop. In the meantime, advances in medicine over the next 20 or 30 years that are likely to affect the man in the street will be of the 'mechanical' type–either physical replacement of injured or diseased pieces, or the application of drugs which act directly on the general presence of a disease. *Tomorrow's World* looks at some of them.

The Cell Stripper

Very little is known about red blood cells–the tiny particles that colour our blood and are so vital to our health. A lack of these cells is called anaemia, and some forms of the disease still baffle doctors. The root of the problem lies in how little is known about the tiny cell itself–a concave disc about a thousandth of an inch across. Until very recently nothing was known about its internal structure–it was, as one haematologist put it, 'like looking at furniture with dust-covers over it'.

Now a new technique has revealed for the first time what a cell looks like inside. It is a method normally used to examine minerals under the tremendous magnification of an electron microscope. The cells are first 'fixed' to stabilise them–rather like tanning a skin to produce leather. The fixing effectively prevents the cells from breaking down under the microscope, which will bombard the specimen with ions. The cell is put in a vacuum, and then a gas, like hydrogen, is fed into the chamber holding the cell. Then power feeding to two electrodes alternates 15 million times a second, producing a rapidly changing positive and negative charge on the surface of the cells. When the surface is negative it pulls the positive ions in the surrounding hydrogen towards it and the ions, in effect, sandblast the surface of the cell, stripping it away atom by atom.

The strange fibrous structure just underneath the skin has yet to be explained. One startling fact has emerged, however: blood cells from people suffering from chronic anaemia were more resistant to the stripping, suggesting perhaps a change in the chemical composition of the skin. Deep within these cells researchers found a tangled and distorted web of filaments instead of the orderly structure of a healthy cell.

This is a first step in this new investigation of the cell structure–but already haematologists hope that the

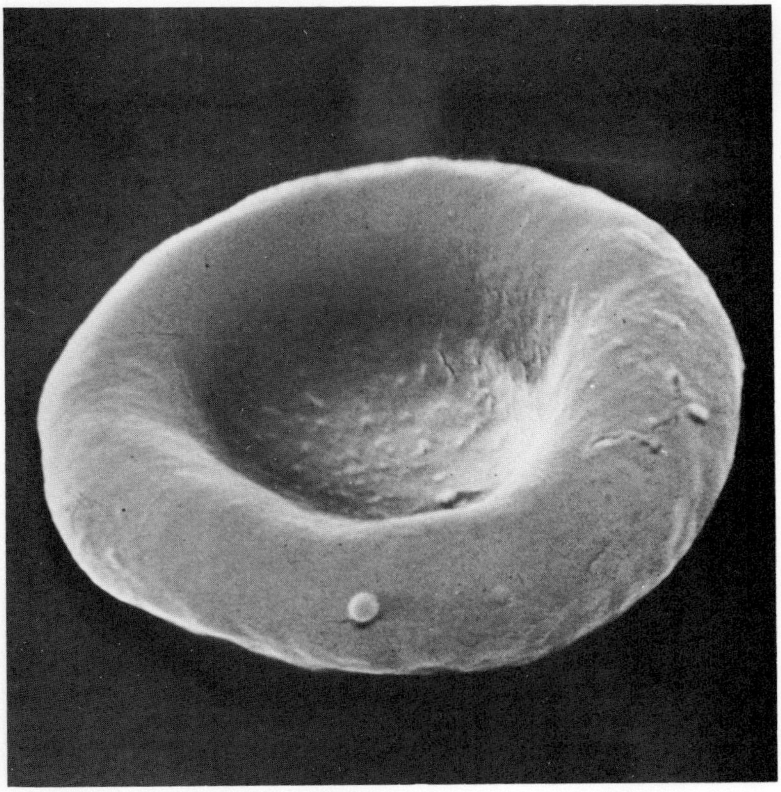

Blood cell—An electron microscope photograph of a healthy red blood cell taken from a live human being. Nothing of the interior structure can be seen (Magnification × 11,500)

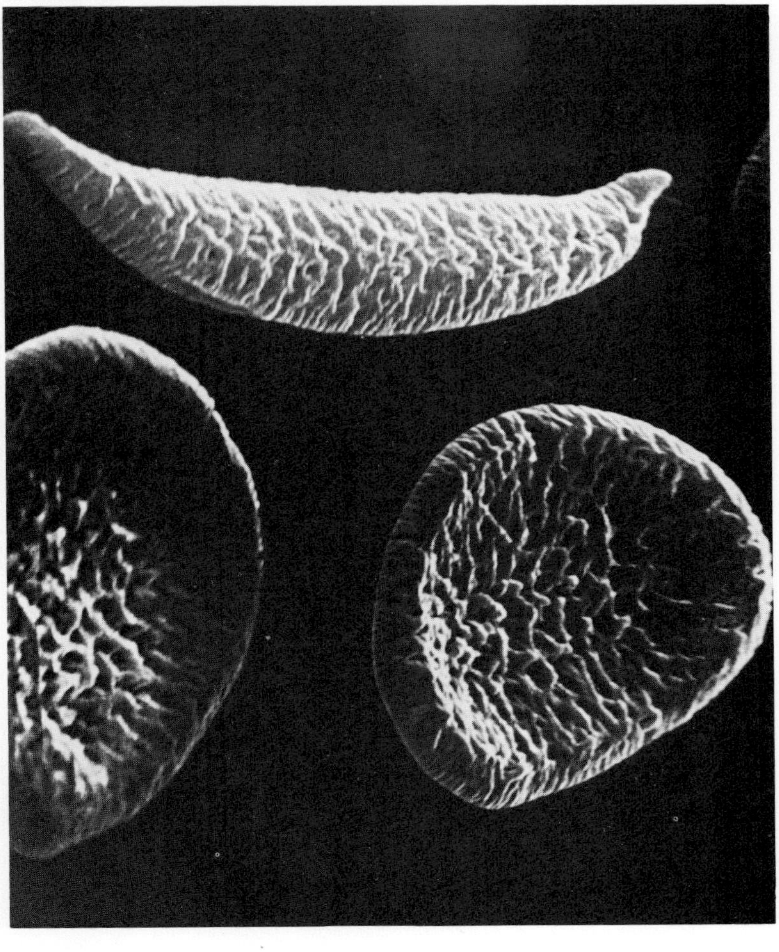

Stripped cells—One of the world's first pictures of red blood cells after being stripped by ion etching process. Positive hydrogen ions have bombarded the outer membrane, revealing the mysterious fibrous structure beneath. These cells are from a subject with sickle-cell anaemia. Researchers are now looking for common patterns or structures beneath the outer membrane of cells affected by the same disease

Computerised doctor—The computer presents a patient in a Multiphasic Health Screening Centre with one of the series of questions through which she can give a complete personal medical history for later comparison and analysis with her medical test results. On the console are buttons for answers: Yes/no/don't know/don't understand

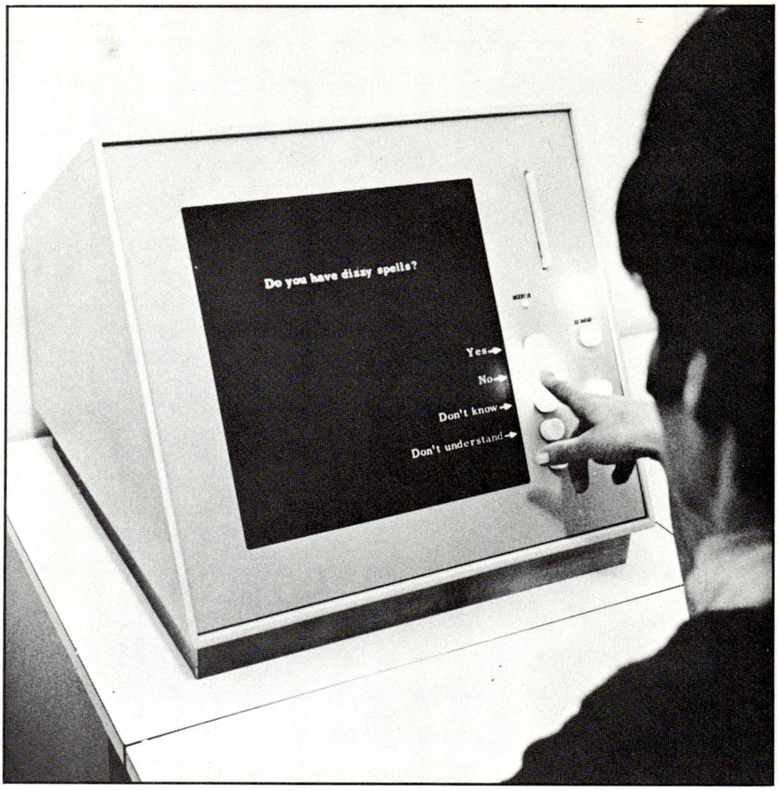

Computerised questions—Part of the latest automated medical examination. The patient answers 'branched' questions to give a medical history to the computer, pressing answer buttons (yes/no/don't know/ don't understand) that move the questions on along a path branched relative to the reply

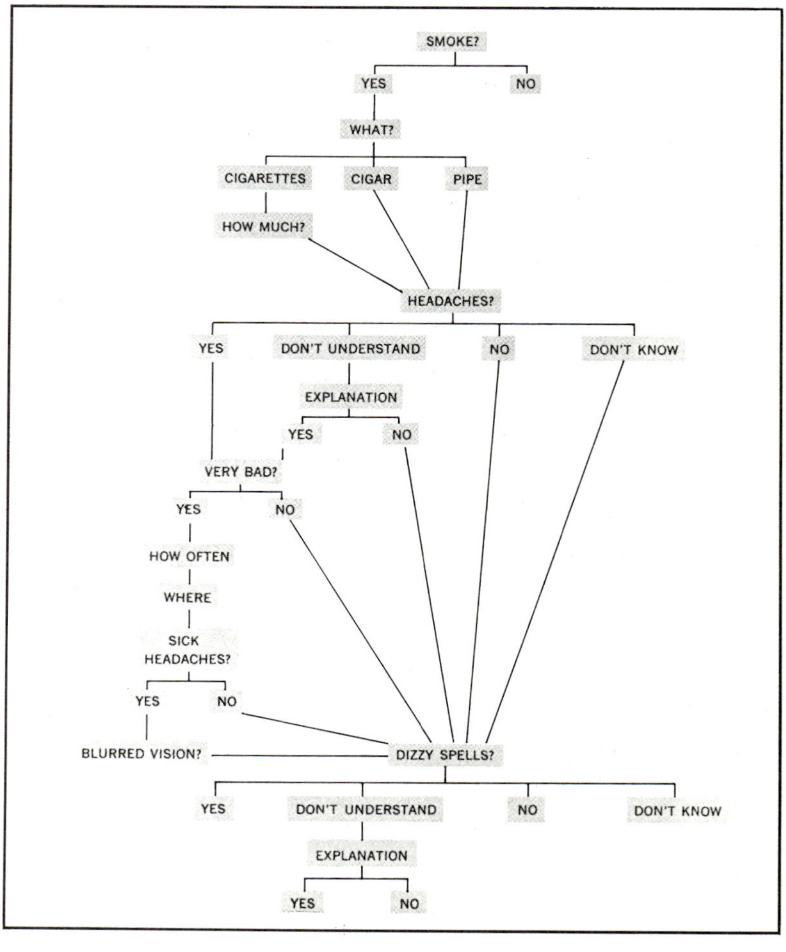

trail will lead them that bit nearer a cure to several blood diseases including anaemia.

Multiphasic Health-screening

Technology is developing ways of dealing with the increasing number of people who are ill, or who may be about to be ill. Health screening is an essential part of preventive medicine, and as we become more aware of the effects on us of our environment, we become more conscious of the fact that, although the conditions in which we live or work may not make us ill, they could be making us less healthy than we ought to be. Preventive medicine through screening aims at stopping a minor condition from developing into an acute illness.

There are centres in America and Britain which may be setting a trend in medical examinations. They do not require the presence of a doctor, for the patient is checked over by machine.

The patient begins at the Medical History Station. In front of him a TV screen flashes a series of questions. Buttons to the side of the screen match all the possible answers which are also on the screen, i.e.,

'Do you have dizzy spells?'
Yes.
No.
Don't know.
Don't understand.

Once the initial question has been answered the machine follows a 'branch' technique in taking the patient's medical history. For instance, if you reveal that you have had pneumonia, questioning checks on possible after-effects. If you did not have pneumonia these questions are by-passed. This system leaves patients to record, in total confidence, their entire personal medical history in about half an hour. The answers are automatically processed–and if indications show that there may be something wrong, later tests will try to diagnose it.

The patient moves on to a sight test. The latest optical test is a tabletop model looking like a circular drum with a viewing hood. It tests for visual acuity far and near, each and both eyes, depth and colour perception. At the same area hearing is tested with signals fed into each or both ears by headphones. When you hear the test tone you press a button which moves the tape recording on to the next tone. When you can no longer hear properly the test is over and along with the results of the eye examination the data is automatically transferred to a processer.

The patient moves on to the anthropometry section. Here a split scale measures not only weight, but weight on each foot and weight shift as the patient balances to stay upright. An arm moves down to the top of the head to measure height. A pincer device for measuring fat folds informs the computer of the ratio of weight to body fat.

A lung-function test is carried out by expelling air into a tube. This automatically checks the amount of air in the lungs and measures it against known norms for the patient's age, height, sex, etc.

During an electrocardiogram examination, in which the heart-beat analysis can either be recorded by the computer or printed out on a strip of paper for examination by a specialist, measurements are also made of blood pressure and pressure inside the eye. X-rays and, in the case of women, gynaecological tests are carried out as well. Blood samples are put into an autochemist which carries out a number of tests. The results, either printed out or fed into the computer, are closely scanned to find evidence connected with observations or symptoms noted at some earlier stage of the examination.

The advantages of this system are that it is quick, simple, and labour-saving. The automatic processing of the data gets you to the actual doctor faster and leaves him free to spend all his available time discussing your health rather than finding out about it. The one automated, computerised, health-testing centre so far in Britain has been crowded since the day it opened in 1970.

Autochemist

One of the tests carried out by the screening system, or by your own doctor if you visit him rather than the computerised screening unit, is a blood analysis. In most cases the doctor will send your blood to a hospital where the tests will be carried out.

Eighty per cent of all work in hospital laboratories is routine checking of blood and urine. It is a time- and money-consuming business, and quite often patients must spend days worrying until they hear the results.

The autochemist will cut that waiting time considerably, as well as reducing labour costs by a very large

Auto analysis—Blood samples move along for 'production line' analysis in an autochemist. One operator can run the system, capable of 24 different tests on 135 specimens in one hour. With this system you could have your blood-test results in hours instead of weeks. 80 per cent of all hospital laboratory work is taken up with blood and urine analysis

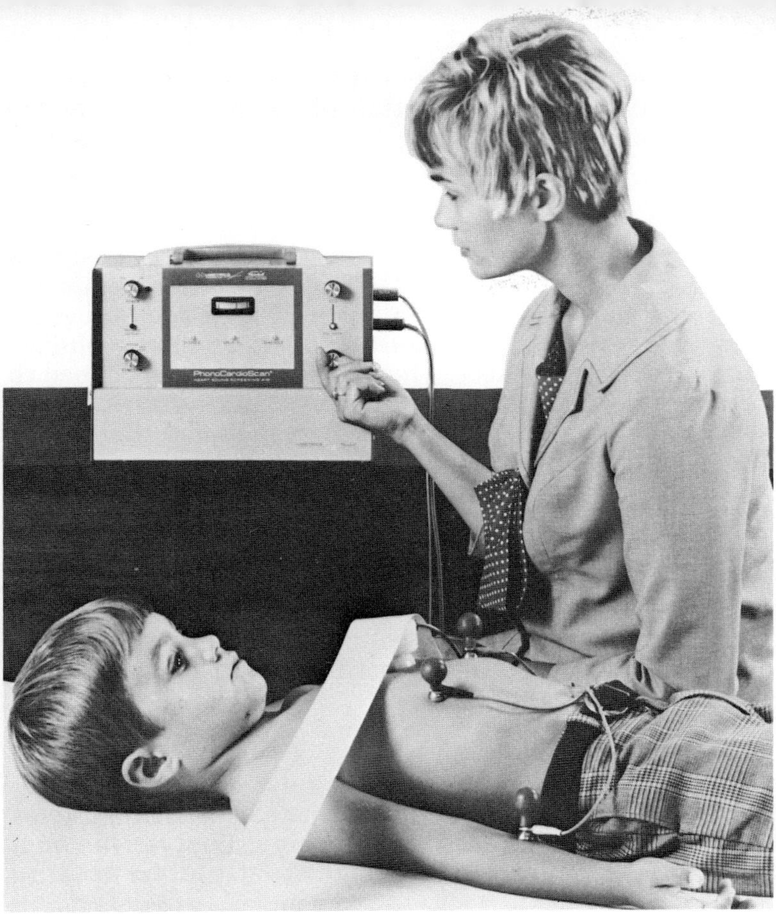

Phonocardioscan—A child in Los Angeles has his heart checked by a medical technician. The Electrocardiogram leads are attached to his wrists and chest while under the bandage a microphone listens to his heart. If the heart rate exceeds programmed normal limits the small computer will warn that the boy should visit a doctor

margin. The blood specimen is put in a centrifuge, and spun to separate the red corpuscles from the plasma in which they float. The plasma is then put into individual phials, marked according to the test required, and each one is tagged with magnetic tape. An electric eye scans the tape and stores the identity figures in a computer before the samples pass under a line of nozzles, each of which sucks out enough for one test. The mini-sample is mixed with a chemical that will turn it a certain colour according to the substance being looked for. Each colour change is registered on another electric eye and the result is fed to the computer. Autochemist can carry out 24 different tests on 135 separate specimens in one hour, and only one operation is needed to keep the machine loaded. At the end of it all, the computer prints out the results automatically.

Many doctors believe that one day automated laboratories like this could screen us all every year and give advance warning of treatment even before we ourselves knew we were going to be ill.

Mass Heart-screening

It may be that you have the kind of heart trouble that is serious enough for your doctor to keep a regular eye on you. In the near future this kind of check-up will be carried out not by your doctor, but by a technician or even by yourself.

One direct result of American space research has been the development of two new instruments that will permit

Children in an American school play Drug Decision, a game introduced as part of an anti-drug educational course. Three players take the part of town mayor, health officer, and Narcotics Bureau Chief. As the game progresses, the players have to remove Drug Pusher counters from the board or get them into clinics or prison. The game's designers hope this introduction of young children to the drug problem will become a regular part of school life. Its critics say it only serves to encourage them to try drugs they might otherwise not have known about

heart-screening on the same scale as mass X-ray screening. An ever-increasing proportion of the population of the United States is suffering from heart conditions and research has now shown that five children in every thousand have heart problems nobody suspected.

The Phonocardioscan is a transistorised computer about the size of a small suitcase, with circuitry designed to analyse each heart-beat in order to differentiate between normal and abnormal sounds. Electrocardiogram leads are fastened to the chest and leg, and microphones are placed on the chest. The computer is programmed to reveal abnormalities in the diastonic and systonic activities and to detect heart sounds shorter than 0.05 seconds. When it does, a light shows in which area the abnormality has been detected.

The Electrocardioanalyser does the same analysis, using the electrical impulses given off by the heart's action.

Both devices can be used by non-specialists, and it has been estimated that eight technicians could screen up to 50,000 people a year, and warn the specialists, in advance, of developing heart-cases.

The Telltale Pill

One day soon, should you, for instance, go to the doctor with persistent stomach trouble, instead of arranging for you to go to hospital for tests, he may simply give you a pill to swallow that will tell him instantly all he needs to know.

The pill will be a fully automatic internal investigator, containing a tiny power source–similar to a mercury battery–a receiver and a transmitter. Inside the capsule there will also be three minute aerials to transmit data as the pill travels through your tubes. Two of these pills have already been produced. One of them has a tiny diaphragm which responds to the slightest changes in pressure in the areas it travels through. As the diaphragm expands or contracts the movement activates pressure-sensitive switches which beam the data out through the transmitter. Temperature changes could be found by using a nickel-alloy material whose magnetic properties change with heat and cold.

A receiving aerial placed near you would pick up the signals and feed them to a computer. If the doctor wanted data from only one particular area, he could even transmit a signal to start the pill working some time after you had swallowed it. And by feeding the signals down a telephone line or over satellite, your tiny pill could be telling its tale to the best specialist in the world if need be. One day, with your box of pills in the medicine cupboard, a visit to the doctor might involve no more than swallowing one at regular intervals and 'broadcasting' your state of health to him from the comfort of your home.

The Bright Red Kiss

There are often reports of people being saved by the 'kiss of life' after accidents or when they have been pulled out of the sea unconscious. And there are just as often reports where the kiss of life failed to work.

Now a tiny, bright red metal container the size of a box of chocolates has been created by a British firm to administer the 'kiss' to patients who have stopped breathing. It is so simple that it can be easily operated by anyone without risk of mistakes. Resuscitation by the 'kiss of life', while effective, is distasteful to many people and it is difficult to keep it up over long periods. The automatic box will carry on indefinitely–and there need never be physical contact with the patient's mouth. There is only one control on it, a small black knob which is set to correspond to the type of patient under treatment–child, adult, or large adult. One end of the box is connected to a supply of oxygen or compressed air and the other is fixed to a mask which goes over the patient's nose and mouth. The life-giving air passes through a valve and is divided into two channels: the main flow goes on to a relay valve, the residue to a pneumatic timing circuit. From the timing circuit a signal passes to the relay valve and can go no further; at the same time air goes to the pneumatic timing unit, and pressure builds up to a critical level until a pulse of the air travels from the timing unit to the relay valve. The valve then opens and the air is pushed into the patient's lungs at a pressure of 55 lb. per square inch. At the end of this cycle another valve opens and the air in the patient's lungs pours out again.

The 'kiss of life' machine with its small cylinder of oxygen or compressed air is designed to be carried around by general medical practitioners. It could also be installed in factories, swimming-baths and other places where it might be needed. It is totally waterproof and works even when submerged. And its entire motive power is provided by the compressed air that the patient will breathe.

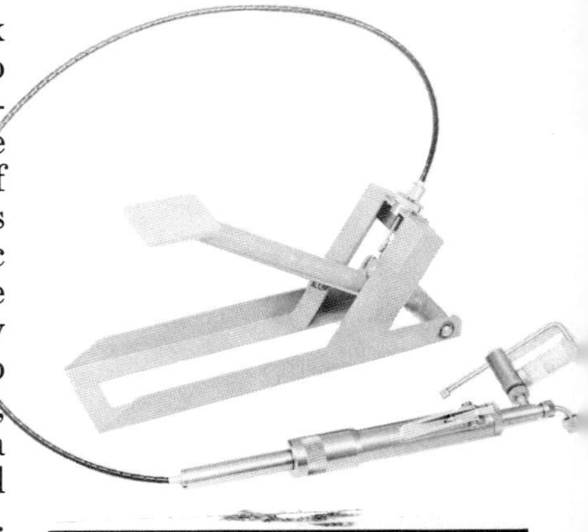

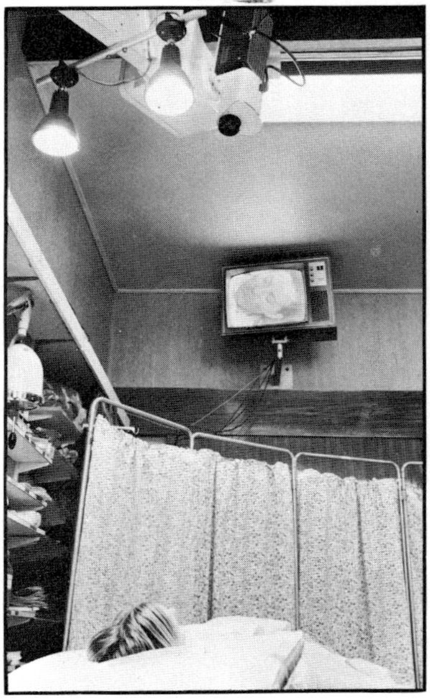

Top No-pain injection—The jet injector ready for use—the foot-pedal activates a plunger in the handle of the injector. This in turn draws the drug in from the capsule supply unit on the head of the injector, atomises it, and streams it out in a fine, high-speed spray that penetrates the skin

Above Monitor surgery—A patient is prepared for surgery by remote control, the TV picture being transmitted to a surgeon several miles away. He can make a minutely detailed examination with the remote-controlled camera above the bed and give the necessary pre-operation orders to the hospital staff

The Jet Injector

Having an injection can be a frightening, even disturbing, experience. The idea of a hollow needle puncturing the skin to convey quantities of fluid inside the body has been known to make strong men faint. But the days of the needle are numbered.

The National Research Development Corporation has now produced a jet injector which can put between 0.1 and 1.3 millilitres of drug under your skin, and you feel nothing. By pressing a foot-pedal, a plunger in the handle of the injector is activated, and compressed air shoots the drug, which is held in a capsule, through the injector nozzle as a high-velocity spray which penetrates the skin. Other nozzles compress the jet to shoot it with greater force for intra-muscular injections. There is no danger of injecting into the bloodstream as the jet is deflected by the elastic walls of blood vessels. It is safe, effective, and painless. And the jet injector can give its pneumatic 'jab' to 800 people an hour.

TV Doctor

In an emergency one of the factors that can determine life or death is the speed with which treatment begins. Often, in accident cases, the examination in the casualty ward indicates that highly specialised surgery is necessary as quickly as possible. But the specialist may be some distance away and crucial time is lost.

Now a new communication development could mean that one day in an emergency the specialist could begin treatment and surgery long before he arrives on the scene. By means of closed-circuit television he will be able to see the patient in the casualty department.

The key to this new system is a television camera on a movable boom above the patient's trolley. It can pan from side to side, tilt up and down, and most important of all, zoom in on the smallest detail.

The casualty doctor on duty can alert the specialist, who may be several miles away. The specialist switches on his monitor set and directs the examination by calling for the necessary close-ups of injuries. When he arrives at a diagnosis he then gives the necessary instructions for preliminary care, and preparation for any operation he thinks necessary. The zoom lens can provide pictures of details as small as the pupil of an eye, so he may even choose to get the patient into the operating theatre and, with the use of the zoom, tell the surgeon on the spot where the first incisions should be made. X-rays, heart-rate, brain activity, and other vital information, can be shown to him. If he is near enough to go to the hospital, by the time he arrives the operation could well have reached the stage where his particular skills are needed.

Eventually, through the use of satellites, an emergency case could benefit from the advice of the world's leading specialists, without the time and expense of bringing them to the bedside or operating table.

Light-Knife Surgery

The surgeon's oldest tool–the razor-sharp scalpel–may one day become a museum piece. For the knife could be replaced by the laser if laser-power can be harnessed effectively. The possible application of continuous wave-gas lasers in surgery could mean that it would be possible to cut through human tissue with very little bleeding. Already an argon laser, producing power in the order of watts, has been used to make continuous incisions in the liver, where bleeding would normally obscure the operative field. As the researchers hoped, it was discovered that the focused beam from the argon laser sealed off blood vessels as they were incised and the result was very much less bleeding. Other experiments have been carried out to use lasers to excise cancerous growths. The potential advantage here is that, because there is no physical contact with the tumour or the healthy tissue around it, the use of a light knife reduces the risk of spreading cancer cells during the operation.

Lasers may also be used to cut through bone and some surgeons believe that a laser knife could be of tremendous advantage in chest and heart surgery because it does not interfere with the delicate electrical activity in the heart as other methods tend to do.

Soapy Blood

While in hospital, during or after the operation, you may well need a blood transfusion, but there is a good chance

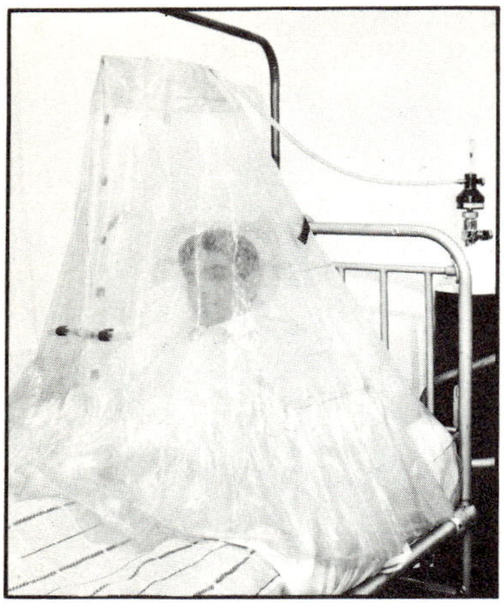

Life-saving tent–The new conical oxygen tent. It is specially designed to give downward oxygen flow over the face before it leaves the tent at the overlapping fringes resting on the bedclothes in front of the patient. Magnetic fasteners down the front entry flap keep oxygen loss to a minimum

that in a few years you will not be given human blood. A milky liquid invented by scientists at the University of Cincinnati Medical Centre may become the very lifeblood of human beings, because researchers have created an artificial blood made from inert, organic oxygen solvents. It has been used to keep dogs alive, and during experiments on animals up to 50 per cent of the total blood-volume has been replaced with the new artificial substance without any apparent complications, because the artificial blood appears to be replaced slowly by normal blood after the transfusion. The synthetic cells are composed of fluro-carbon liquids and make up 30–40 per cent of the whitish blood substitute. These oily fluro-carbon liquids are able to dissolve tremendous amounts of oxygen, about twenty times as much as water and even more than normal blood plasma.

The fluro-carbon is mixed with salt water to which traces of other essential ingredients, such as glucose, are added. A detergent or surface-active agent is added as the mixture is subjected to ultrasonic waves to convert it into a milky emulsion containing millions of tiny particles of fluro-carbon, each about one-tenth the size of human red blood cells. Researchers had to screen over 100 different detergents to find one that did not destroy normal blood cells. But before their artificial blood is used on humans a great deal more work still has to be done, though there is reason to believe that under certain circumstances artificial red blood cells may be even better than normal blood. Research has already shown, for example, that oxygen tensions in the brain as high as those obtained in a hyperbaric chamber under several atmospheres of pressure can be achieved with artificial blood in an animal breathing pure oxygen. Because of the small size of the synthetic cells the scientists believe that they will get through more easily than normal blood to partially clogged or restricted capillaries in oxygen-starved tissue. Other advantages are the indestructibility and relatively inexpensive cost of the artificial blood cells when manufactured in quantity. Estimates suggest that a pint of artificial blood would be a quarter the price of a pint of human whole blood. There is also one other big advantage–the lack of risk from blood-borne infections such as hepatitis and veneral disease.

If the synthetic blood is perfected for human use it could be used widely: there is no need for typing and cross-matching, since no reaction occurs to the inert substances from which it is made. But the main questions yet to be answered are what effect the artificial blood has on the human body and vice versa. Work goes on to find out.

Oxygen Pup Tent

Oxygen tents–which in most cases merely provide an oxygen environment to be breathed in through the patient's mouth and nose–are at present bigger than they need be. The result is that the oxygen is not being used efficiently, since under normal working conditions it is difficult to ensure that none escapes.

A new mini-tent has now been produced that should solve these problems. Conical in shape and made of transparent PVC, it hangs down over the bed and tucks in below the mattress at the back. In front, multi-layer fringes conform to the irregular contours of a patient and his bedclothes. These fringes are the only place where oxygen-escape can occur, since a vertical opening down the front, for use during treatment, is kept closed by magnetic 'buttons'.

The gas inlet at the top ensures that any oxygen entering the tent must pass the patient's face before reaching the only outlet possible, the first fringe. The great advantage of the new tent is that it uses the oxygen very efficiently, and above all it takes only 5 minutes in an emergency for effectively high oxygen concentrations to be built up.

The Burn Camera and Skin-Bank

Every week about 2000 people burn themselves so badly that they need hospital treatment. Ahead of them can be weeks of painful waiting while their bodies struggle to replace damaged tissue. And until recently it took that time to find out which were the areas so badly burned that a skin-graft was needed.

At East Grinstead a new technique has been developed that will cut out the waiting period and increase the chances of recovery. When patients arrive an infra-red camera is used to find out immediately which areas of damaged flesh are cold, and therefore dead and beyond recovery. A crystal in the heart of the camera is kept at −200°C in a bath of liquid nitrogen. This crystal turns the heat emanating from the skin into electrical charges that can be converted into pictures on a TV screen which show the temperature pattern of the area of skin under examination. The hotter the burnt area, the better the chance of recovery.

Once the graft has been decided on, the damaged layer has to be removed and replaced with new skin. Until recently the body rejected the grafted skin after a few days and it had to be replaced again, but now skin can be tissue-typed according to 16 separate factors, and in most cases will match so closely with the patient's own skin that rejection does not occur for up to 90 days.

The skin is kept in the hospital's skin-bank, frozen solid. The new method of freezing in an 'environmental freezing device' works so well that the process may one day be used to bank larger organs for transplants. It is a delicate business. The skin, removed from corpses, is mounted on a stainless-steel frame. Liquid nitrogen (at −196°C) is pumped round it to lower the temperature by 1 degree a minute. Then for almost an hour the rate is increased to 4 degrees a minute. Towards the end of the 70 minutes that the process takes, the temperature is dropping 10 degrees a minute. The rock-solid frozen skin is then ready to go into the bank vault where it will last for several years. When it is needed, warm water brings it back to life ready for the graft.

The Swinging Patient

One day soon disabled patients may literally get taken for a ride. A new device will enable the tiniest nurse to

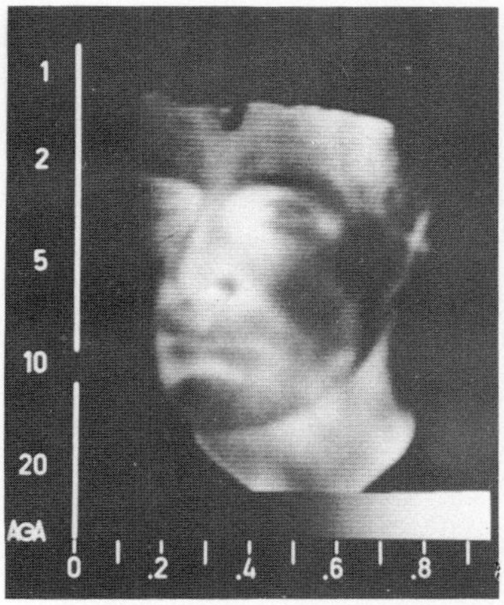

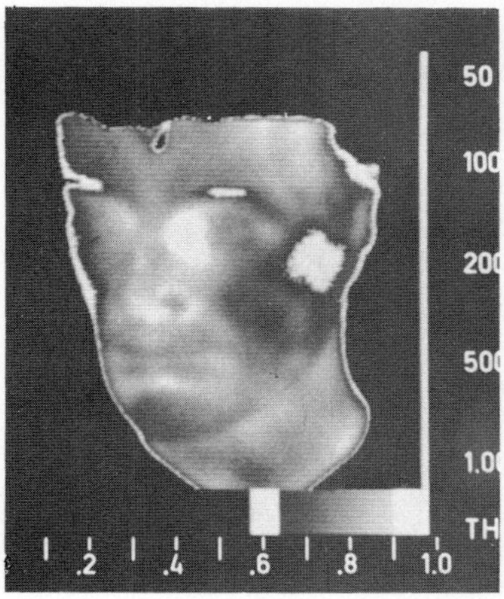

Heat pictures—Thermal photographs of a healthy face. *Top* The light areas of the face show warm sectors, dark areas are cooler. *Above* Shows isothermic mapping according to specific temperature. Eyebrows and part of cheek and nose show 6°C, inner part of eye shows $9\frac{1}{2}$°C

move the biggest patient anywhere without having to call in orderlies or trolleys.

Ambulift carries the patient in a dual sling–one part supporting the back, the other under the hips and thighs. There is also a stretcher attachment for hoisting people who, for reasons of treatment, have to stay supine. The sling fits on to a holding arm which can then be winched up to a horizontal position where it locks into place, with the patient swinging safely in mid-air. A system of chain-drives in the vertical mast allows the nurse to wind the patient up by rotating the lifting handle with one finger. Castors on the base permit the swinging patient to be taken anywhere and to be lowered gently on to another bed, or even into the bath.

Electrodrive—A new wheelchair-cum-runabout. Battery-powered, with a range of 20 miles at up to 10 mph, the chair telescopes open for entry and closes round the driver. It can be driven indoors as well as on the road—as long as you have a driving licence

The Runabout Wheelchair

A new Swiss idea may solve the problem of getting about in a wheelchair outside one's own home. The Electrodrive combines the wheelchair and the motorcar. It is driven by 26-volt batteries powering a special traction compound motor; five speeds are obtainable from $1\frac{1}{2}$ to 10 mph, forward and reverse. Drum brakes are fitted to the front wheels; drive and steering are through the single rear wheel.

There are adjustable front and back seats, lights and a horn. A telescopic seat moves outwards for side-entry and head- and leg-rests are also adjustable. There are only two main controls (operated by either one or two hands) for steering and movement and an extra handbrake on the side cuts off the electric power when applied, so it is impossible to drive away with the handbrake on.

The Electrodrive is powerful enough to drive up on to a kerb, and its suspension is designed to absorb the impact. It has a turning circle of $33\frac{1}{2}$ inches, and its range on the open road is 20 miles. All you need is a driving licence and you can drive from your bedroom straight to the shops–assuming you are sleeping on the ground floor.

The Artificial Brain

A great deal of research is going into replacements for injured, diseased or worn-out parts of the body. One of the most exciting projects is investigating what part the brain plays in controlling the state of the body, in the hope that one day doctors will be able to send 'messages' through the nervous system to cause necessary changes in the body's behaviour. For this reason, the United States Navy is studying the possibility of a computer with the ability to sense its surroundings and respond intelligently to spoken commands. If development work is successful the robot computer could be operated by anyone without the need for special technical training.

The system is not the conventional digital computer, which basically adds up numbers at great speed in a sequential process, but rather utilises a process based on the workings of the human brain. This means that, just like the human brain, the robot could perform several different procedures simultaneously. The design of the computer is based on the living nervous system, whose building-blocks are neurons or nerve-cells. The machine's

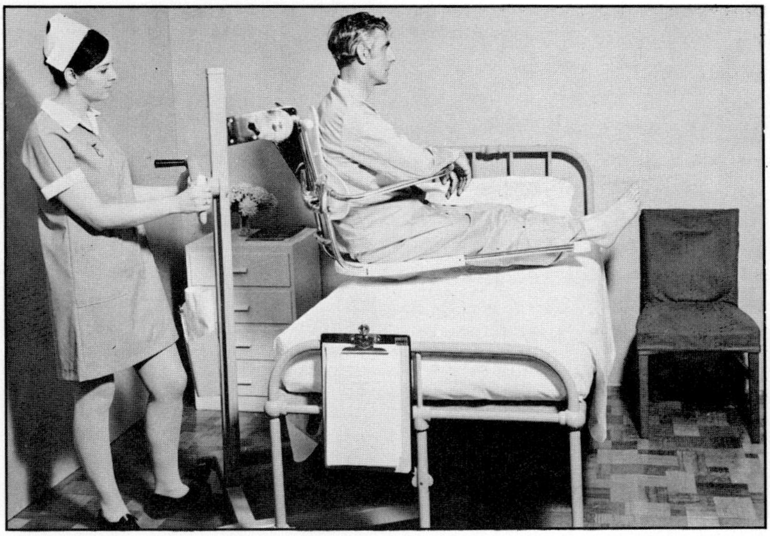

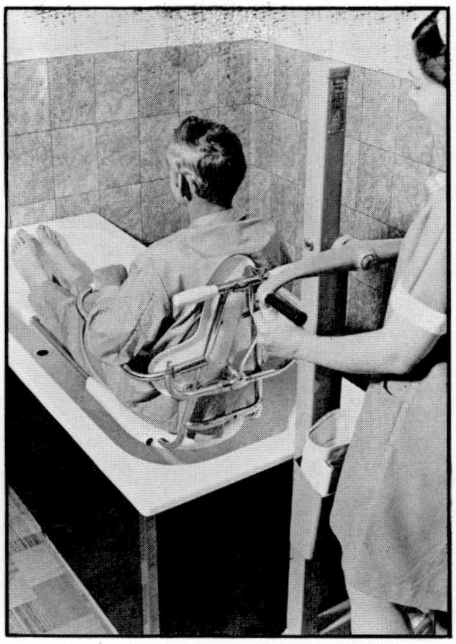

Swinging patient—A partly paralysed or weakened patient swings in a new lifting device designed to move hospitalised or bedridden people with minimum effort. The winching handle can, through a gearing system in the mast, be moved with a fingertip

artificial neurons—'neuromimes'—are electronic circuits that can to some extent mimic the information processes of neurons, and when wired together into a complicated network form the rudimentary brain of the computer robot, which the scientists call an 'operant conditionable' machine. This means that the machine can be trained by modifying its responses through 'punishment' or 'reward' signals. When a message is given to the network some of the neuromimes fire in response. If the response is incorrect, or one not desired by the trainer, he can 'punish' the neuromimes that send the unwanted responses by suppressing or preventing them from firing. This 'punishment' has the effect of persuading the network to try to find a new way to solve the problem. At the same time the group of neuromimes which react correctly are 'rewarded' by the trainer who strengthens their signals. This process is repeated over and over again, like an animal trainer working in a circus, until the robot produces the correct answer or response every time. Once trained to perform various tasks the machine would execute them faithfully when given the appropriate command.

The US Navy has high hopes for their latest recruit because aboard ship it could be used to help in a variety of tasks from navigating to preparing the meals. The scientists at Santa Monica, California, who are developing the robot computer hope that by the time their work is finished it will not only perform in response to voice commands but will also answer back with its own electronic vocal cords. They are planning to create a brain consisting of 3000 neuromimes that could shed light on the workings of the human brain as well as earning its keep on board ship.

Artificial Mouth

Dentists have known for years that saliva contains millions of bacteria which grow in a creamy film over teeth. Even apparently healthy teeth are victims of this harmful layer which dentists call 'plaque'. It takes the sugar from the food you eat and converts it to acid which even attacks dentures.

In order to study plaque, researchers in London have developed an artificial mouth where, under controlled conditions, they can create this damaging substance and chart its effects on human teeth. The artificial mouth consists of two interconnecting glass chambers: the lower incubation chamber contains teeth, sections of enamel or dentine, dentures and filling materials; the upper chamber contains artificial saliva–a mixture of electrolytes, mucine vitamins, amino-acids, and urea at concentrations within the range of normal mixed human saliva. Bacteria are sometimes added together with a mixture of 95 per cent air and 5 per cent carbon dioxide. The liquid from the upper chamber is dripped on to the teeth planted in plastic gums in the artificial mouth at a rate that simulates three meals a day with snacks. The saliva may be programmed to imitate oral secretions, even to including vital rest-periods for dehydration. After only a week at blood temperature the scientists are able to produce thick, artificial plaque which is similar in many respects to natural plaque. By varying the diet it is possible to produce almost exactly the same kind of plaque that would be found in a natural mouth. Further research should help to solve the widespread problem of tooth decay.

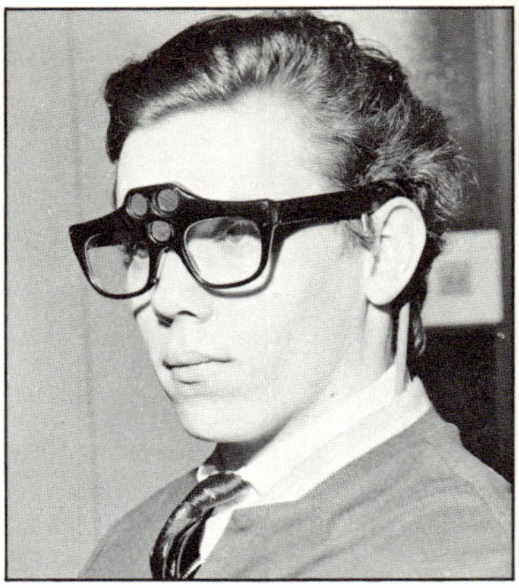

Sound spectacles—Ian Pivac, the first blind man in Britain to test the new aid. The three disks between the lenses are transducers radiating and receiving signals from a cone-shaped area covering a 60° arc in front of the wearer. 'Echoes' received from obstacles are passed down the cable behind the ear to an amplifier worn on the chest, then re-routed to the tiny ear-piece. Mr Pivac can 'see' up to 20 feet in front of him

Sound Pictures

Many people who are blind develop an acute sense of hearing, and are able to recognise subtle changes in sound that sighted people cannot even hear. The latest development by the National Research Development Corporation makes use of this fact. It is a pair of sound-spectacles for the blind. Three disks are mounted in the centre of an ordinary spectacle frame; these are transducers radiating and receiving invisible ultrasonic waves in a 60° beam 'shining' ahead of the wearer. The echoes received from objects hit by the ultrasonic wave are converted to audible signals in the lightweight electronic control-box carried at the waist. The sounds vary in pitch and loudness as the wearer approaches obstacles, and the blind person hears them through miniature earphones fitted to the sideframe of the glasses.

Ian Pivac, the first man to wear them, says that just being aware of what is up to 20 feet in front of him produces an enormous sense of confidence. And from practice Pivac can now tell the difference between an iron fence and a brick wall, a lamp-post and a pillar-box. The sonic spectacles have yet to go into commercial production.

Artificial Hand

Many people who have had a hand amputated still have healthy muscles in the stump. The essential factor in using a new artificial hand, based on a principle called myoelectric control, is that any muscle that is left should still be capable of generating its own electrical signals. These tiny electrical charges are the 'messages' produced by a muscle when it contracts.

This latest artificial hand is fitted with a complex receiver and control-unit working in response to these signals. Since the signals are produced randomly, the hand mechanism compensates for the fluctuation in the signal and reduces the 'power' to a point where steady manipulation at an acceptable speed is possible, through tiny motors driving each 'joint' individually and in concert with the rest of the hand.

The prototype hand has been developed by the West Hendon Hospital, St Thomas's Hospital and the Atomic Weapons Research Establishment, and is already sophisticated enough to be of practical use.

Myoelectric control—The new artificial hand, operated by a receiver/transmitter unit in the hand: Signals from the wearer's own arm muscles are translated into mechanical hand and finger movements, through tiny motors driving jointed areas in the device

Artificial Arms

Rudimentary artificial arms for accident victims and for children suffering from congenital deformities have been available for many years. But the construction of an easily fitted, easily operated pair of artificial arms capable of performing all the tasks of the natural arm is still beyond the capabilities of medical engineering. Many functions of the human arm will probably prove impossible to reproduce, but one researcher, Alex Godden at the Department of Engineering Science, Oxford University, believes that it should be possible to create an artificial hydraulic arm that is capable of making 60 per cent of the movements of a living arm. He believes this is feasible by concentrating on systems that will control the hand position rather than each joint in turn.

Basically the arm he is working on consists of an assembly of plastic components representing the skeleton of the human arm, with built-in plastic actuators to do the work of the muscles. The fleshy part of the artificial arms and hands is made from resilient urethane foam and the whole system is powered by hydraulics housed in a metal container. The 'jacket' to which the arms and the power-supply are attached will be made from nylon or light-weight stainless steel. At the moment Mr Godden's main problem is the interface between the patient's shoulders and the false arms. He believes that somehow, vestigial muscles must be able to operate the unit's shoulder, elbow and wrist joints. The tiny electrical currents generated by these muscles were once considered as a means of control but the main difficulty here was that current miniature hydraulic valves were far too big for this application. So after careful research it was decided to utilise the bone-movement method of control.

Research is now going on to reduce friction between all the moving parts in the artificial arm, because the smoother the action the less power will be needed to operate it. One of the biggest problems was to find a suitable hydraulic fluid that was not poisonous or inflammable, and silicone fluid was finally selected. Mr Godden says, 'Our aim is to develop a reliable and versatile unit which will operate in the most adverse conditions without maintenance for 9 months. Should any part fail, these moulded plastic components can be replaced quickly and cheaply. The principle can be extended to other prostheses to ease the lot of the disabled, but the arms are not yet at the stage where we can fit and forget.' When development work is complete it is expected that the new artificial arms will cost about £120 a pair, compared with £400 for the conventional equipment available now. And they will be much lighter too: at 4½lb. they will weigh considerably less than any other kind currently available–including your own.

The Toe Thumb

A recent operation has shown that people who lose fingers or thumbs may be walking on the answer to the problem of disfigurement and handicap, for a new operation can graft your big toe on to your hand to replace a missing thumb.

The toe is attached to the thumb stump by the main blood vessels and the nerves. Some weeks after the operation a tendon is grafted in so that the new 'thumb' can be flexed. Although the surgeons involved draw no conclusions from the one operation of this kind they have performed so far, it seems likely that their new technique will solve a lot of problems for people with hand injuries.

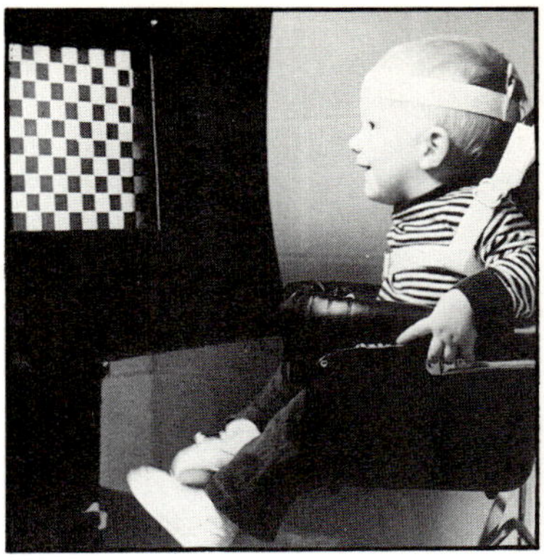

You need never speak to the optician during this eye test. This child's brain will produce signals relative to the clarity with which he sees the flashing chequerboard, or any other test object. Electrodes on the back of his head pick up the signals and a computer translates them into a print-out which tells the optician in what way the eyesight is defective. All the patient has to do is keep his eyes open

Portable Paunch

One day in the not too distant future, you may carry your artificial gut around with you in a plastic bag. The bag will provide a system of putting enough nutrient into the blood system to stay alive if your digestive processes do not work or if for some reason you cannot eat.

The plastic bag contains the food in solution, a small pump, and a battery power-pack. A plastic tube connects with the necessary vessels in the body. Dr B. H. Scribner of the University of Washington has designed the bag for long-term intravenous feeding. At night the wearer would 'plug in' to a home infusion unit that would add amino-acids, electrolyte solutions and other foods as needed. The plastic bag's main job would be to get you through the day or the effects of an operation. So far the device has been tested successfully for periods up to 3 weeks.

Electric Babies

A corset with electrodes built into it could soon be helping thousands of mothers to give birth to their babies. It has been designed by the NRDC in collaboration with University College Hospital, London, and is meant to be used to induce labour in pregnant women. The idea is that the electrodes in the corset pass low-voltage pulses to nerve endings in the expectant mother's abdominal area, which stimulate natural muscular contractions. The researchers have had to overcome many problems and have had to create special electrodes containing their own supply of jelly to ensure good electrical contact between the electrodes and the skin. Special low-weight electronic circuits, powered by rechargeable batteries, have been developed to produce the positive and negative pulses at predetermined intervals. The original corsets were tested only in hospitals but it is expected that as an extension of the initial clinical trials, these new units will soon be used by expectant mothers at home.

Baby Warmer

One of the most critical moments in life is just after birth, especially a premature one. The danger comes from heat-loss. A baby can lose heat rapidly in its wet state, and if surgery is necessary the problem of keeping the baby warm enough to stay alive can become acute, since under anaesthesia the thermostatic reflexes no longer work, muscles relax, and heat production is decreased. A new child-care system is aimed at solving this problem.

The Warmer heat source has two features that make it unique. The heating unit radiates ultra-long wavelengths of infra-red energy. These wavelengths are 10 microns long, the range at which human skin blocks off only half the heat. Thus, unlike other devices in the visible spectrum range (2 microns), the Warmer can heat a baby with warm rays that actually penetrate deep into the muscles. Secondly, the unit has been designed to throw out heat in concentric circles so that it envelops the baby rather than going out in straight lines, as with normal units, to warm the nearest point of the baby.

The amount of heat being given off by the Warmer is controlled by a thermister placed on the baby's skin which changes the *level* of heating so as to maintain the required temperature. When the temperature sensed by the thermister falls below 34°C (93·2°F), visible and audible alarms are activated. Such a situation would occur if the heating output was too low, or if the thermister became dislodged. The only time when the heat is switched off is if the baby's temperature rises above 100°F–and again alarms are triggered.

This new system looks rather like a cot with a canopy fixed above it. It may soon be saving infant lives, either during that first 20 minutes of heat-loss after birth, or under anaesthesia.

Germ-free Babies

In most cases a normal baby begins soon after birth to build up immunity to the infection-organisms surrounding it. But now and again babies are born without this ability –and until recently many of them died because of it.

A new technique gives doctors time to make sure that, in cases of doubt, a baby is not immunologically deficient, before letting it into contact with normal air. Such a birth takes place in complete sterility. A sterile operating-tent made of clear plastic, with air filters to remove bacteria, is used for the birth itself. Using gloves in the wall of the tent, the child is then passed through a tunnel to an isolator. There it is surrounded by tubes filled with warm water to prevent heat-loss. The tunnel is clamped shut and the totally germ-free baby can be taken to a larger isolation unit for treatment. Until it is ready to live normally, it breathes sterile air, eats sterile food, and is handled with sterile gloves. Already there are mothers in Britain who have cause to be thankful to this new technique for keeping their babies alive while congenital deficiencies are being corrected.

Medical Crystal Ball

Recently a team of experts carried out an exhaustive analysis on general trends in medicine and related technology. Their object was to examine what medicine was likely to be doing over the next 20 to 30 years. Each expert made forecasts for his own field and every member of the group studied each other's statements. This gave rise to a kind of cross-fertilisation of ideas, and forecasts were changed or modified in the light of knowledge gained from reading other reports. What follows is a selection of the group's final report.

Antibiotics/There are likely to be many small changes and a few big ones–but by the 1980s there will be a new generation of specific antibacterial substances. Mass-screening techniques will have advanced to the point where early diagnosis will often point to the use of a specific antibiotic in order to control the development of a disease almost before it starts.

The problem of antibiotic resistance is likely to grow, although within 10 years most people will have their own particular body reactions to antibiotics noted–much as a blood group is noted today–and this information will permit selective personal use of antibiotics.

Virus Infections/Almost the whole battle against virus infections has so far been carried out with vaccines. The next 5 or 10 years will continue the same way. Measles is likely to be eradicated within the next 5 years, and perhaps also rubella. Before the end of this decade a mumps vaccine should be developed and there is a good chance that influenza may also be countered. With the approach of an epidemic, there will be nation-wide treatment available well in advance. Some scientists believe that by the 1990s effective broad-spectrum antiviral compounds may be used to give protection against several diseases in one treatment.

The main problem to overcome will be to find drugs that will act against the virus within the living cells and yet leave the cells' metabolism untouched. Application of such drugs will therefore depend on investigation into cell structure and behaviour which is only just beginning today.

Cancers/Forecasts are that at best the problem of cancers will be difficult to solve. The less optimistic forecasters believe that, as existing cancers are dealt with, new ones will emerge to present a continuing problem. The main difficulty will be to deal with the multiplicity of causes and types of cancer, and to isolate and identify the many points in the relevant cells where the very complex processes of malfunction that produce cancer can occur. Some scientists feel that accidental discoveries are more likely to find answers than planned research. Others feel that by the 1980s major organ-transplants will provide the solution–with medical effort concentrating on dealing with secondary cancers in their early stages. This 'pessimistic' group also feel that no single 'wonder'-cure is to be found; if anything there will be a number of anti-cancer compounds to deal with specific cancers.

The more optimistic predict that studies of cell metabolism and genetics will show how the various substances that cause cancer fit into the system. They suggest that once we have this knowledge we will be able to prevent cancerous activity from happening by altering the process

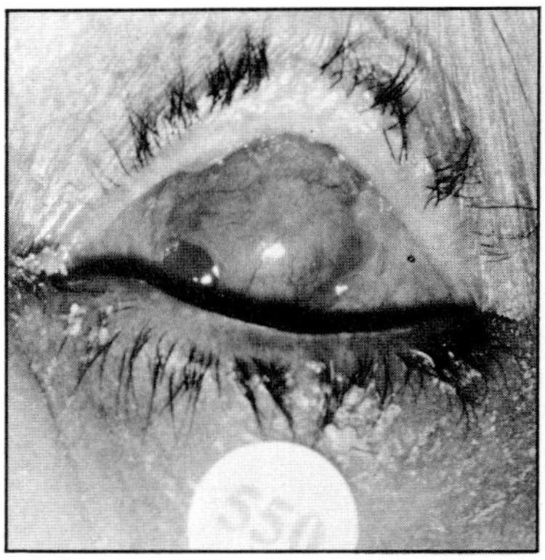

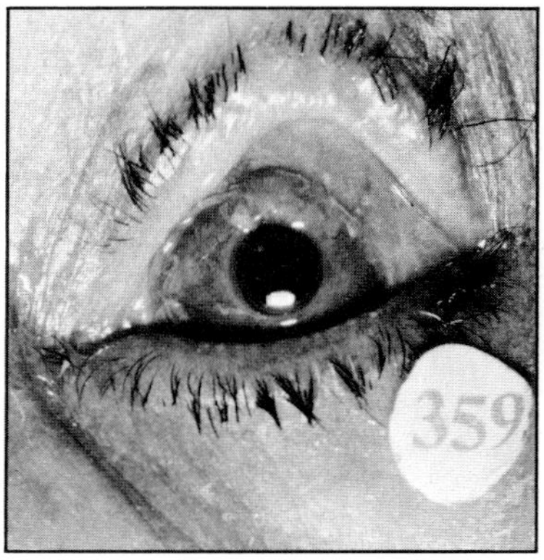

These two photographs show the result of a remarkable new technique used to insert a transparent section into a diseased lens, iris and cornea of the eye. The eye *(top)* was injured 35 years ago. After removing the lens, the cornea is strengthened with a graft of corneal tissue and a transparent plastic tube is inserted. Eventually a new, specially corrected optical lens moulded in plastic is inserted. The patient *(above)* can now literally see through the artificial window in his eye

itself. At best this group foresee 70 per cent of cancers being controlled by 1990. But all agree that this particular field is the most difficult to forecast.

Heart Disease/In the prevention of coronary heart disease five abnormalities must be considered: fats in the blood, high blood-pressure, diabetes, obesity and cigarette smoking.

By 1975 the effectiveness of cholesterol-lowering medicine will have been explored, and there should be medicines to decrease the body's absorption and synthesis of blood-fats. A definitive answer to the effect of dietary changes on blood-fats should be known by 1980. But whatever method is used, mass-screening, probably concentrating on young adults and children, will have identified those with the problem. Safe and effective medicines for the control of high blood-pressure will be available by 1980–and routine screening will be widespread, probably given at yearly intervals either by trained technicians or by the patient himself.

The connection between diabetes and coronary heart-disease is already established and by 1975 the effectiveness of anti-diabetic treatment will have proved its value in preventing the occurrence of heart disease. From that point on, preventive measures will be taken against diabetes. In all probability educational methods to improve dietary habits in those with obesity will be less effective than medicines which will 'burn off' excess calories without causing harmful side-effects. These medicines should be available in the near future, as work is already well advanced on them.

Cigarette smoking is expected to be a continuing problem. All forecasts agree that much stronger evidence connecting smoking and heart disease will be found by 1975–but that in high-risk subjects there will still be half today's number with the smoking habit.

Treatment of heart disease will be aided by discoveries of drugs to be taken as soon as the attack occurs–most people die in the first hour after coronary thrombosis–or even when a symptom suggesting an attack occurs. By 1990 it is reckoned that a heart attack will be foreseen a matter of days before it occurs. Studies in the acceleration of blood in the heart, and great advances in cardiology, will give warnings that will minimise the chance of an unexpected attack. High-risk subjects may get automated predictive examinations every 3 months to see how the heart is performing. Other medicines will be developed to strengthen the contraction of the heart muscle while not affecting the rest of the arterial system. By 1990 artificial hearts should be perfected, with parallel advances in the field of immunology producing drugs to prevent rejection of the new organ by the host body.

Mental Disorders/Many forecasts predict an increase in nervous disorders. This will be partly due to the recognition that many diseases are essentially neurotic in origin, and more generally that overpopulation and the spread of technology will cause an increase in both anxiety-states and depression. One expert believes that by 1990 nearly every individual will be taking drugs to counteract mental disorders, and that such drugs will become as socially acceptable as aspirin is today.

Research on the subject will concentrate on the biochemistry of the changes that occur with mental disorder. Powerful anti-depressants will be available by 1975; 'mood-normalising' compounds will be ready by 1980; and by 1985 drugs will 'stabilise' those people whose nervous states are caused by environmental and personality factors. Considerable progress is expected in the treatment and care of schizophrenics by 1985, including perhaps specifically-oriented drugs.

By 1985 there will be drug control of psychoses originating in genetic or biochemical abnormalities. Genetic counselling–advising prospective parents not to have children if their family histories show that they would be likely to have an abnormal child–should reduce the number of certain types of subnormality, and better care after birth will reduce the harmful effects of nutritional deficiencies. By the end of this decade there will be medicines capable of altering mood or quieting anxiety without any of the side-effects produced by today's drugs. And by 1990 there will almost certainly be drugs which could be used to blunt curiosity and initiative while inducing an unassailable sense of security and well-being *whatever* the real external situation may be.

Artificial Replacements/By 1990 it should be possible to replace almost every part of the body with artificial devices except the brain and the spinal cord. In the case of some internal organs many of the problems of rejection may be solved by not actually implanting replacement systems but placing them outside. At the moment it looks as if advances in microminiaturisation of power sources could reduce the size of some replacement packs to a size where they could be worn inconspicuously.

Replacements will be available by 1990 for heart-valves, possibly even the entire heart system, blood vessels, joints and bones. Lung and liver transplants should be commonplace. Electrical devices will be implanted to give control of muscles in cases where natural control has failed. Microsurgery will mean in some cases curing internal disorders, like stomach ulcers, by manipulation of related parts of the nervous system.

Anaesthetics/Short-acting anaesthetics will have improved by 1975 to the stage where they will work rapidly, and recovery from them will be almost immediate. Later, recovery may be as quick as waking up from natural sleep. By 1980 anaesthesia will be electrical, not chemical, allowing the patient to remain conscious if necessary, while only certain areas are anaesthetised. Chemical developments will effect local control on muscle-tone, secretion or oxygen needs, so that surgical shock will be greatly reduced and post-operative recovery made that much swifter. At the same time anaesthetised patients will probably be monitored automatically, permitting one anaesthetist to handle several patients at once. And by 1990 the forecast is for anaesthetics inducing hibernation of weeks or months if necessary. You may go to the operating theatre for a critical operation and electrodes will touch you, so that you fall asleep, to wake again weeks later with the incision healed and your body almost completely recovered. Subjectively, it will have all happened in a moment, as you 'slept'.

Perhaps the most dramatic change ahead will be that by 1990 the social use of medicine will be accepted just as the use of tea, coffee, cigarettes and alcohol have been. As health education advances, people will turn more and more to drugs which do the same to them as alcohol and cigarettes, but more safely. The gradual acceptance of drugs to stimulate the learning processes, to improve co-ordination and physical ability, to counter stress and increase awareness of surroundings will all have a considerable influence.

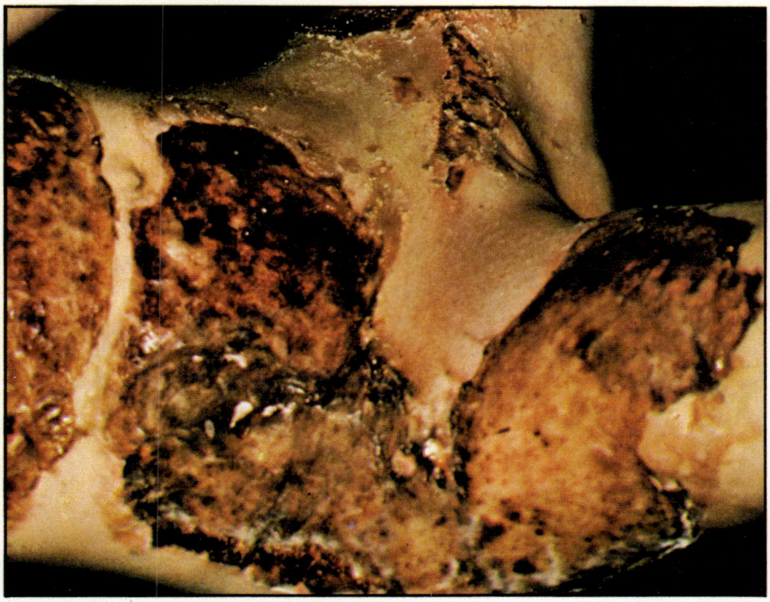

Above The body's micro-climate—one of the first photographs to show the thin layer of air that moves constantly upwards over the body. The tiny 'wind' is generated by body heat, carrying in it microscopic particles of skin, hair and bacteria. Parallel beams of light deflected by the 'wind' revealed its presence, and its germ-carrying potential, to researchers at the National Institute of Medical Research. In the picture it looks like a thin layer of light-coloured steam

Left Heat in colour—The abdomen and upper thighs of a child badly burned in a fire. In the isotherm picture of the same areas the darker the colour, the worse the burn, because it indicates dead skin (see page 115)

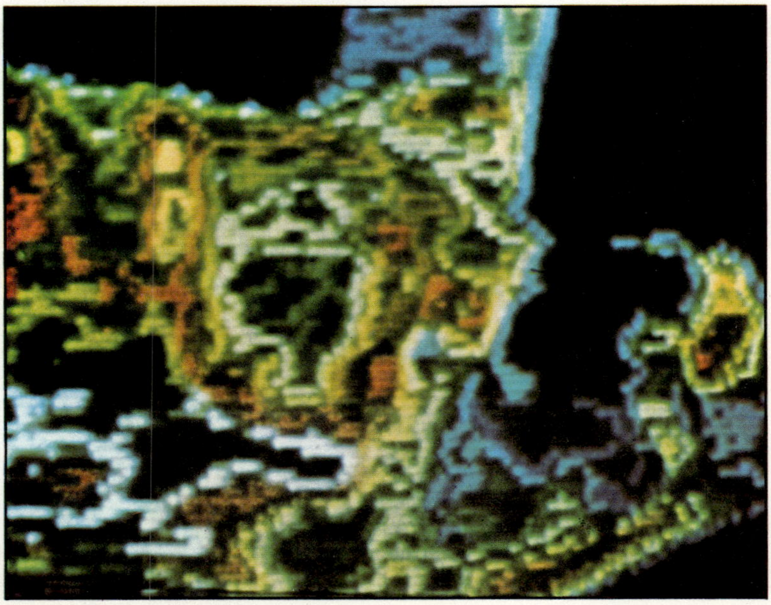

Apollo Saturn V moonrocket moves on giant crawler towards launch-pad. At the end of the top gantry arm is the sterile White Room, where the crew enter the ship. Above Command Module Launch Escape Tower rocket used in launch emergency to lift CM up and away

Saturn V: 363 ft, 3-stage, 33 ft diameter, 6,262,500 lb. (fuelled). Speed reachec for orbit 17,450 mph

The Saturn V clears the launch pad bound for the Moon. 4,000 gallons of water are pumped over the base of the pad in the first minute to protect it from the rocket exhaust. Note the immediate pitch-over from the vertical to ensure that the rising rocket does not foul the support gantry. The five first-stage rockets develop 7½ million lb. thrust, using 15 tons fuel per second. As Apollo clears launch-tower control is passed from Cape Kennedy to Mission Control Centre, Houston

Beyond Tomorrow

Apollo/Skylab/Spacebase/ The Space Shuttle/The Lifting Body/Planetary/ Automatic Satellites/ Propulsion/

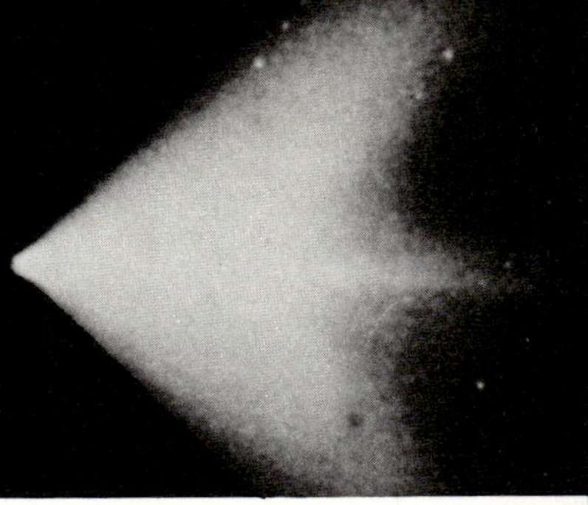

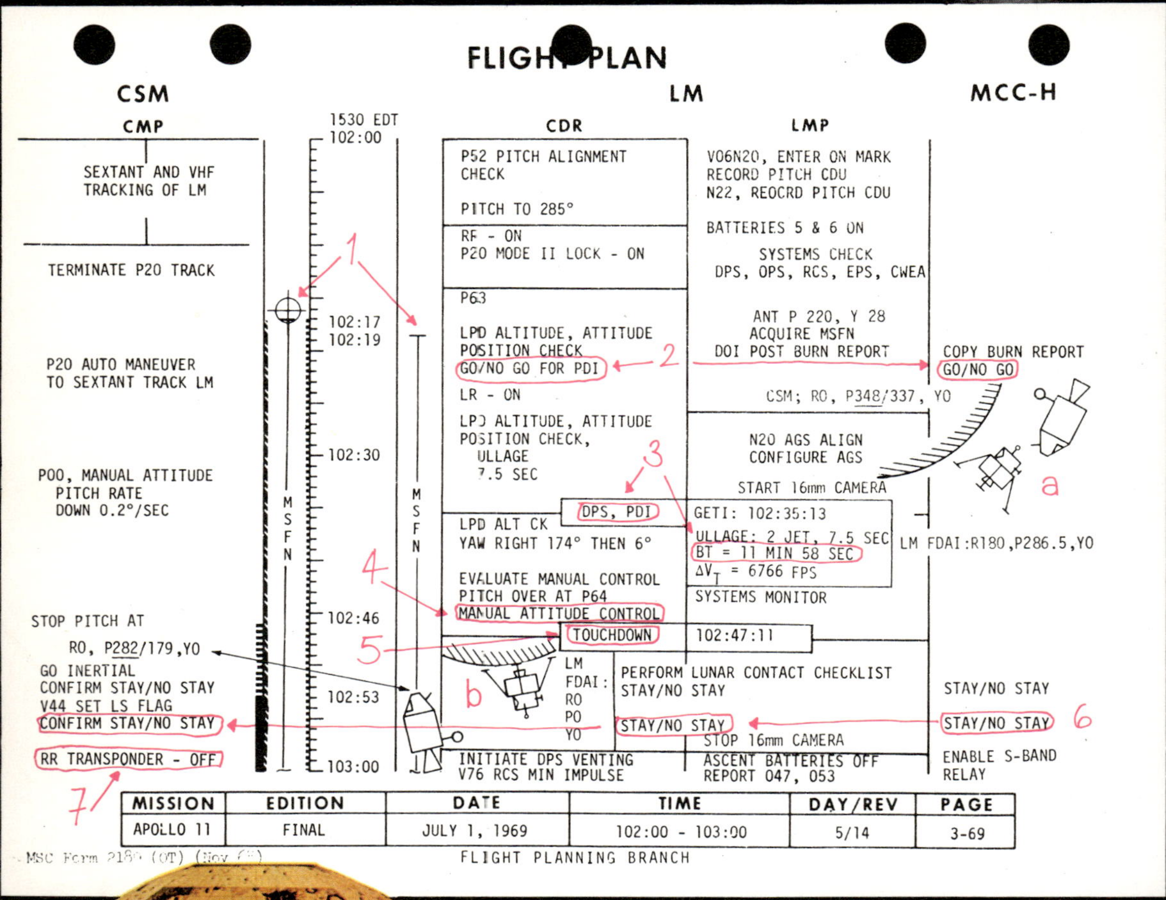

FLIGHT PLAN

CSM		LM		MCC-H
CMP	1530 EDT	**CDR**	**LMP**	
SEXTANT AND VHF TRACKING OF LM	102:00	P52 PITCH ALIGNMENT CHECK PITCH TO 285°	V06N20, ENTER ON MARK RECORD PITCH CDU N22, REOCRD PITCH CDU BATTERIES 5 & 6 ON	
TERMINATE P20 TRACK		RF - ON P20 MODE II LOCK - ON	SYSTEMS CHECK DPS, OPS, RCS, EPS, CWEA	
P20 AUTO MANEUVER TO SEXTANT TRACK LM	102:17 102:19	P63 LPD ALTITUDE, ATTITUDE POSITION CHECK GO/NO GO FOR PDI LR - ON	ANT P 220, Y 28 ACQUIRE MSFN DOI POST BURN REPORT CSM; RO, P348/337, YO	COPY BURN REPORT GO/NO GO
P00, MANUAL ATTITUDE PITCH RATE DOWN 0.2°/SEC	102:30	LPD ALTITUDE, ATTITUDE POSITION CHECK, ULLAGE 7.5 SEC DPS, PDI	N20 AGS ALIGN CONFIGURE AGS START 16mm CAMERA GETI: 102:35:13 ULLAGE: 2 JET, 7.5 SEC BT = 11 MIN 58 SEC ΔV_T = 6766 FPS	LM FDAI:R180,P286.5,YO
STOP PITCH AT RO, P282/179,YO	102:46	LPD ALT CK YAW RIGHT 174° THEN 6° EVALUATE MANUAL CONTROL PITCH OVER AT P64 MANUAL ATTITUDE CONTROL TOUCHDOWN	SYSTEMS MONITOR 102:47:11	
GO INERTIAL CONFIRM STAY/NO STAY V44 SET LS FLAG CONFIRM STAY/NO STAY RR TRANSPONDER - OFF	102:53 103:00	LM FDAI: RO PO YO INITIATE DPS VENTING V76 RCS MIN IMPULSE	PERFORM LUNAR CONTACT CHECKLIST STAY/NO STAY STAY/NO STAY STOP 16mm CAMERA ASCENT BATTERIES OFF REPORT 047, 053	STAY/NO STAY STAY/NO STAY ENABLE S-BAND RELAY

MISSION	EDITION	DATE	TIME	DAY/REV	PAGE
APOLLO 11	FINAL	JULY 1, 1969	102:00 - 103:00	5/14	3-69

MSC Form 2180 (OT) (Nov 68) FLIGHT PLANNING BRANCH

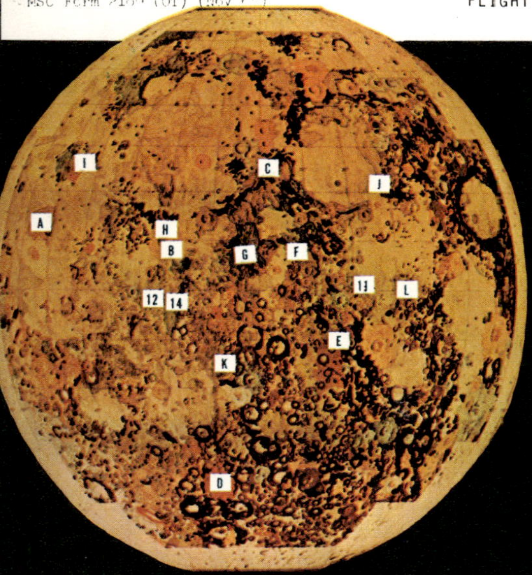

Top The journey begins—the Apollo engine firing to leave Earth orbit for the Moon, photographed by Hawaii observatory tracking-camera

Centre Apollo 11 Flightplan. Touchdown page. Carried by crew on Moon landing, 20 July 1969.
1-hr period divided vertically in minutes. GET—Ground Elapsed Time (from Mission start); CSM—Command and Service Module; LM—Lunar Module; CMP—Command Module Pilot; CDR—Commander; LMP—Lunar Module Pilot; MCC-H—Mission Control Centre, Houston.

Main activities:
1 At 102:17 the CSM emerges from behind the moon into radio contact with Manned Space Flight Network; 2 minutes later LM makes contact. Contact maintained through rest of page (see MSFN vertical lines)
2 At 102:21 Armstrong receives 'go' for PDI (Powered Descent Initiation) from MCC-H
3 At 102:35:13 Armstrong fires DPS (Descent Propulsion System) for PDI. Firing time (BT) 11 minutes 58 seconds
4 At 102:46 Armstrong takes manual control
5 At 102:47:11 touchdown
6 MCC-H makes second and final check on LM's ability to stay on Moon before CSM disappears behind Moon. Information relayed to Armstrong and Aldrin, and to Collins
7 Collins turns off rendezvous radar transponder. This would have sent signals to the LM for automatic rendezvous in event of emergency blast-off immediately following touchdown
a) & b) relative positions (not to scale) of spacecraft before PDI and after touchdown

Bottom Geologic Map of Moon with Candidate Future Landing Sites

Landing Sites
Apollo
11 Sea of Tranquillity
12 Ocean of Storms
14 Fra Mauro

Candidate Sites
A Marius Hills
B Copernicus Peaks
C Hadley-Apennines
D Tycho Rim
E Descartes
F Hyginus
G Rima Bode
H Copernicus (Interior North Wall)
I Aristarchus
J Littrow
K Davy Crater Chain
L Censorinus

At 4.18 pm US Eastern Daylight Time on 20 July 1969, 102 hours and 47 minutes after lift-off from Cape Kennedy, Neil Armstrong piloted the Apollo 11 Lunar Module 'Eagle' to touchdown in the lunar dust on the Sea of Tranquillity. This successful journey to the surface of another planet brought the rest of the solar system within the reach of our technology. It is ironic that the message they carried on board, 'We came in peace for all mankind', should have reached the Moon on a spacecraft that might not have made the voyage for a good many years had the world been at peace. For the mission was purely political in origin, the answer to a real or imagined need for a prestige victory over the Soviet Union. Be that as it may, mankind had taken the first step on its incredible journey to the stars.

Our solar system covers a pitifully tiny area in the outer reaches of our galaxy. If you look up at the Milky Way from a spot in the Southern Hemisphere you are looking through the spiral arms of that galaxy towards its centre. We know that as the galaxy formed, our solar system was one of the last to come into existence, far out on the edge. So in terms of age Earth is considerably younger than the stars and planets close to the centre. It is amost certain, therefore, that thousands, maybe millions, of other intelligent civilisations exist somewhere out there and with our new-found, though infant, abilities, we have it within our power to begin the search for them. It will be a long and frustrating journey—and one that may end almost before it begins. Perhaps our first contact with an alien intelligence will produce a 'culture shock' that will send us back across the light-years to our own solar system again, to decay and become extinct, much as some of the primitive societies on Earth reacted before the onslaught of a Western civilisation with which they could not live. But the journey will be made—it has already begun.

Between now and the end of the century, automatic spacecraft will visit all the planets in the solar system. Man will land on Mars, and the technology used to investigate outer space will also be turned to advantage on and in orbit round the Earth. This chapter examines the various activities, that are planned by the National Aeronautics and Space Administration of the United States. Repeated attempts to find out what specific plans are being made in the Soviet Union have met with failure. It is, however, highly likely that the exploration of space will make similar technological demands on both countries. So it is fairly safe to assume that the technology and systems used by the Russians will be, to all intents and purposes, similar to those used by NASA. The only question remaining is 'who does what first?'—and we have no means of knowing that.

Apollo

The rest of the Apollo missions to the Moon will be undertaken in the same basic vehicles that have been used since Apollo 9: the three-stage, 363-foot-high Saturn V (SV) launch vehicle, the Command and Service Module (CSM) and the Lunar Module (LM). Apart from the fact that they have proved adequate for the job, the hardware for a total of nine manned systems has already been paid for and is either already used, ready to go, or in the final phases of preparation. One extra launch vehicle and spacecraft, planned as Apollo 20, has already been diverted for use on the orbiting laboratory programme in 1972.

Future plans call for modification in order to increase time spent by the astronauts on the lunar surface from 35 to 54 hours; an increase in the payload taken down to the lunar surface from 350 lb. to 1000 lb.; an increase in lunar walks from 2–4-hour periods to 3–6-hour periods; an increase in lunar-sample return from 100 lb. to 175 lb. In addition to all this the entire mission will lengthen from 10·7 days to 16 days and the CSM will carry out more scientific investigations in orbit round the Moon. It will also include an EVA (Extra Vehicular Activity) by a crewman outside the CSM to retrieve instruments.

Command and Service Module/The familiar shape of the CSM will not change, nor will the modifications mentioned above be introduced before Apollo 15, which will mark the start of the J series, or the ALEM (Advanced Lunar Exploration Missions). In orbit round the Moon on these J missions the CSM will monitor alpha-, gamma- and X-ray emissions from the lunar surface in order to explore the homogeneity and similarity of the surface and subsurface, as well as the characteristics of the Moon's tenuous atmosphere. A panoramic camera will take pictures of areas 2½ miles by 2½ miles, from a height of 10 miles, showing everything larger than 2 foot in size. There will also be a mapping camera on board taking high-quality shots of the surface simultaneously with pictures of the positions of the stars relative to the camera. Through the correlation of these two pictures cartographers will learn the exact location and dimensions of the area mapped. A laser altimeter will record the height of the CSM to the inch above the surface, and show even the most minute changes in orbit caused by any change in the gravitational pull of the Moon. A mass spectrometer and magnetometer will also be carried, and on one of the next missions the CSM will launch its own tiny satellite for other experiments in lunar orbit. All these automatic experiments will be housed in one quadrant of the Service Module. The film cassettes from the cameras will be retrieved by a crewman and brought back to Earth.

The Lunar Module/The biggest modification to the LM from Apollo 15 on will be the inclusion of a Lunar Rover. It will be carried in the quadrant to the right of the LM exit ladder, in a folded-up position. As the astronaut pulls on a latch at the top of the ladder the Rover will swing out to a 45° position. There it will be unfolded by stages and lowered to the surface.

The Rover will weigh 400 lb. and be capable of travelling at 10 mph for a range of 70 miles. But for safety reasons it will travel no farther than 3 miles from the LM, because even on a perfectly flat area the horizon is only 4–5 miles away, and line of sight could be easily lost with the LM. Powered by two independent batteries, the Rover is driven with two controls, set in a console between the two seats at the front of the vehicle. One stick applies forward acceleration when pushed forward, braking when pushed backward. The steering stick is simply pushed in the direction the driver wishes to go. A small computer, which senses which way the Rover is moving and measures the number of wheel revolutions, will work out the vehicle's position at all times. The Rover will also carry up to 170 lb. of equipment and samples. On a 3-mile radius from a circular track round the LM it will cover about 20 miles of surface. The astronauts will be able to stop to make investigations whenever they like. With three trips on each landing, the crew will cover much more lunar surface than previously possible, and with much less effort. This saving in effort, and a modified oxygen supply system, will allow a total of up to 18 hours outside the spacecraft.

The Sites/If nothing prevents the completion of the Apollo flights, NASA plan to visit no less than six of the following areas of the Moon:
The Sea of Tranquillity (plains).
The Ocean of Storms (plains).
Fra Mauro (uplands).
Littrow (crater on the edge of a plain).
The Davy Crater chain.
Censorinus (crater).
Marius Hills (the mysterious 'domes' about 600 feet high, and the 1200-mile-long ridges).
Copernicus (crater 55 miles wide, 12,000 feet deep. The landing point is in the crater by its central, 1500-foot peaks).
Copernicus (inner north wall).
Hadley-Apennines (a 600-foot-deep, mile-wide 'rille' or canyon 50 miles long, at the foot of the mountains).
Tycho (the rim of the giant crater).
Descartes (crater).
Hyginus (canyon or 'rille').
Aristarchus (crater).

The certain thing about the Apollo manned lunar missions is that unless the technicians who have been laid off are given their jobs back to begin launch-vehicle and spacecraft construction again, there will be no American lunar landings after the present number of rockets are used up. When that time arrives NASA may well decide to wait until the nuclear engine is perfected and in flight before continuing with other plans to send men up to explore the surface of the Moon.

Skylab

Just as in the past every generation of space vehicle, manned or unmanned, has in its later stages provided impetus and experience for the next generation, so will Apollo and the systems primarily developed to land men on the Moon. And it is in this next generation of manned flight that we on Earth can expect, at last, some imme-

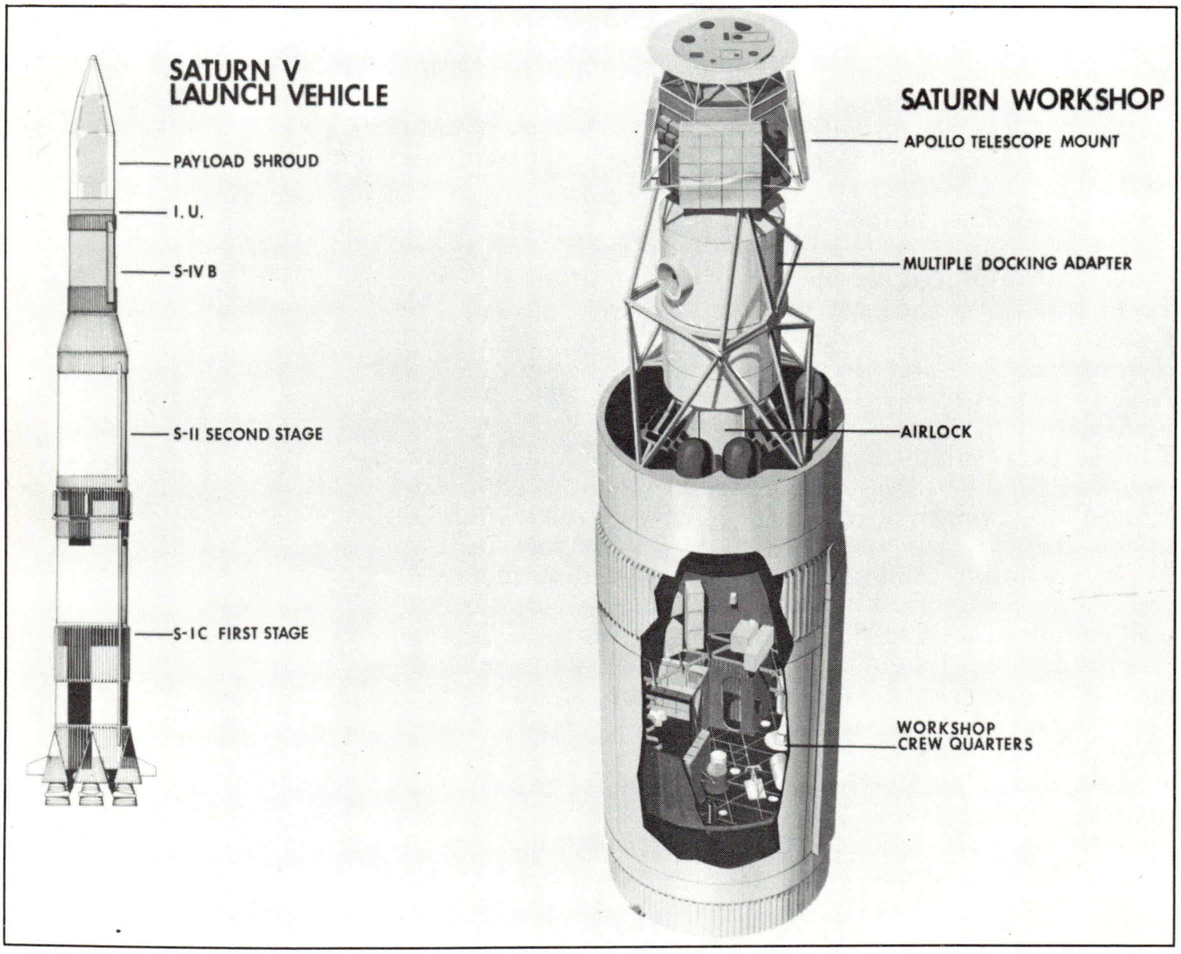
SATURN V
LAUNCH VEHICLE
PAYLOAD SHROUD
I. U.
S-IV B
S-II SECOND STAGE
S-IC FIRST STAGE
SATURN WORKSHOP
APOLLO TELESCOPE MOUNT
MULTIPLE DOCKING ADAPTER
AIRLOCK
WORKSHOP
CREW QUARTERS

diate return for the money spent over the past 12 years.

For in 1972 the first orbiting space station is due to blast off. Originally known as Apollo Applications 1–now renamed Skylab–the station will consist of the top stage of the Saturn V (the stage known as the S4B). Before launch the giant empty tank will have been rigged out with living-quarters and an experimental 'workshop' area, and the engines and associated machinery will have been removed. The Skylab will provide 10,000 cubic feet of living and working space for three astronauts, who will stay on board for periods of up to 56 days.

Next to the S4B will be an attachment called the Multiple Docking Adaptor (MDA). The MDA contains all the controls for the onboard telescope, and has an optical-quality window for Earth-viewing. At its outer end, and to one side, there will be two docking ports to enable the nose of the Apollo spacecraft to ease gently into dock on arrival. The spacecraft docking hatch is then removed, the port hatch swung open, and the astronauts float through into the MDA. A few feet away down at the end of the 5-foot-high MDA cylinder they will find the Airlock Module. Working on the same principle as the submarine airlock the module permits entry to the station without having to depressurise the entire structure. Apart from permitting access to the station, the Airlock Module also houses all the environmental control, power, communications and altitude-control systems for the Skylab.

Since the Skylab will be floating in orbit 235 miles above the Earth's surface, the crew will be living and working in zero gravity. For this reason, running the entire length of the station is a fireman's pole. After the astronauts have removed and stowed their spacesuits in the Airlock Module, they will move into the station along the pole, using it as a sort of guide-rail.

Work Scheme/Two-thirds of Skylab is taken up by the workshop. There the first crew to go on board will begin work on fifty separate experiments divided among five areas: medicine, engineering, science, technology and defence.

The medical tests will include investigations into the effects on the body of prolonged weightlessness, the changes in bone density, red blood-cell changes, and any alteration in heart rhythms and body fluids.

Engineering experiments will look mainly at what kind of structure is easiest to work and live in at zero gravity. Tiny control thrusters will be used for personal manoeuvring when astronauts are floating in space outside the station.

Part of the scientific programme will investigate the effects of zero gravity on the daily (or circadian) body-rhythms of insect and plant life and researchers will also concentrate on galaxy-mapping by X-rays, as well as an intensive examination of the Earth's atmosphere and the Sun itself.

In the zero-gravity environment on board the crew will have one major piece of equipment to test and use when objects being worked on need to stay firmly anchored to a working surface without floating away. It is called an

Top left Skylab cluster orbiting Earth in 1972. Right to left: The main workshop area, inside an empty S4B rocket, with solar panels extended to convert sunlight to electricity; the Airlock Module, housed inside the left-hand end of the S4B (the AM carries the Skylab Central Control Console); the cylindrical Multiple Docking Adapter (one port shown, one with spaceship docked); behind the MDA the Telescope Mount, with its own solar panels. Skylab will orbit about 250 miles up

Bottom left Skylab at launch. The right-hand figure shows Skylab with the telescope mount in its folded position on top. Once in orbit the solar panels open and the telescope swings right through 90 degrees to lock into its final position

Above top The workshop structure before modification–the giant S4B (normally the third stage on the Moon rocket) being removed from storage. The engine nozzle on the left will be removed, and the interior fitted out with crew quarters and experimental equipment. The S4B is 58′ 7″ long, 21′ 8″ in diameter

Above The spaceship dock–final work being carried out on the Multiple Docking Adapter. The two technicians are working on the ports into which the nose of Apollo spaceships will fit on arrival from Earth. Once the hatches are opened the crew will enter Skylab through these narrow tunnels

electrostatic zero-gravity workbench. Consisting of a metal-surfaced, 3-foot-square table-top and an object over it resembling a reading-lamp, the device is the first practical application of the science-fiction writers' dream–the force field. The 'lamp' will emit a stream of ions (electrically charged atoms given off when certain materials are subjected to a considerable electric charge of over 10,000 volts). The ions hit the metal bench-top, setting up a positive and negative static electricity force field which holds the object down on to the surface of the table.

All the experimental equipment inside the workshop will be placed on the 'walls', since in zero g the crew will be able to stand in any position, once their special shoes (called Dutch shoes since they look like metal clogs) clip into the metal mesh that will act as flooring by each work area. The living-quarters will consist of an exercising area, sleeping-quarters with strapdown bed covers (to prevent the sleeper from sleep-floating), a kitchen, shower-room and 'waste management section' (lavatories).

Power for Skylab will come from two enormous solar panel arrays. These 'wings' will be covered with small black tiles which generate electrical power when the Sun's light hits them. A smaller set of solar panels will

The first spacebase—an artist's concept of the first cylindrical spacebase module showing the five decks and the central connecting tunnel. The module will carry a crew of twelve. New crews, fuel, equipment and supplies will be delivered by spacecraft which will dock at one of the circular ports standing out from the module's surface

also power the Apollo Telescope Mount fixed on top of the MDA. Free from the degrading effects of the Earth's atmosphere, the telescope will be principally used to make an intensive survey of the Sun.

Mission plans at the moment extend through four launches. The first will take the S4B up into orbit on top of the first two stages of a Saturn V, launched from Complex 39 at Cape Kennedy. Once in orbit all systems will be activated from Earth, and Skylab will be pressurised with an oxygen/nitrogen atmosphere. One day later, on board an Apollo Command and Service Module, the three-man crew will lift off from Complex 34 on a Saturn 1B. The ship will coast up through four orbits, rendezvous, and dock at the MDA, where it will be powered down to minimum status to await the crew's return to Earth some 29 days later. If the de-orbit burn and re-entry go as planned splashdown should be in the West Atlantic. About three months after the first crew launch, the second Skylab crew will lift off, this time for a period lasting up to 56 days. Priority on this session will be given to work on the telescope. Splashdown will again be in the West Atlantic. Then three more months will pass before a third crew will leave for another 56 days in space. One of the members of this crew will be a doctor. Further Skylab missions beyond 1973 have yet to be defined. One more is probable.

Spacebase 1

In 1977, five years after Skylab goes up, the first spacebase will be put into orbit about 260 miles out. The launch vehicle will be the two-stage Saturn V known as INT-21. It will be a modest beginning, following on and using all the information gained by the five years in Skylab. Consisting of a single cylindrical module about 33 feet in diameter and 40–50 feet long, it will house twelve people and will be built to last up to ten years. The crew will probably be changed every three to six months.

The module will be totally self-reliant. The command and control systems will be on board, as will life-support systems, but plans will not be finalised until the Skylab results are evaluated sometime in 1975. Five basic problems should already have been solved when they launch Spacebase 1: the bodily effects of zero gravity on a crew for periods of up to 56 days, how efficiently the crew can work in zero gravity, how habitable the module is in space, what kinds of power and control systems work best, and what particular supply and maintenance problems arise.

There will be five decks: two for living, eating, sleeping and controlling the station; one for housing supplies and cargo; and one for subsystems such as life-support machinery. The fifth deck will house the laboratory.

Two large solar panel wings will provide all the necessary electrical power. Around the outside will be airlock components through which new crews and cargo can enter the module. These airlocks would also serve as lock-on sections for the addition of other modules.

Once Spacebase 1 is established, plans will go ahead for the gradual build-up of the base to a large space station

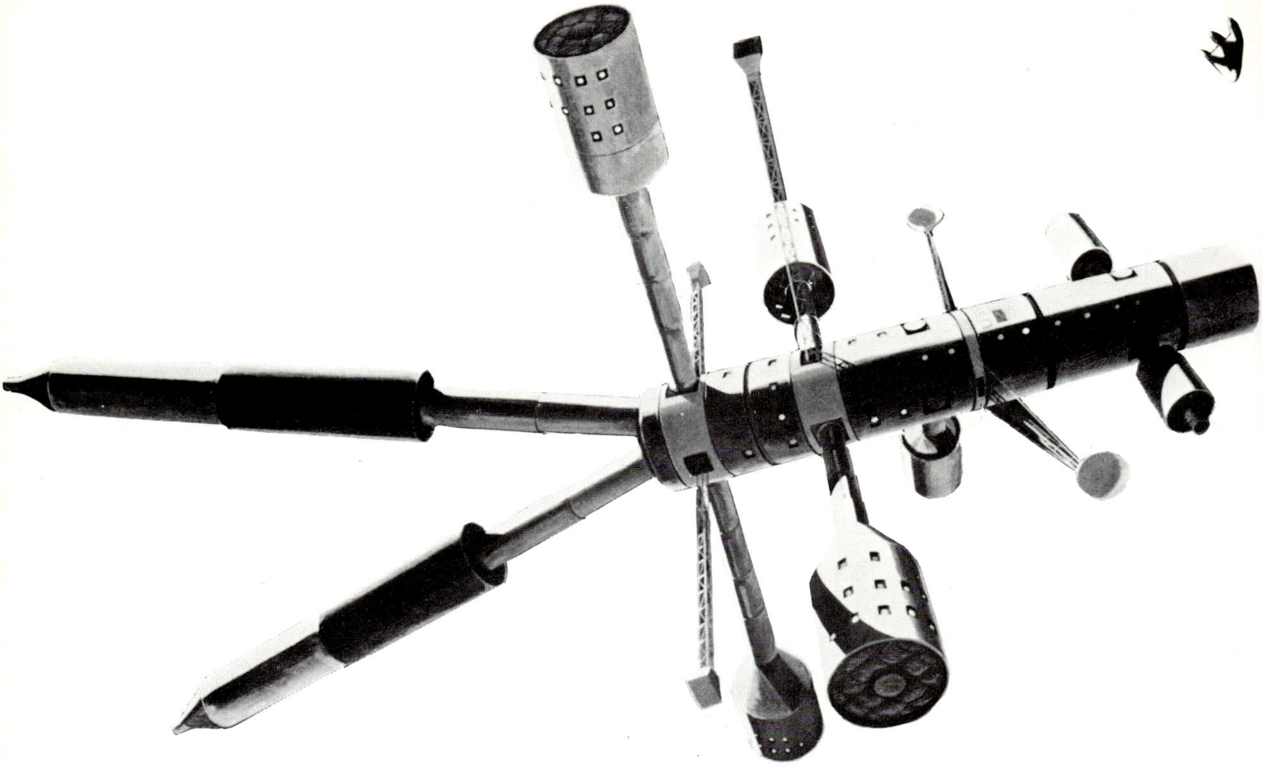

Above Spacebase 1990—an artist's concept of what the final spacebase in orbit round the Earth will look like towards the end of the century. Made up of multiple modules, which can be detached for special autonomous experimental work, the entire structure will be spun slowly in order to create artificial gravity conditions. In the distance, upper right, a space shuttle arrives from Earth carrying equipment and personnel for the station

Right Two impressions of crew quarters and control room in any one of the modules making up the spacebase cluster. Note the restraining arm curving over the bed to keep the sleeping astronaut in position should the module be in a weightless condition

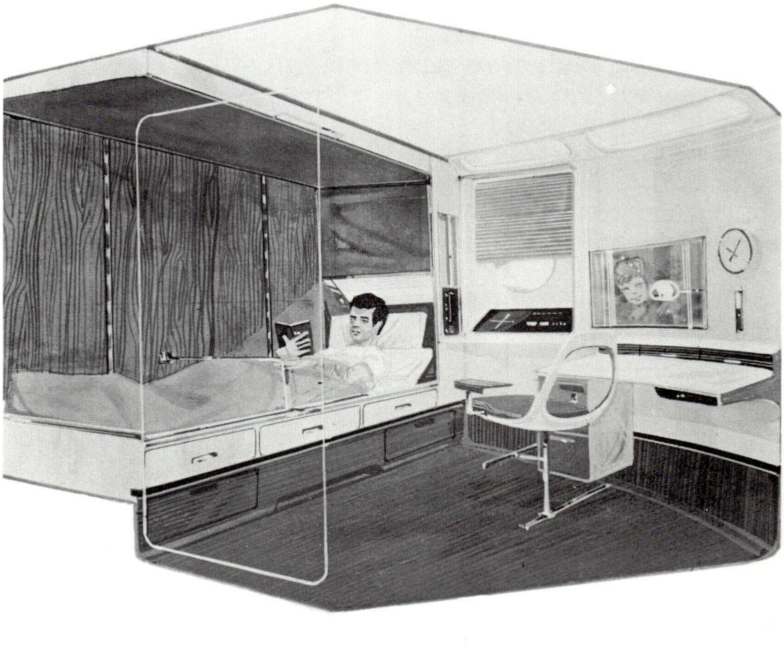

housing up to 100 people by 1980. This will be achieved simply by adding extra modules. By this time enough will be known to permit the introduction of non-specialist personnel and eventually the only qualification for living on board might be average good health. The 100-man station will serve many purposes. First of all the experimental modules will provide specialised environments for work demanding particular conditions. For this reason modules will be detachable so that they can if necessary float alongside the main station and avoid problems such as vibration, atmospheric contamination, and the kind of interference that comes of being part of a larger station.

New personnel will arrive in advance of their specialist modules in order to supervise the arrival and attachment of the modules. Most of the spacebase will have artificial gravity and experiments indicate that in all probability the station will be spun at about four revolutions a minute in order to give about $\frac{3}{10}$g. This will permit workbench activity under almost normal conditions, and make it easy for people like you and me to go on board without having to go through training for weightlessness.

Life on board will be much closer to that on Earth than today's astronauts have to put up with. Each deck, for instance, will have a ceiling height of over 6 feet. Those parts of the base dealing with Earth-observation studies will rotate slowly so as to maintain a view of the surface at all times through their portholes. Other areas will move so as to null out the local artificial gravity in order to use zero gravity for specific experimental purposes. It is likely that those parts slowly spinning to develop the partial gravity will have no window ports because of the considerable disorientation that could be caused by views of tumbling stars, Earth and Moon.

Food in the base will also be much nearer to normal Earth food. Although wrapped longlife snacks of the Apollo variety will be stored, there will be microwave ovens cooking entire TV dinner-type meals on trays–in partial gravity even the crumbs will stay on the plate!

At the moment plans are for a lavatory unit which will aid the partial gravity situation by slight negative pressure in the container, so that to a certain extent solid body wastes will be sucked out of the body to avoid any possibility of escape and contamination of the rest of the cubicle. Taking a shower will be an unusual experience. In partial gravity the water droplets will probably not fall properly. So, instead, you will pat yourself all over with a soap-impregnated sponge attached to a hose. The hose will then deliver water to wash away the soap and at the same time a vacuum tube inside the hose will suck away the soapy water as it is used.

The job of the spacebase, as defined at the moment, covers many areas:

Earth observation (weather, geology, crop analysis, etc.).
Astronomy.
Engineering investigation in vacuum and zero gravity conditions.
Manned mission training centre.
Repair of automatic orbiting satellites.

The flexibility provided by the spacebase can be best appreciated in regard to the last type of work. Repair crews living on the base can go out to fix an automatic satellite, or bring it to the base for adjustments to be made, or bring it in and send it home to Earth for a major overhaul.

The base will also be used as an assembly and departure point for interplanetary ships and their crews, as headquarters for scientists who work in the specialist modules floating in orbit nearby or farther out in space.

Shuttle designs—an artist's impression of the two principal shuttle designs under consideration. The delta wing is shown firing engines to move from one point to another in orbit, and on re-entry, nose up, into the atmosphere. The more conventional-looking winged shuttle is more likely to be accepted. The orbiter fires engines to leave its initial position, fixed on top of the booster vehicle. The booster returns for a jet-powered landing at an airfield. The orbiter re-enters the atmosphere, nose up, to take the heat along its belly

The Space Shuttle

The modular space station in orbit round the Earth in the 1980s with its large crew of scientists, and eventually non-specialists like you and me, is an exciting prospect. But unless spacecraft technology changes radically it will be orbiting the Earth uselessly because it will be empty most of the time. The present-day Apollo launch vehicles use most of their 6 million lb. of fuel heaving three men and not much else into space. But the fastest you could get another three men up there after the first lot had gone, on present-day launch schedules, even in an emergency, is a month from the decision to blast off. And even if you accept the idea of a 50-man space station being used only by three men at a time, there is then the relatively important problem of getting them back. At the moment with Apollo launch technology this is impossible. It takes three men to fly Apollo–and there's no room for anybody else.

That is why the space shuttle is now under development. Although at this moment the design of the shuttle has not been finalised, the first test launch is programmed for 1974 from Cape Kennedy; and the first operational launch in 1977. The space shuttle will answer the three requirements set up by the American President's Space Task Force: economy, reusability, and commonality.

At present it costs $1000 to put 100 lb. in low Earth orbit (100 miles up). The launch of a Saturn V costs $1½ million and all but the top 22 feet (the Command Module) of the 363-foot beast is either dumped into the ocean on the way up or left floating uselessly in space.

The space shuttle will weigh just over half the Saturn V–3½ million lb. The fuel systems will be essentially the same (liquid hydrogen and liquid oxygen), but unlike the Saturn V, the entire space shuttle will come back to Earth. The basic designs being looked at at the moment are seen opposite. All of them consist of a two-stage vehicle. In a typical mission the first stage lifts off the pad and up to 200,000 feet. At that point, travelling at 7000 mph, the lower, or booster, stage separates and continues rising to about 230,000 feet when it pulls over and drops to about 50,000 feet, 250 miles downrange of launch. Then it turns back and lands under power at the launch site. This manoeuvre will be executed either by remote control or by a pilot. Meanwhile, the orbiter stage boosts up to orbit. The first orbit would be elliptical, beginning when the spacecraft had reached 45 miles and swinging out to 100 miles on the other side of the Earth. It would then gradually circularise its orbit to

Below Re-entry experiment—a special photograph taken in the NASA Ames Research Center laboratories' wind tunnel, showing a space shuttle aerodynamic model being subjected to the effects of hypersonic re-entry to Earth's atmosphere. The bright areas show those parts of the surface most affected by heat generated as the shuttle plunges through the atmosphere. As the nose takes the greatest punishment, these tests led to the design of a shuttle that would take most of the initial heat tilted up to expose the broad belly-section to the airflow

100 miles and from that point coast up to the height of the space station.

After rendezvousing and docking with the space station and transferring its 14 passengers and 50,000 lb. of cargo, it would then be ready to make the return journey with any passengers and crew due to make the trip. It dives towards the upper edges of the atmosphere, and using small directional jets the pilots place it in a tail-first position. The main engines then open up to slow the shuttle down to entry speed and the craft spins round and re-enters the Earth's atmosphere at 400,000 feet, nose forward. The re-entry manoeuvre takes place with the nose tilted well up through about 60 degrees. This causes most of the intense heat on re-entry (up to 3000°F.) to affect the broad belly of the shuttle. At a lower altitude, when the heat begins to drop off, the pilots push the nose of the shuttle down and from there on they fly back to a landing like a conventional aeroplane.

Unlike the Apollo Command Module, the shuttle pilots would be able to choose which hemisphere they wished to land in, aim for the appropriate re-entry corridor and, once in the atmosphere, select any landing runway over 10,000 feet in length within a range of 1725 miles from the point of re-entry. That in effect means, for example, coming in over the Californian coast and landing at virtually any airfield in the United States. Landing speed would not be in excess of 250 mph.

If the shuttle is to be used effectively and efficiently, plans are that any one shuttle will last 10 years and be capable of 100 launches during its life. In extreme cases of emergency a shuttle could be fuelled, manned and launched within two hours of a decision to go. The fuel technology already exists. By the time of the first launches at the end of the 1970s, crew training will be at a point where the shuttle will require not much more pilot ability than is needed for a conventional aeroplane today –so manning a shuttle at two hours' notice should not present crew-readiness problems.

The requirement is for a shuttle that can, if necessary, be flown by one man, although normally there would also be a co-pilot. The pilots will be almost totally self-reliant once off the pad, since plans are for virtually total control-autonomy on board. At present the Apollo spacecraft has three times as many switches and dials as a normal airliner, and there are 400 people on Earth directly connected with the control of the spacecraft at all times. The shuttle will do away with most of that. The pilot will have a television screen presentation in front of him. On it he will see all the essential information minute by minute throughout the flight, provided automatically by the on-board computer. He will have a computer keyboard which will allow him to override the computer in order to ask for specific information at any time. When he has received what he asked for and signalled acceptance to the computer, the screen will start to flash regular automatic information once more.

There will be no 'dedicated' instruments–those that show only one piece of information, like speed or height or fuel state. This will minimise pilot effort in flying the shuttle. To see where he's going there will be portholes for general visibility and closed-circuit television. The television is needed because as the shuttle uses up its fuel its centre of gravity will shift and the shuttle will tilt. Portholes useful at a fully-fuelled stage could end up showing the wrong view at half-fuel stage, and so on. This is particularly critical during rendezvous and docking with the space station. For this reason a forward-looking television camera will provide the pilot with a 'window' giving forward visibility at all times.

Behind the control cockpit there will be room for up to 14 passengers. The shuttle is designed to take a maximum of 4g, but normally it will not exceed 3g. That means that any healthy human being will be able to fly without specialised training. You will sit in ordinary airliner seats with seatbelts, so the problem of weightlessness will occur only during the last few minutes as you transfer from the shuttle to the space station. The interior of the passenger cabin will be pressurised as in an airliner. No spacesuits will be worn.

The cargo compartment will be approximately 15 feet across by 60 feet in length, and carry up to 50,000 lb. This particular size has been chosen because one of the shuttle's most important jobs will be to tour between satellites carrying out repairs. And the maximum size of a satellite that might have to be taken up and positioned as a replacement for an unrepairable satellite would be 15 by 60 feet.

The Lifting Body

Although most of the prospective designs for the shuttle being prepared by private companies in the United States are still on paper, one is not. It is called the Lifting Body–and present indications are that the shuttle will possibly closely resemble it when it lifts off the launch-pad sometime in 1974.

At the moment there are three lifting bodies being flown at Edwards Air Force Base, in the Mojave Desert in California. There, at the NASA Flight Research Centre, more than 60 missions have been flown with the HL-10, the M2-F2, and the X-24. All of them have one thing in common–they're wingless.

At NASA's Ames Research Centre in California ten

The lifting bodies–the three experimental bodies (left to right: X-24A, MS-F3, HL-10) wait for take-off on the dry lake at Edwards Air Force Base in the Mojave Desert, California. The tests are conducted by the NASA Flight Research Center

years ago it was discovered that it was possible for a roughly delta-shaped body to produce enough aerodynamic lift for it not to fall quite like a stone. The shape flies–but only just. Falling through the air at speeds up to 800 mph there is just enough low pressure produced over the top of the shape for it to behave as the wing of an aircraft does, so allowing the lifting body to go through a 'controlled fall'.

During landing the pilots of the lifting bodies have to carry out what they call a 'very hairy manoeuvre'. Controlled falling from 90,000 feet gives just enough lift to provide forward motion–but not much. For every 1 foot forward, they fall 4 feet. This is scarcely adequate forward motion for anything other than a suicidal landing attempt. So, at a few hundred feet above the runway they tip the nose of the lifting body forward and dive it at the ground. At the very last minute they pull the nose up very sharply. The high speed that has built up increases the lifting capability of their wingless aircraft just sufficiently for it to shoot forward long enough to get the wheels down and touch down under full control.

These test flights (or 'drops') are carried out by a very quiet-mannered veteran NASA test pilot called Bill Dana. He and his lifting body are slung under the wing of a B–52 which takes them up to 45,000 feet. The pilot releases himself and drops away fast. A small hydrogen-peroxide engine opens up to lift him at 1200 mph to over 90,000 feet.

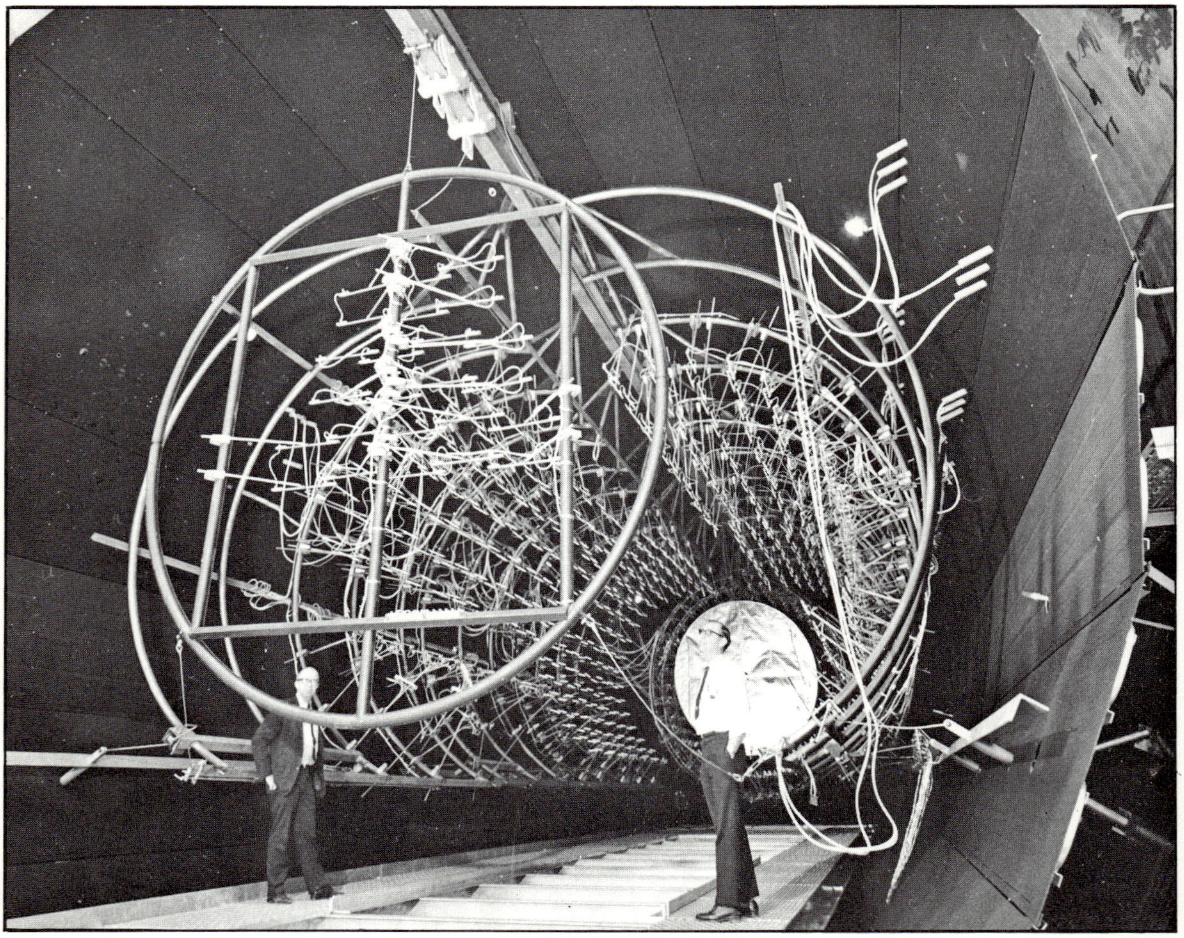

Spacecraft cooker–the Lockheed Space Simulator. Spacecraft are placed inside the cylindrical web of equipment. Special lamps match the solar radiation to be encountered by the spacecraft in flight. In the 30-ft-wide, 66-ft-deep chamber the heat test is carried out under space vacuum conditions

Then the fuel runs out and, as Dana says, 'the fun begins'. He and the team at Edwards have proved that the lifting body is as easy to manoeuvre as a conventional aircraft, as flexible in terms of range as will be necessary for the space shuttle, and extremely stable. 'I think we could take a passenger all the way from 90,000 feet to touchdown and not spill his coffee,' Bill Dana says. And if he can do that now there is not much doubt that in ten years the coffee-drinker coming back in from the space station to a landing at London Airport could be you or me.

Planetary

The arrival at the spacebase of a huge nuclear-ion-propelled interplanetary ship with its crew and cargo on its return from the first manned landing on Mars is not expected to occur before the mid-1980s. But exploratory flights have already begun to map out the journey and investigate the hazards faced by men bound for Mars, Venus, Mercury and the outer planets. Both America and Russia are still receiving data from deep-space probes. In spite of fly-bys of Mars and Venus and one Russian probe (Venera 5) which actually descended into the Venusian atmosphere and presumably liquefied in the tremendous heat, no spacecraft has, as yet, managed to orbit Mars, much less visit Mercury or any of the outer planets. But by 1980 all this should have been done, and man should have received the first information back from Pluto, the outermost planet in our solar system, swinging around the Sun on the very edge of the unimaginably vast ocean of interstellar space that separates our solar system from the nearest one to us, Barnard's star, 5½ light-years away.

Mars/The only detailed information we have on Mars has come from those spacecraft that have already flown by the planet. They told us little. Mars has an atmosphere, in all probability composed of 90 per cent carbon dioxide with a little nitrogen and oxygen in the upper atmosphere. Television cameras found little trace of cloud, although it looks as if there is a tenuous layer of haze several miles above the surface. The surface itself is heavily cratered, with vast featureless plains. There are polar icecaps, though it looks as if the ice is solid carbon dioxide. The famous canals are merely an accidental line-up of rows of craters. But most important of all, the spacecraft did not tell us enough to know if life exists there or not.

The next flight should do so. It is a dual unmanned mission: Mariner 71 A and B. As soon as they are successfully on their way they will be called Mariners 8 and 9. Each one will weigh about 2200 lb. and will be launched by an Atlas-Centaur from Cape Kennedy. Both will carry a retro-engine with 970 lb. of fuel, and be capable of five course-change manoeuvres on the way and a long firing at the other end to slow down by 3400 mph in order to go into Martian orbit. Launch will occur sometime between early May and late June 1971, eight days apart. The Atlas fires for lift-off. After its fuel runs out, the Centaur separates and puts the Mariner carried in its nose cone on course for Mars. It then separates after the shroud panels protecting Mariner have been jettisoned. As soon as Mariner is alone it spreads its solar panels and immediately locks its navigational sensors on to the Sun and the star Canopus. These provide reference points should a course correction be necessary. Nearly six months later, on 16 November, Mariner 8 will fire retro and enter Martian orbit–elliptical, 1000 miles high at low point, swinging out on the other side to 10,500 miles. Each orbit will take 12 hours. On 24 November Mariner 9 arrives and goes into a 1000- by 27,000-mile orbit lasting 32.8 hours.

Both spacecraft will operate for 90 days. Mariner 8 is set in such an orbit that the planet moves around slightly faster than it does, so each orbit covers a slightly different area of surface each time. In this way, by the end of the mission 70 per cent of the surface of Mars will have been examined. Mariner 9's orbit is set to ensure that it will cover specific areas every four days so as to observe any changes that occur throughout the entire period.

Scientific experiments on board each spacecraft will include:

1 A wide-angle and narrow-angle television camera. The camera will see objects 330 feet across, and although the pictures will not provide sufficient evidence of the existence of life, they will enhance our knowledge of whether or not it is likely. The greatest mystery that should be cleared up is the seasonal 'wave of darkness', a change in the colouring of the northern hemisphere which some scientists have speculated might be plant-life activity.

2 An infra-red radiometer (IRR). This will measure soil temperature and reveal any unusual hot spots that might result from volcanic activity.

3 An ultra-violet spectrometer (UVS). This will analyse the upper atmosphere.

4 An infra-red interferometer spectrometer (IRIS). This will analyse the lower atmosphere together with the surface in terms particularly of heat and composition. The importance of this experiment lies in the fact that if there is life it will affect the atmosphere above it just as, for example, a forest on Earth will cause warmth and updraughts in the air immediately overhead.

All data and pictures will be transmitted back to Earth by two antennae.

The Viking Mission/The Mariner missions have been planned to lay the ground for the first unmanned landing on Mars–the Viking Mission. Two Vikings will lift off on Titan-Centaur launch vehicles from Cape Kennedy. Each will consist of an Orbiter carrying a Lander in a specially sterilised condition inside a structure looking much like an oyster shell. The spacecraft will each weigh 7020 lb., over three times Mariner's weight.

The Vikings will leave about 30 days apart in late 1975 and arrive in early 1976. Because of the extra weight, the retro-fire to go into Martian orbit will take no less than 45 minutes. The first orbiter will map general sites on the surface as commanded, and then select the site for its lander to head for. About 9 days after arrival it will position itself and jettison the lander. Just before separation the lander jettisons the top half of its oyster-shell bio-shield and 8 jets around the edge of the lower shell manoeuvre it into de-orbit position. Four jets then fire the de-orbit burn. Shell forward, the lander burns down through the atmosphere, tiny gyros on board ready to keep it level in the event of turbulence. The parachute goes out at $5\frac{3}{4}$ minutes after entry, 14,500 feet up; it jerks the lander out of its shell and is released. At 3900 feet three jets on the lander open up to slow it down and at 10 feet as they cut out, the lander drops on to the surface of Mars.

As it touches, the orbiter photographs the area for reference and slows down to Mars rotation speed in order to remain hovering over the spot where the lander is, so that it can act as a relay station to send data back to Earth. For a few days most of the lander's information relates to the kind of place it is in–and depending on how fortunate the site has been, Mission Control will redirect the orbit of Viking 2 to its dropping zone. A total of 30 days after the arrival of the first spacecraft, Viking 2 swings into its now rearranged orbit and drops its lander. Both orbiters are then free, if so desired, to wander off above the planet on tasks radioed from Earth. While they do so the landers go on working and recording data. When the orbiter returns within range it sends a wake-up signal to its lander's transmitters and collects what has been learnt since it was last there.

The landers are designed to work for 90 days. Unknown meteorological conditions on the Martian surface might deprive any solar-power system of the sunlight it needs to make electricity, so each lander will have its own power source; two radioactive thermoelectric generators (RTG).

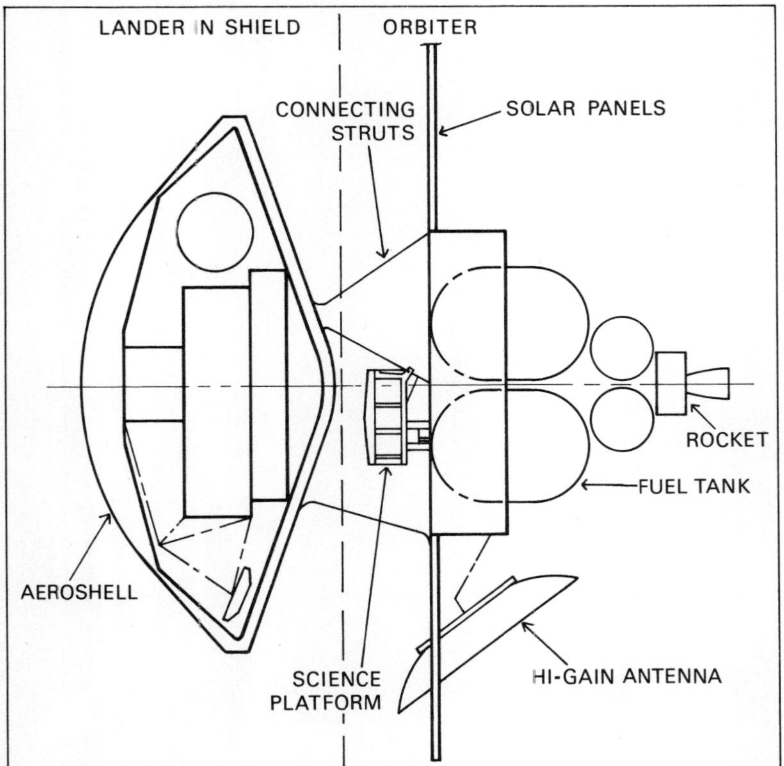

Viking—a cross-section drawing of the spacecraft. Once in orbit round Mars explosive bolts separate the Lander for Martian atmosphere entry, aero-shield forward

Dressing for Space—what the astronaut wears under his spacesuit. This clothing is worn only for periods outside the spacecraft. At other times the crew wear conventional coveralls

1 The basic garment—special underpants made of super-absorbent material for dealing with solid waste matter. Ointment is used to protect the skin from contact with the waste matter and to avoid drying or irritation

2 The garment looking like long underwear contains dozens of tiny tubes which carry water and glycol pumped from the backpack through the large transparent connecting tube to the right of the zip. The fluid is heated or cooled to maintain the body at a comfortable temperature, whatever the conditions

3 The small boxes at the waist contain bio-instrumentation units

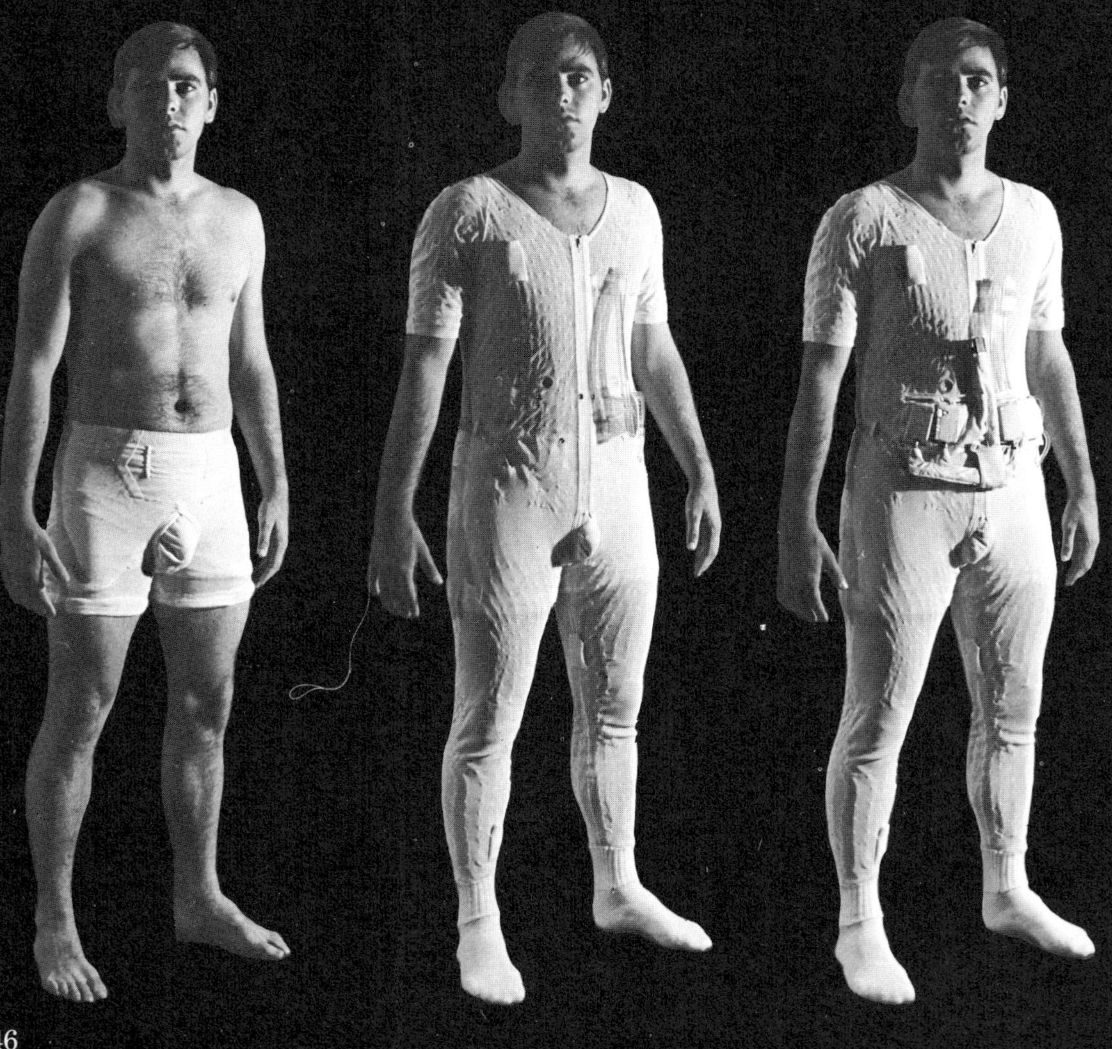

4 The V-shaped belt worn over the boxes is a urine container

5 & **6** The outer suit provides protection against low pressure, micrometeorites, and the intense heat and cold of the lunar surface

6 Life support systems and radio are housed in the backpack

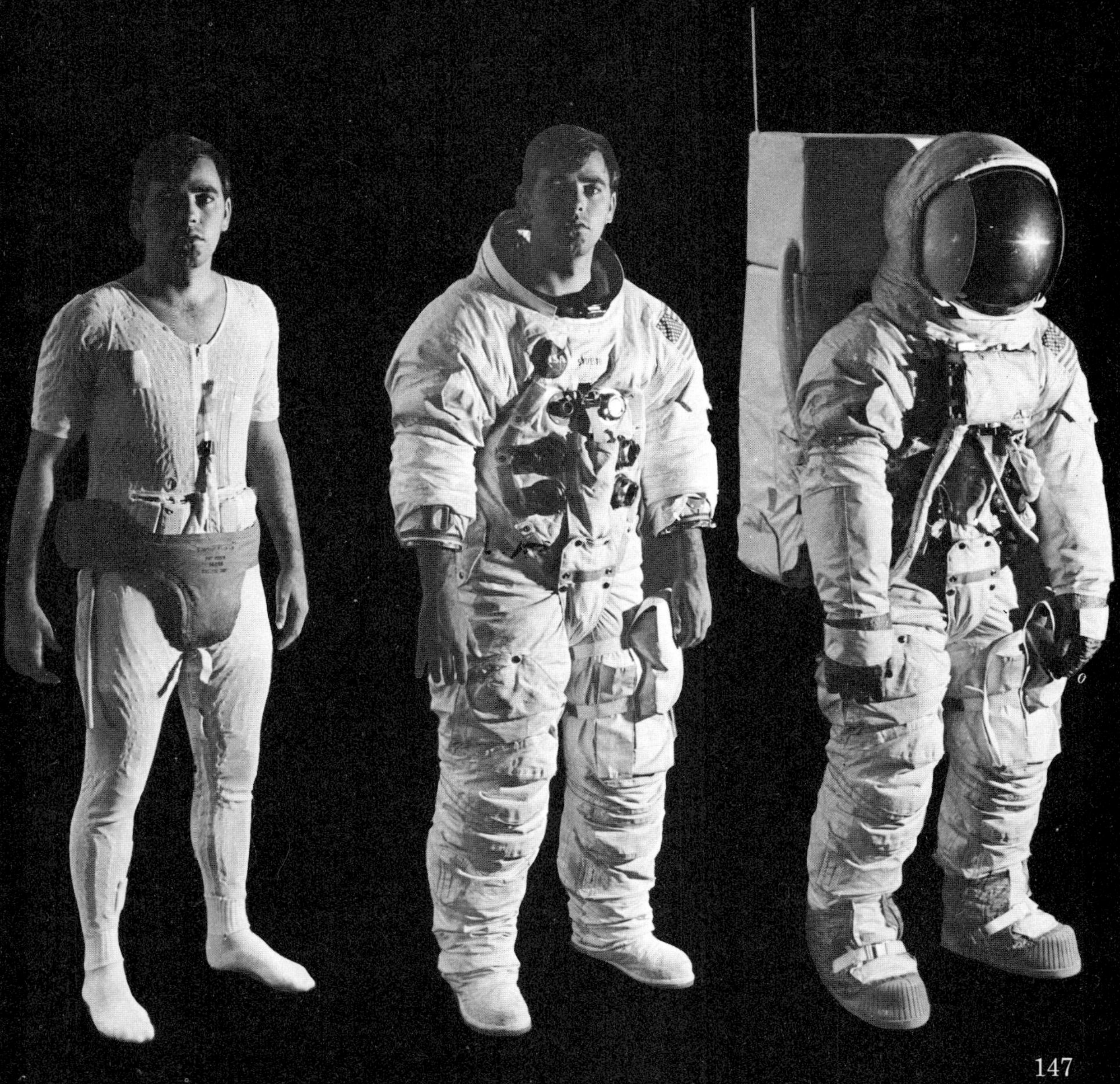

These tiny nuclear power stations will each produce 50 watts continuously throughout the period, aided by small back-up batteries for peak power-load periods.

The lander will be three-cornered, standing on three footpads. It will carry cameras with movable mirrors to photograph all round the spacecraft and a flexible arm to scoop up samples. The instruments carried in the area between the legs are:

1 A mass spectrometer and probes–to analyse the atmosphere on the way down.

2 A device (as yet undecided) to find water. One system suggested is to heat a sample of soil and look for steam. The device will also check on humidity in the atmosphere.

3 Meteorological instruments to measure the wind, temperature and air pressure around the lander.

4 A seismometer–to check for Mars-quakes.

5 An ultra-violet photometer–to measure the amount of ultra-violet rays from the Sun arriving at the surface, to discover to what extent Mars is protected by its atmosphere.

6 A gas chromatograph and mass spectrometer to analyse the soil: a special oven will heat a sample of soil and as it vaporises the gases are separated by the chromatograph and identified by the spectrometer. This test will also reveal the presence of organic materials in the soil.

And the most important test of all:

7 The life detector, an integrated system that will probably conduct four different tests for life:

a Feed a sample, gathered from the surface of Mars, on some radioactive carbon dioxide. After waiting, test the sample. If it is radioactive, the sample has absorbed the carbon dioxide and is alive.

b A double-check on the previous system: look for radioactive CO_2 being given off by the sample after it has been subjected to the original radioactive carbon dioxide. Its presence would indicate life.

c Feed the sample with a growth-medium containing radioactive traces. If the sample is living its metabolism should absorb, and then give off radioactive waste products. The big problem here is finding the right food so as not to end up, for example, trying to feed 'milk' to a 'tree'.

d Put the sample into food. If it is alive and its cells multiply, they will thicken the consistency of the food and measurements will reveal that less light will pass through it.

The presence of life-forms on Mars is of critical importance. For if life exists on two planets within one solar system, then the statistical probability of it existing elsewhere in the galaxy, and of even having evolved into intelligent beings, is enormous. If we find life on Mars in 1975 it will be almost certainly final proof that we are not alone in the universe.

Venus and Mercury/In October or November 1973 a Mariner-class spacecraft will lift off and head inwards towards the Sun. As with other Mariner launches, the lift-off will be on board an Atlas-Centaur at Cape Kennedy. Solar panels deployed, navigation sensors locked on the Sun and Canopus, Mariner 73 will head first for the planet of Venus.

Little is known about Venus. The planet is similar to Earth in size, density, and gravitational force. But its atmosphere is very different: at the surface the atmospheric pressure is 100 times greater than ours, and the temperature is almost 900°F.

Mariner 73 will take only a quick look at Venus, since its goal is tiny Mercury, the closest planet in the solar system to the Sun. It is going via Venus to save fuel, because if it went direct it would constantly have to fire retro-rockets to slow it down so that it could fall out of one orbit round the Sun into another. By passing Venus it will only use fuel to slow down enough to fall in as far as Venus. Then, as it goes by, the pull of gravity from Venus will tend to slow it down further. This in turn will cause it to fall towards Mercury without extra use of fuel. This 'gravity-assist' will mean tremendous savings in retro-fire fuel for the rest of the trip, but it is a most complicated calculation.

Almost nothing is known about Mercury. It is a little larger than the Moon, it is made of very high-density material, and has little, if any, atmosphere. Earth-based astronomical investigations have been thwarted by its proximity to the Sun. The spacecraft will take thousands of pictures of Mercury's surface. Nobody knows what it will look like. The temperature characteristics at the surface will be checked, and sensors will look for an atmosphere to analyse. Its magnetic field, if any, will be

The Mariner 1973 Venus/Mercury spacecraft

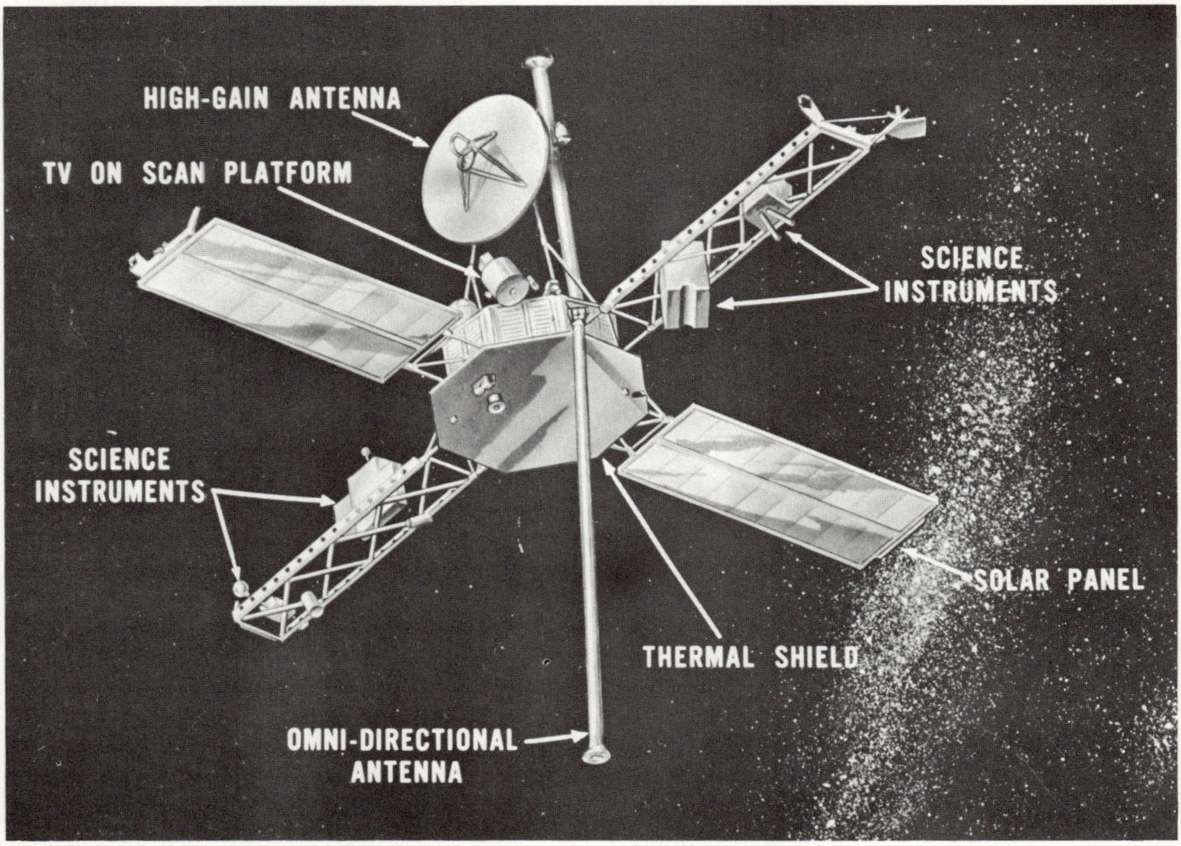

Jupiter—a telescope photo of the giant planet, 318 times Earth's mass. It rotates over twice as fast as Earth. The atmosphere is largely hydrogen. The black spot is a silhouette of one of the twelve moons and the grey oval shape, bottom left, is the mysterious giant Red Spot, slowly moving round the planet

measured, as well as what happens to the solar wind around it. This will be the first visit to Mercury.

Jupiter and the Outer Planets/Nothing man-made has ever gone beyond the orbit of Mars. In February 1972 the first of two automatic spacecraft, Pioneer F, will do so, heading out past Mars through the asteroid belt and on, to fly by Jupiter, the most unusual of all the Sun's planets. Pioneer G will follow 13 months later in 1973. Launch will be on the Atlas-Centaur vehicle from Cape Kennedy. The Pioneer missions will investigate interplanetary space beyond Mars; examine the asteroid belt both scientifically and as a potential hazard to space flight; and explore Jupiter and its surroundings.

The voyage will take nearly two years and will include only two main mid-course corrections, as well as minor thrusting, if needed, to avoid collision while in the asteroid belt. As the Pioneers head out away from the Sun, they will become the first unmanned spacecraft to have to do without solar panels as a source of power, since the Sun's light is too weak at that distance.

Power will be supplied by two Radioisotope Thermoelectric Generators (RTGs), set out on booms extending away from the spacecraft's instrument package in order to avoid radioactive disturbance. The central instrument package fits against the largest component on board–the main antenna, a large dish. The size of this antenna is necessary because of the tremendous distance over which data must be sent and commands received. As the spacecraft nears Jupiter, it will take 90 minutes for a radio signal (travelling near the speed of light) to make the journey to Earth from Pioneer and back. For this reason memory circuits are also being built into the control-systems so that the spacecraft can act on two commands received some time apart.

The spacecraft will weigh 550 lb. and will carry 13 experiments which will examine:

How dense the asteroid belt is.
The slowly-decreasing effect of the Sun at that distance.
How much cosmic radiation is entering the solar system from galactic space.
The magnetic properties of Jupiter.
The contents of its atmosphere.
The cloud structure.
The heat given off by the planet's surface.
The planet's 12 moons, one of which, Io, may have an atmosphere.

As the Pioneers swing in from below Jupiter's equator and on, leaving it to the north-east, cameras will take thousands of pictures of the planet's surface. Pioneers F and G will tell us much about Jupiter and about the strange red spot that moves constantly within its atmosphere (mostly composed of hydrogen). One of the theories put forward is that the red spot is made up of organic dyes that may hold the molecular structures that are the basic building blocks of life as we know it. But F and G will also prepare the way for the greatest interplanetary

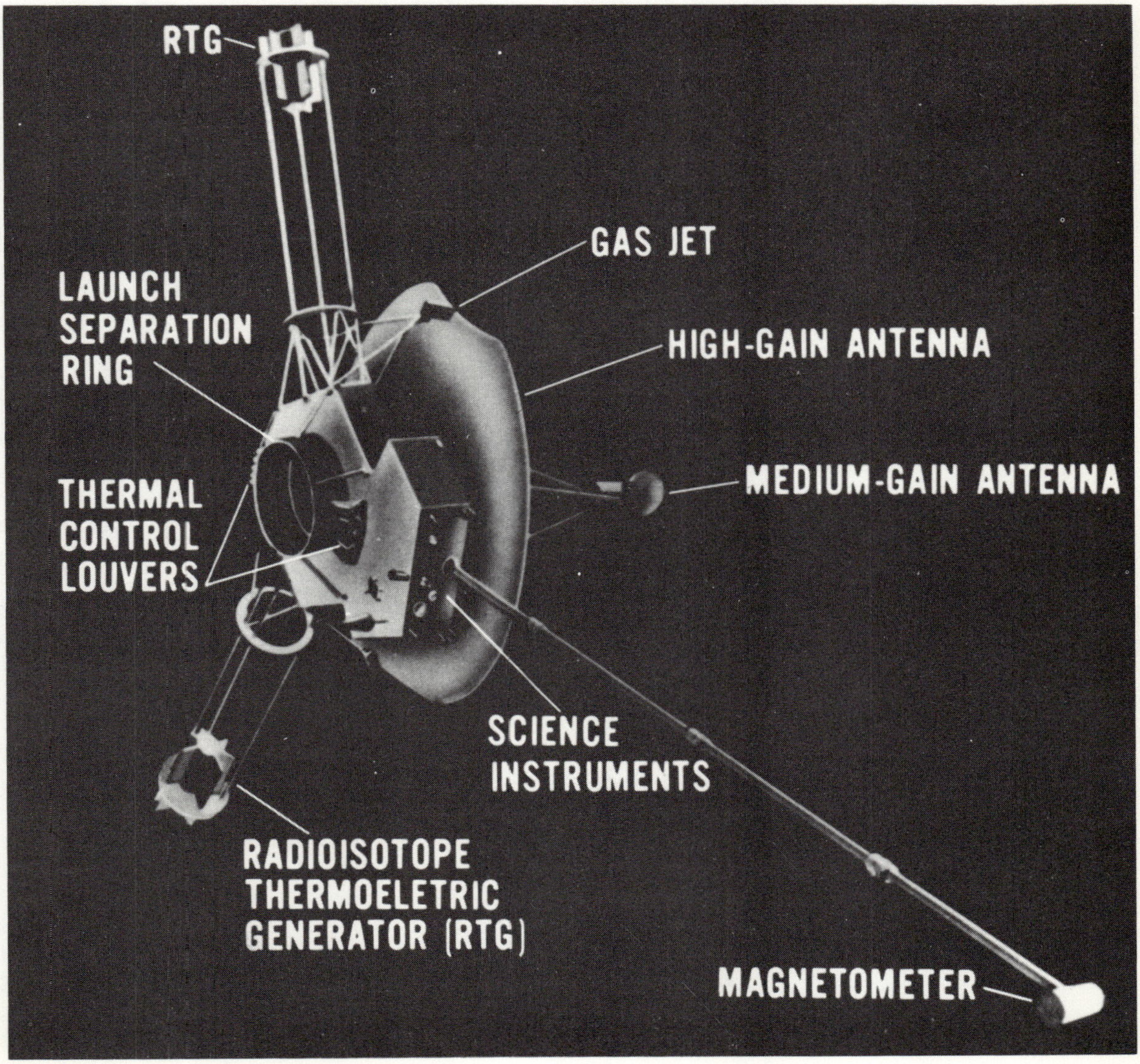

The Pioneer F/G spacecraft—due for launch 1972 and 1973. The two vehicles will explore the asteroid belt and Jupiter

adventure in the history of man–the trip to the outermost planet of all on the very edge of the solar system–a trip known as the Grand Tour.

The Grand Tour/Jupiter holds the key to the exploration of the planets beyond–Saturn, Uranus, Neptune and Pluto. Although each of these planets could be reached by a direct flight, such a flight to Pluto, for example, would take 31 years. It is therefore necessary to use what is called the 'planetary carom' technique. This involves aiming the unmanned spacecraft right towards the giant Jupiter, the first of the outer planets. By judging the trajectory very carefully the spacecraft would be dragged in towards Jupiter by its enormous pull of gravity. Accelerating tremendously, and then just missing the planet, it would, in effect, be flung out towards the edge of the solar system.

As it happens, just as we are ready technologically to undertake this journey, we have been aided by a tremendous piece of natural luck. Jupiter and the outer planets, on their varying orbits, are in alignment only once every 179 years. And between 1977 and 1979 this alignment is such that two spacecraft could move out to Jupiter, and be flung towards Saturn and Pluto, or towards Uranus and then Neptune. In both cases the planetary carom technique will also be used at the second planet following Jupiter, to speed the spacecraft on towards its final destination and then into interstellar space. This technique will put Pluto within nine years' flight instead of thirty-one.

The navigational problems are going to be considerable. Particularly after what is called 'planetary encounter', the deflected trajectory of the spacecraft towards a planet must be judged with extreme accuracy. For this reason scientists at the Pasadena Jet Propulsion Laboratory making the preparations for the Grand Tour, are concentrating on producing a planetary-approach television. A camera sensitive to differing light-levels will scan the skies as it approaches the planet which is to give it its 'gravity assist', line up the necessary stars, and inform the computer on board. The computer will make the necessary short firings of the directional thrusters in order to refine the trajectory.

The principal problem, however, will be the length of the journey. The spacecraft will have to pass long periods of operation without Earth-based coverage. So equipment is necessary to collect and store interplanetary information before relaying it on to Earth at an operationally convenient time. In addition, the spacecraft is literally going to have to look after itself: a malfunction signal will take so long to get to Earth, that by the time correcting commands have arrived back the breakdown of the particular unit in trouble might have gone too far for repair.

For this reason the outer-planetary spacecraft will carry STAR. Fans of Stanley Kubrick's film, *2001*, will recognise an infant Hal already in the making, for STAR means 'Self-Testing and Repairing Computer'. Basically, checks for malfunctioning would be by recognising that a

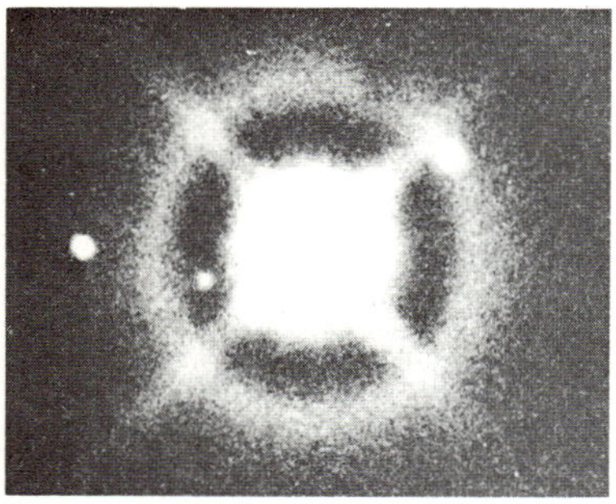

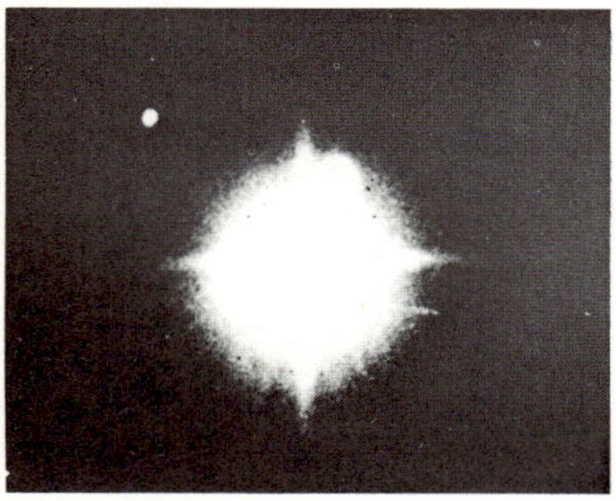

Top Saturn—third planet out from Earth after Mars and Jupiter. The mystery 'rings' may be composed of particles of ice, dust covered with ice, or polymerised formaldehyde. There are ten moons, one similar to Mercury

Centre Uranus—fourth out from Earth, the planet has a dense gaseous atmosphere. The telescope photograph has been over-exposed to show three of the five moons, but even if it were not, we should still be unable to see structure over the distance

Bottom Neptune—fifth out from Earth. Photo over-exposed to show one of the two moons, one of which is similar to Earth's moon. The planet is similar to Uranus in composition

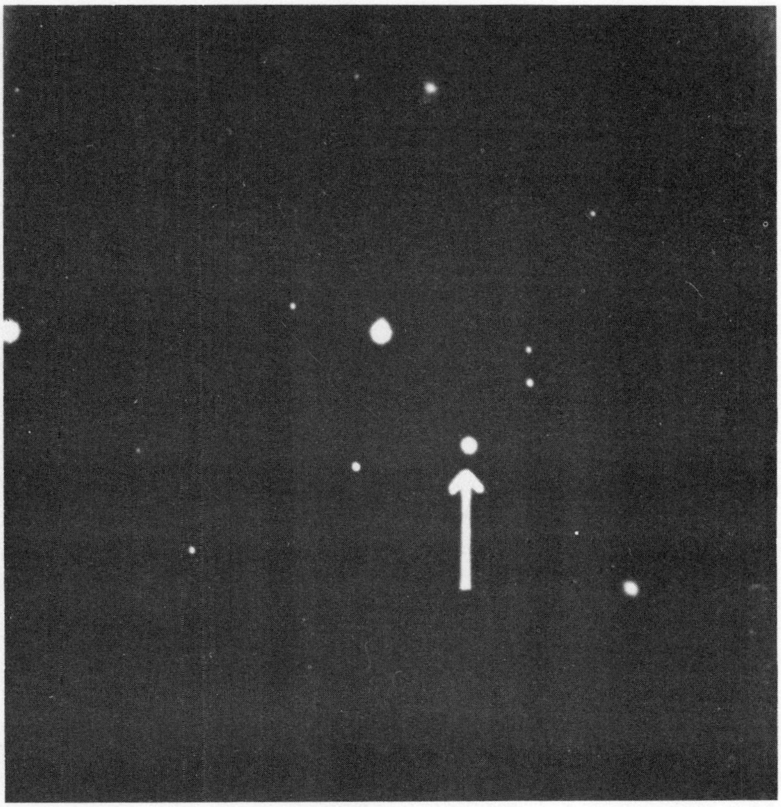

Pluto—the planet on the edge of the Solar System. No planet has been so far detected farther out. The best photo possible shows it no larger than the starfield beyond it. Size uncertain, probably about the same as Mars. It orbits the Sun in a different plane to all other planets, and may once have been one of Neptune's moons

'message' from a sub-unit is 'damaged', i.e. that the pattern, the way it is sending the message, is wrong. The computer would then switch in the back-up sub-unit. With this facility, the mission-planners can risk programming the spacecraft's memory and letting it correct its own trajectory anywhere on the journey.

The spacecraft will, of course, leave the Sun's light and heat far behind. So, as with the Mars Viking project, the power source will be nuclear–the radioisotope thermoelectric generator (RTG). Apollo 12 took one to the second Moon-landing to test it out.

The RTG, the antenna and STAR have dictated the spacecraft's design. The RTG dare not be placed too near the scientific instruments. Several of these will be taking measurements of tiny particles in interplanetary space and would be upset by RTG radiation. The antenna has to be at least 14 feet across if the spacecraft is ever to communicate with Earth. And STAR has to be on board. The resultant design of the Thermoelectric Outer Planet Spacecraft (TOPS) is very simple. The antenna (unfolded in space) is fixed onto a handlebar about 30 feet long. At each end of the handlebar there is an RTG, and slung in the centre behind the antenna is a science platform carrying STAR, two TV cameras, instrumentation, and about 100 lb. of scientific experiments to measure the radiation and magnetic properties of the planets, to find out what their atmospheres are made of, to take thousands of pictures of each planet, and indeed anything else of interest, since TOPS will be going where no one has ever gone before. Little is known of Saturn, even less of Uranus and Neptune–and almost nothing about Pluto.

And beyond Pluto, for $5\frac{1}{2}$ light-years there's absolutely nothing. By 1986 TOPS could be heading out across those $5\frac{1}{2}$ light-years of interstellar space. There is no hope that we will ever get signals back from TOPS as it speeds farther and farther out. But some unknown intelligence deep in interstellar space might.

Automatic Satellites

Nobody knows what effect our civilisation is having on the Earth and its atmosphere or how much needs to be done to correct such effects. Nobody knows how many resources are left for us to take from the land and the sea or, in many instances, where to find those resources. Nobody can predict with accuracy every change in the weather and what effect it will have, nor how our national food supplies will in consequence fluctuate from season to season on a worldwide scale. We live in a period of technological transition where, on many occasions, it seems that we are not masters of our own environment. If Man's efforts in orbit round the planet are successful, the next 20 years should see many of these problems solved by the use of manned and unmanned satellites.

Meteorology/The aim is not only to predict–but also to establish a scientific basis for future weather-control. A certain degree of success has already been reached by seeding clouds, but you have to know in advance what the natural cause of the weather is before you attempt any wide-scale alterations. The first step is 24-hours-a-day global meteorological information.

This information comes in three forms:

1 Pictures of global cloud-cover–for periodical checking of the clouds and tracking their formation and movement, and the dissipation of storms. This is the kind of information needed for the 24–36-hour weather forecast.

2 Continuous viewing of the atmosphere, making it possible to locate and follow short-term disturbances, which may in turn give rise to large-scale changes. This is also useful in providing the short 0–12-hour forecast.

3 Global analysis of the atmosphere structure–factors like temperature, wind, water-vapour. This will eventually provide a three-dimensional model of how the atmosphere may in most cases behave, and through the model permit long-term, 2–3-week forecasts and eventually, many years from now, some kind of control.

The Improved Tiros Operational Satellite (ITOS) will provide cloud pictures in two ways. On board will be two Advanced Vidicon Camera Systems (AVCS). From its 1000-mile polar orbit AVCS will take a series of wide-view high-detail television pictures of Earth and its cover, store the information on tape and transmit it to a ground station on command. This tape can be played back if necessary at speed so that meteorologists can actually see the last lot of weather 'happening'. Also on board will be the Automatic Picture Transmission (APT). This is a camera that takes 11 wide-angle pictures of the Earth and cloud as it passes over the daylight area (48 minutes in every hour and 36 minutes). Both these systems operate in

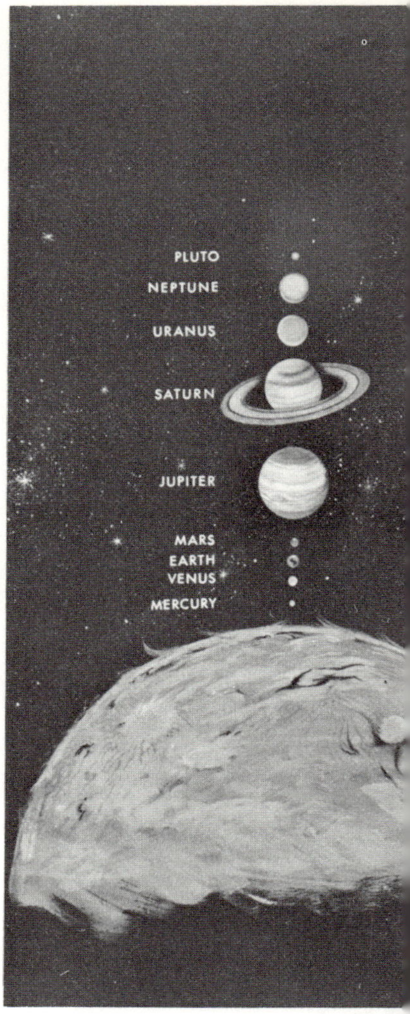

The Solar System. The left-hand drawing shows the planets' relative sizes, Jupiter being the biggest. Stretching across 9000 million miles of space, it holds 9 planets, 32 known moons, and thousands of asteroids, most of which are strung in orbit immediately beyond Mars. It is possible that more planets and moons may be discovered in the next decade. Two comet orbits are shown. So far no spacecraft has ventured beyond the orbit of Mars. Beyond the Solar System the nearest sun with planets around it, Barnard's Star, lies 5 light-years (38,220,000,000,000 miles) away, or about a million times the distance from Earth to the edge of our system

Nimbus D—the present work-horse of the meteorology satellite system orbits nearly 700 miles up, circling Earth every 107 minutes, viewing the entire planet on a polar (i.e. north-south) orbit twice a day. Nimbus weighs 1366 lb. is 10 ft high and 11 ft wide. Each Nimbus is designed to operate for 6 months

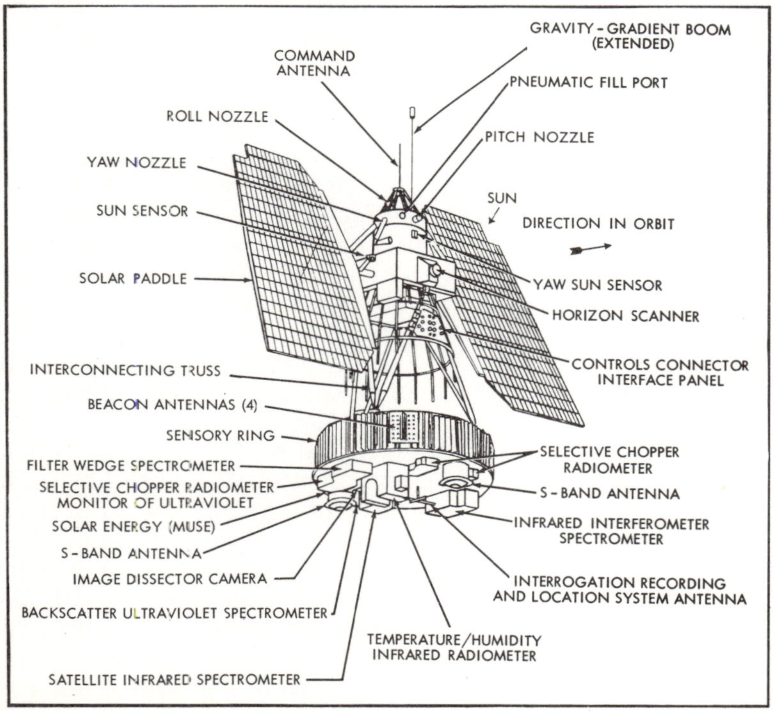
GRAVITY - GRADIENT BOOM (EXTENDED)
COMMAND ANTENNA
PNEUMATIC FILL PORT
ROLL NOZZLE
PITCH NOZZLE
YAW NOZZLE
SUN
SUN SENSOR
DIRECTION IN ORBIT
SOLAR PADDLE
YAW SUN SENSOR
HORIZON SCANNER
INTERCONNECTING TRUSS
CONTROLS CONNECTOR INTERFACE PANEL
BEACON ANTENNAS (4)
SENSORY RING
FILTER WEDGE SPECTROMETER
SELECTIVE CHOPPER RADIOMETER
SELECTIVE CHOPPER RADIOMETER
MONITOR OF ULTRAVIOLET
S - BAND ANTENNA
SOLAR ENERGY (MUSE)
INFRARED INTERFEROMETER SPECTROMETER
S - BAND ANTENNA
IMAGE DISSECTOR CAMERA
INTERROGATION RECORDING AND LOCATION SYSTEM ANTENNA
BACKSCATTER ULTRAVIOLET SPECTROMETER
TEMPERATURE/HUMIDITY INFRARED RADIOMETER
SATELLITE INFRARED SPECTROMETER

ITOS—A scientist completes final checks on the meteorological satellite's main structure. The black disc (top centre) is the communications antenna. On either side the four camera lenses can be seen. The shiny wrapping (bottom) covers and insulates the electronic command and operating circuits from the intense heat and cold of space. ITOS will provide world-wide coverage of the weather, under the direction of NASA's Goddard Space Flight Center

daylight; night coverage will be possible by the use of the Scanning Radiometer (SR) which provides a map of the heat given off by ground, sea and cloud-tops.

Continuous viewing of the atmosphere will be provided by four satellites in geosynchronous orbit. The orbit is set so that the satellite moves around at the same speed as the Earth rotates and is therefore always looking at the same hemisphere. The MET-ATS (Meteorological Applications Technology Satellites) due to go up in 1972 and 1973 will provide full Earth pictures every 20 minutes, night and day, to permit global cloud-movement to be monitored. Eventually the transmission of the Earth's cloud-cover would be continuous so that warnings of winds, rain and storm could be given with an accuracy of within two miles.

But it is the continuous analysis of the atmosphere-structure that will eventually enable us to control the weather. Nimbus 4, launched this year, is the prototype of future satellites that will form the Global Atmosphere Research Programme. Nimbus is butterfly-shaped. Power is derived, as usual, from solar panels. The job is to measure on a global scale the vertical profiles, from ground-level to the outer edge of the atmosphere at 400,000 feet, the changing temperature, ozone and water-vapour content. Temperature is measured by the Selective Chopper Radiometer (SCR) capable of measuring along a strip 7 miles long by 70 miles wide. The Filter Wedge Spectrometer (FWS) measures the amount of water-vapour present. The Backscatter Ultra-violet Spectrometer (BUVS) will measure the ultra-violet radiation that is backscattered into space by the atmosphere. This will indicate the levels of ozone present in the air, since one is related to the other.

By 1980 there will be, then, four geosynchronous, two polar-orbit, and several other satellites up there. Between them they will provide for totally accurate 2-week forecasting. A saving in lives, property and crops should be one of the most profitable results of the programme.

Navigation and Communications/Air Traffic Control over sparsely populated areas and oceans is an extremely difficult business. Since, for example, on an intercontinental route it is left to the navigator on board to work out where he is, allowances are made for error and aircraft are separated by at least 10 minutes to allow for any mistakes. This obviously cuts down on the potential density of air traffic in any given area. Experiments have already shown that Air Traffic Control satellites will probably enhance the use of airspace by almost five or six times.

A central station on the ground will transmit a coded signal to two Air Traffic Control satellites, both of which in turn broadcast it to every vehicle in the area. Such an area might be, for example, the Atlantic Ocean. The coded signal is an interrogation signal asking, in effect, 'Aircraft XYZ, where are you?' The transponder in the nose of aircraft XYZ reacts to its own code and sends out a recognition signal which is picked up by the two satellites and retransmitted to the Earth station. A simple measure-

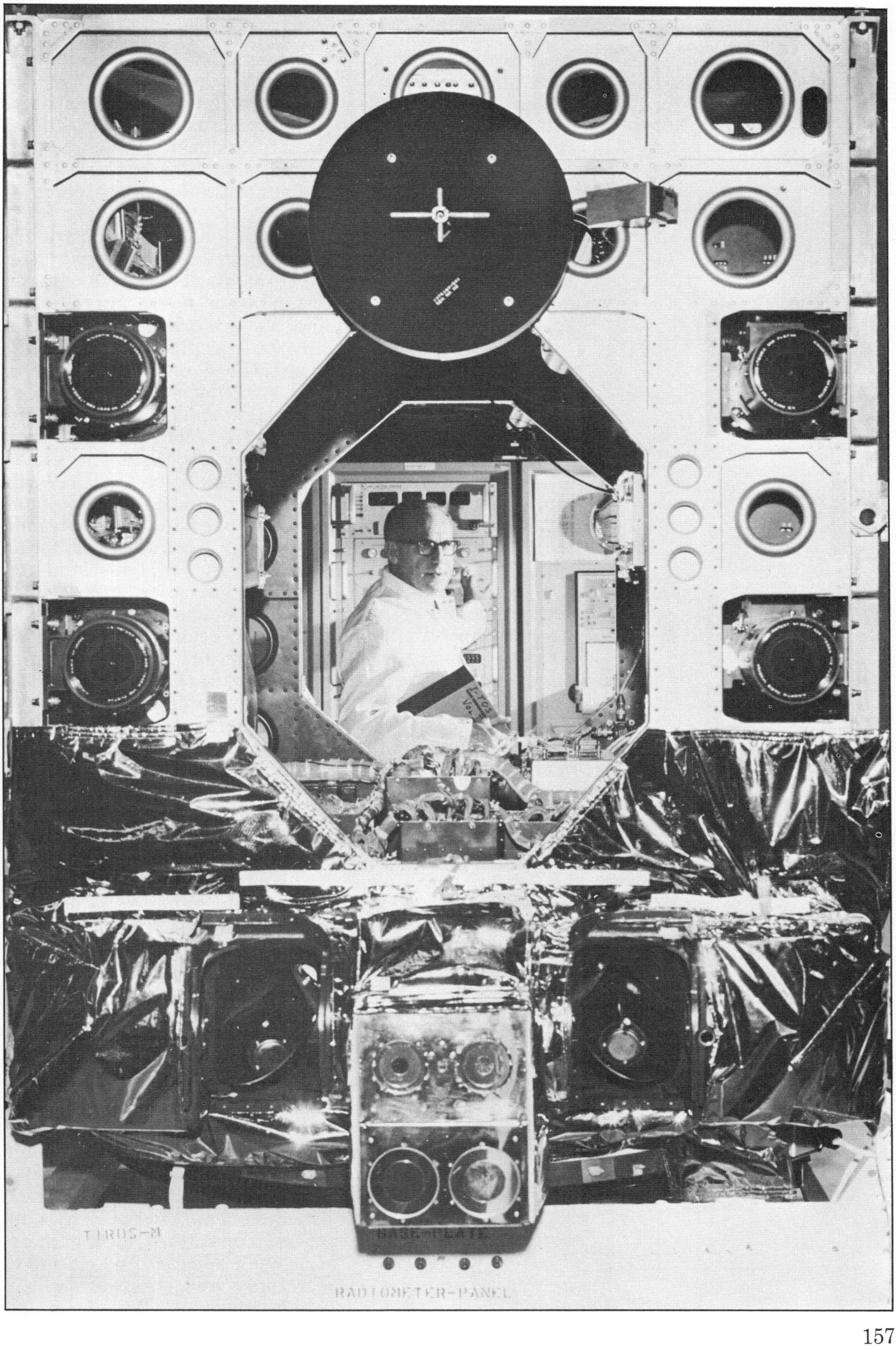
TIROS-M
BASE-PLATE
RADIOMETER-PANEL

ment of how long it took the signal to reach the two satellites from the plane and then back to the station gives the position of the aircraft. At the moment tests have placed aircraft to within 3 miles–better than the present system works–but eventually spot identification will be achieved to plus or minus a number of feet. This system will work as well for seacraft and land-based vehicles. The great advantage is that it permits interrogation of very large areas and continuous monitoring of every vehicle in the zone.

Communications Satellites will aid Traffic Controllers in talking to vehicles when necessary. At the moment the Earth is already covered for transmission by a set of three geosynchronous satellites, placed over the Pacific, Atlantic and Indian Oceans. Their working life is about 5 years, after which they have to be replaced. This should no longer be true once the spacebase and shuttle are up, in the late 1970s, since repairs will be carried out where possible from space, to save money. The only problem with the Communications Satellites, whether the large intercontinental ones, or the many local satellites that will be orbiting 10 years from now, is what traffic they can handle. At the moment they take phone calls, television transmissions and computer hook-ups from one continent to another. The cost is considerable since space on board is limited and only so many sets of circuitry can be packed in. But it looks as if the laser beam will solve the problem. With a giant, many-sided mirror in space, a laser beam can be bounced off each side, each beam simultaneously carrying hundreds of thousands of communications channels. It has already been done in the laboratory. By 1980 it will be happening in space–and we will be witnessing an exchange between continents and races that will make the so-called information explosion of the '60s look like a Women's Institute tea-party.

Earth Resources/The information we get back from all the non-military satellites and manned spacecraft that go up (except for Communications Satellites which move the information to and fro) comes back to Earth to be sucked dry of any potential it might have for helping us here on our planet. Exploration of the Moon and the planets, the Sun and interplanetary space will one day tell us how we came into being. But as we move towards that knowledge, as we learn more and more, we find information that can have more immediately profitable applications, especially in that all-important field of keeping ourselves alive.

Apart from the technological spin-off from man's efforts in space that is being used in industry, medicine, food production, etc., a wide-ranging programme is just about to begin which is expected to bring benefit just as soon as it starts. It is called the Earth Resources Survey–and its eventual overall aim is to examine every inch of the surface of the Earth and the oceans, in order to map new sources of raw material and to maintain the resources we already have, so as to provide adequate world food-supply, satisfactory water-supply and availability, an adequate supply of mineral resources, efficient use of

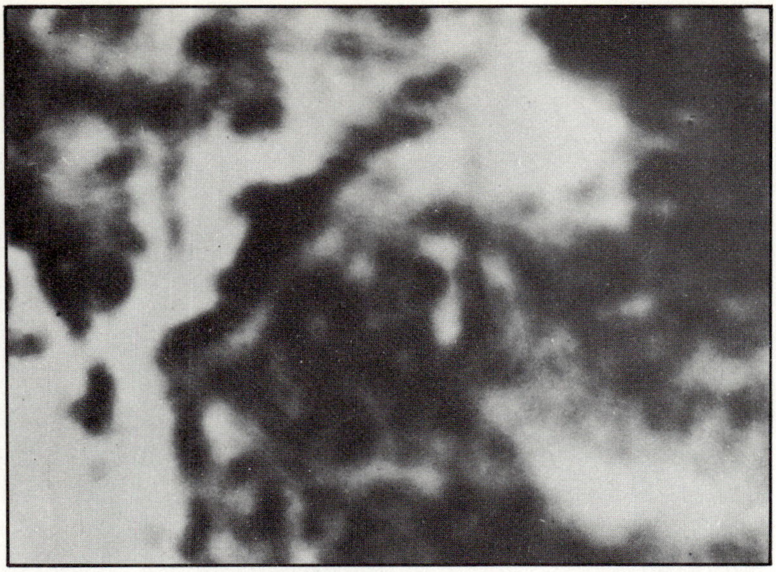

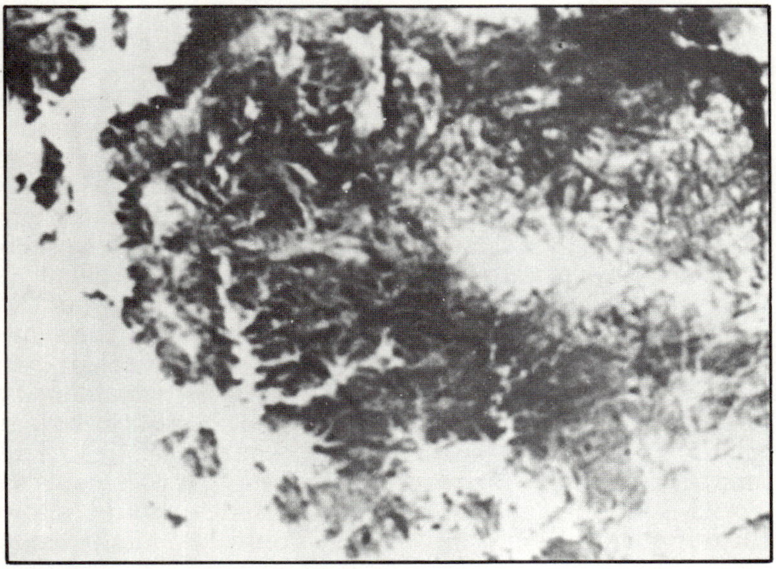

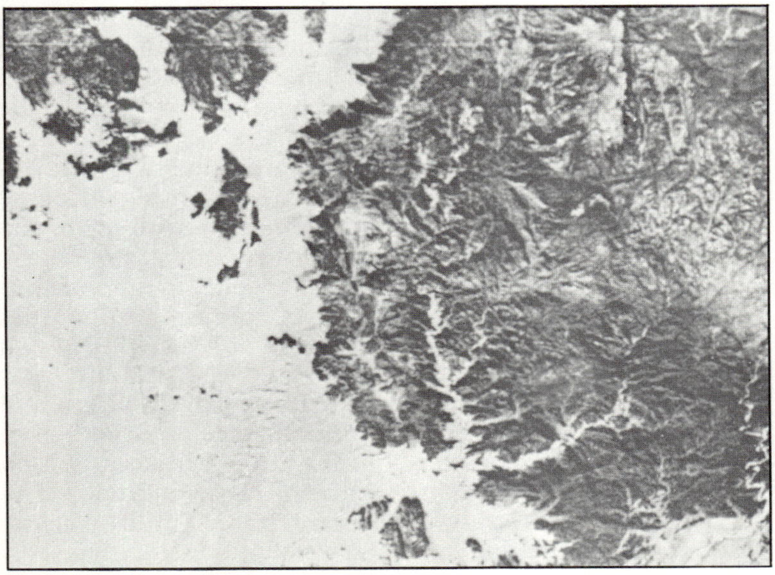

What you can see from space—three pictures taken of an area in Saudi Arabia using a camera at different heights. As it comes close it sees more. *Top* The photo showing features bigger than 1.4 kilometres in size reveals only vague light and shade forms. *Centre* showing features down to 100 metres in size, ground details begin to emerge. *Bottom* feature detail down to 10 metres in size reveals mountain slopes, river-beds, rock outcrops and an idea of height difference from the shadows thrown by the sunlight

Orbiting information centre—an artist's concept of the Earth Surveys Module floating above Earth at the end of the century. Connected by radio to computers on Earth, it will flash pictures and analysis data on demand to geologists, farmers, oilmen, underwater explorers—anyone who needs it in order to get advance information on how their crops are doing, where to look for oil, or gold, or other minerals, where to lay a new road that will not suffer subsidence. The module could be steered automatically to move over the area you want analysed, and would return to a spacebase for repair and maintenance

land, well-planned urban development and control of pollution. Clearly much of the work of the ERTS (Earth Resources Technology Satellite) will be aided by the MET-ATS (Meteorological Applications Technology Satellite).

It was discovered by successive manned missions since Mercury that the higher you go the deeper you see–below the surface of the oceans, and to a certain extent, below the earth itself. A series of photographic experiments carried out on Apollo 9 by McDivitt, Scott and Schweickart produced extremely promising results. It was seen that the use of a multispectral camera (i.e. several cameras carrying different-coloured filters taking a picture of the same spot) revealed different information about the same place. For example, a certain green will reveal underwater information, the reds and near infra-reds identify crops and specific soil features, and near infra-reds reveal plant disease and surface-water characteristics. The implications are far-reaching indeed. New resources could be found because their presence would show through the filter; plant disease could be caught early because the plant's move from good health to the first hint of disease will show up long before it could even be detected from ground-level. Ground stresses are revealed in many cases through heat changes at the surface. The appearance of a long thin line of heat across the surface could well give long-term warning of a developing fault and permit evacuation, or indeed help prevent the development of a town in that direction. The integration of information on crops, soil and general features in one area with an analysis of snow cover and atmospheric changes in another could reveal weeks in advance if farmers and low-lying towns should prepare for flooding. A fishing fleet need only tell the satellite what kind of fish it is looking for to be told where to go. The list is endless.

The first steps will be taken in 1972 with the launch of ERTS-A and B, carrying multispectral scanners and cameras, as well as on-board data-processing which converts the pictures to transmittable signals. The objectives of A and B will be to provide 1:1 million scale, photographic images of the Earth, plot geological features, produce plots of land-use, plot coastal areas, snow and ice cover, and obtain a record of changes that occur throughout the year. Both satellites will be 496 miles up on polar orbit and Sun-synchronous. That means that the cameras will cover a swath 100 miles wide going north to south and staying with the Sun, i.e. letting the Earth rotate under them.

Propulsion

In all the plans for future space-exploration, whether they involve a lonely Pioneer satellite on its way out to the very edge of the solar system, or shuttling men and materials back and forth on the Earth-Moon-Mars trajectories, the most important factor is fuel. At the moment that is what costs the most after the initial design and research on the vehicle. And with the amount of activity planned for the 1980s fuel costs would be prohibitive if we were to go on using today's chemical rockets. A new fuel technology is needed.

The latest, and possibly last, chemical fuel to be used as a producer of main thrust in order to kick a manned mission out of Earth orbit into deep space is a mixture of liquid oxygen and liquid hydrogen. When the engine turns on, the two fuels–kept in super-cool liquid state in order to store more in the tanks–are released. They become gaseous as their temperature rises, and then they are ignited. Ignition raises the temperature of the fuel, the gases expand and rush out of the rocket engine-nozzle at high speed, pushing the spacecraft in the opposite direction. There are two crippling limitations to this type of propulsion:

a The velocity at which the gases leave the nozzle is limited by the heat produced by the ignition temperature of the gases themselves;

b that heat will only push the expanding gas molecules at a certain speed. The hydrogen could go faster because it is the lightest gas known and can be pushed around more easily, but it is held back by the sluggish oxygen molecules with which it is mixed–and the molecular convoy, so to speak, can go only at the speed of the slowest ship.

The result of these factors is that, since high speed cannot be produced by a short sharp burst of gases at very high exhaust velocity, it becomes a long process, building the speed up gradually with a long firing of the engine. That in turn uses up the fuel faster, so you have to carry a great deal. Over $\frac{7}{8}$ of the total weight of a Saturn V is fuel, all of which burns off in less than 20 minutes.

The task, then, is to build a rocket engine that can work on hydrogen alone. It is the fastest gas around when heated and will produce twice as much thrust as any other mixture of gases. So the second task is to find a way to heat it without involving it in speed-reducing ignition with a slower gas like oxygen. NERVA seems to have solved both problems.

NERVA–The Nuclear Rocket/The Nuclear Engine for Rocket Vehicle Applications consists very simply of a tiny nuclear reactor which gives the hydrogen the heat it

NERVA—the Nuclear Engine. Hydrogen enters from the fuel tanks (top right), driving a turbine pump which then moves the hydrogen through the cylindrical shell of the reactor to cool it. The fuel returns to enter the reactor at the top end of the cylinder, heats as it passes through the radioactive core, and exits through the nozzle. The reactor contro'-rod tops can be seen around the top end of the reactor cylinder

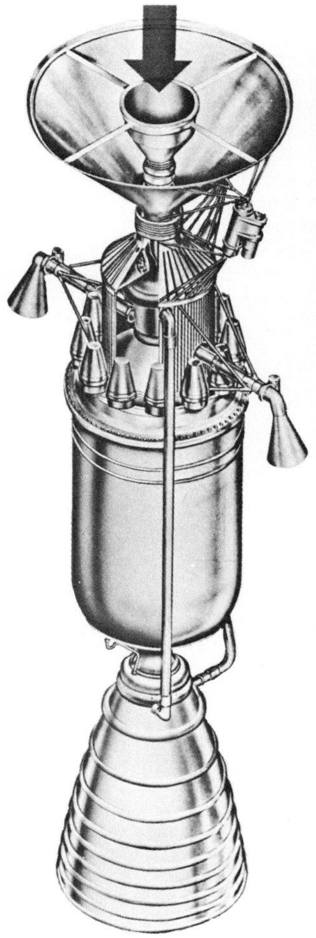

NERVA under test. *Top* shows the latest experimental engine (XE) on the test stand at the Nuclear Rocket Development Station in the Nevada Desert. Data from these tests will aid the final flight-engine construction

Top right illustrates the hydrogen flow through the engine

Bottom right In orbit 1990—an artist's concept of manned-flight activity in Earth orbit near the end of the century. An interplanetary nuclear tug (centre) returning from outer space arrives above Earth. The astronaut crew carry out minor repairs on the nose, as automatic grapples take on an extra fuel-tank brought up from Earth on board the shuttle (bottom right). Already the shuttle's directional nose-jets are firing to move it away

needs. Pound for pound of fuel it produces twice as much thrust as any engine flying today.

As with today's chemical engines, the hydrogen is stored in liquid state. In operation the fuel flows from the tank down and round the outside of the reactor and the engine-nozzle in order to keep them cool during firing. It then flows back up around the reactor and passes through many tiny channels cut through the radioactive fuel element in the reactor itself. The tremendous heat of the reactor core turns the hydrogen almost instantaneously into high-temperature gas, which is expelled through the engine-nozzle to produce thrust.

The requirement is that NERVA should be able to come to full power in one minute, and that it be controllable. Throttling is the most delicate task of all, since it involves extremely precise control over the nuclear reaction itself. Around the radioactive graphite core are placed many rods running its whole length. These rods can be rotated. One side of the rods absorbs the neutrons given off by the nuclear fuel, the other side reflects them back into the core. If the neutrons are absorbed, nothing happens. If they are reflected back into the centre, they collide with other neutrons being produced by the radioactive core, and this creates tremendous heat. The less of the reflecting side of the control rod facing inwards, the more neutrons are absorbed and the lower the temperature of the core. So to build up thrust slowly, for example, you simply run the hydrogen through the core and slowly revolve the control rods until the side facing inwards is all neutron-reflecting, when you have reached maximum heat and therefore maximum thrust.

In outward appearance a nuclear-powered rocket will look much like the upper stage of a Saturn V. The main feature will be the huge cylindrical tank for the liquid hydrogen. When the first test flight occurs in 1978, a two-stage Saturn V (known as the INT-21) will take the third, nuclear stage, to orbit. The nuclear engine will be used only in space, since the stream of hydrogen gas leaving the nozzle will be radioactive. The crew compartment will not suffer from the Apollo capsule design limitations, since, as the space shuttle will also be in operation at that time, the nuclear space capsule will never have to re-enter Earth's atmosphere and therefore will not have to conform to aerodynamic design laws.

Eventually replacement hydrogen will be flown up on other Saturns, and reactor replacements will go via the shuttles. Once in space the nuclear ship will never return to Earth. Nor will it ever approach any orbiting spacebase because of the radioactivity. At a discreet distance it will stop engines and a shuttle will be sent out from the base to pick up the cargo of men and materials. For the nuclear crew the isolation will be rather like living on a lighthouse down on Earth.

Beyond the 1980s comes a totally new form of propulsion—the ion engine. The first one is already in orbit round the Earth. It's an infant as yet, producing only one thousandth of a pound thrust, but eventually it will double the thrust capacity of even the nuclear rocket. It acts as an accelerator. A material such as Caesium is heated; the heating produces charged particles which fly off at speeds up to five times the highest speed attained by hot hydrogen. These particles, when electromagnetically directed, in effect kick the hydrogen atoms along faster. The particles themselves can also act as thrust producers, though on a smaller scale. Their advantage is that minute amounts of material give off the particles continuously and in enough quantity to increase the speed of an interplanetary vehicle very gradually. In this way over a long voyage the speed of such a spacecraft could be built up to what was previously an impossibly high velocity. At present this looks like the only propulsion system likely to take a manned-flight to the outer planets by the end of the century.

Beyond the outer planets lie the unimaginable vastnesses of galactic and inter-galactic space. There is no known propulsion system, based on any known physical laws, that will permit us to cross those frontiers within the span of one man's lifetime.

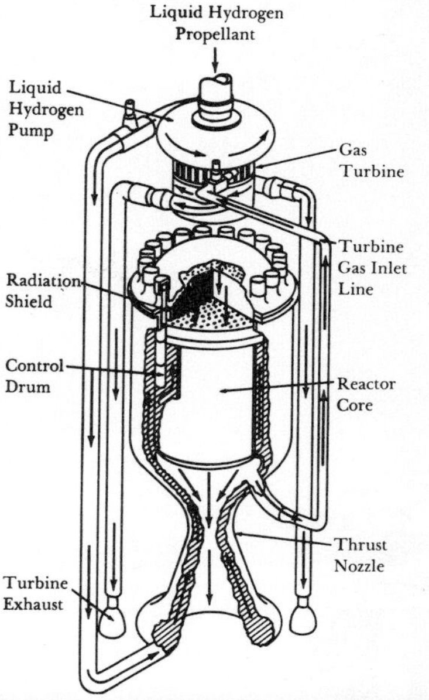

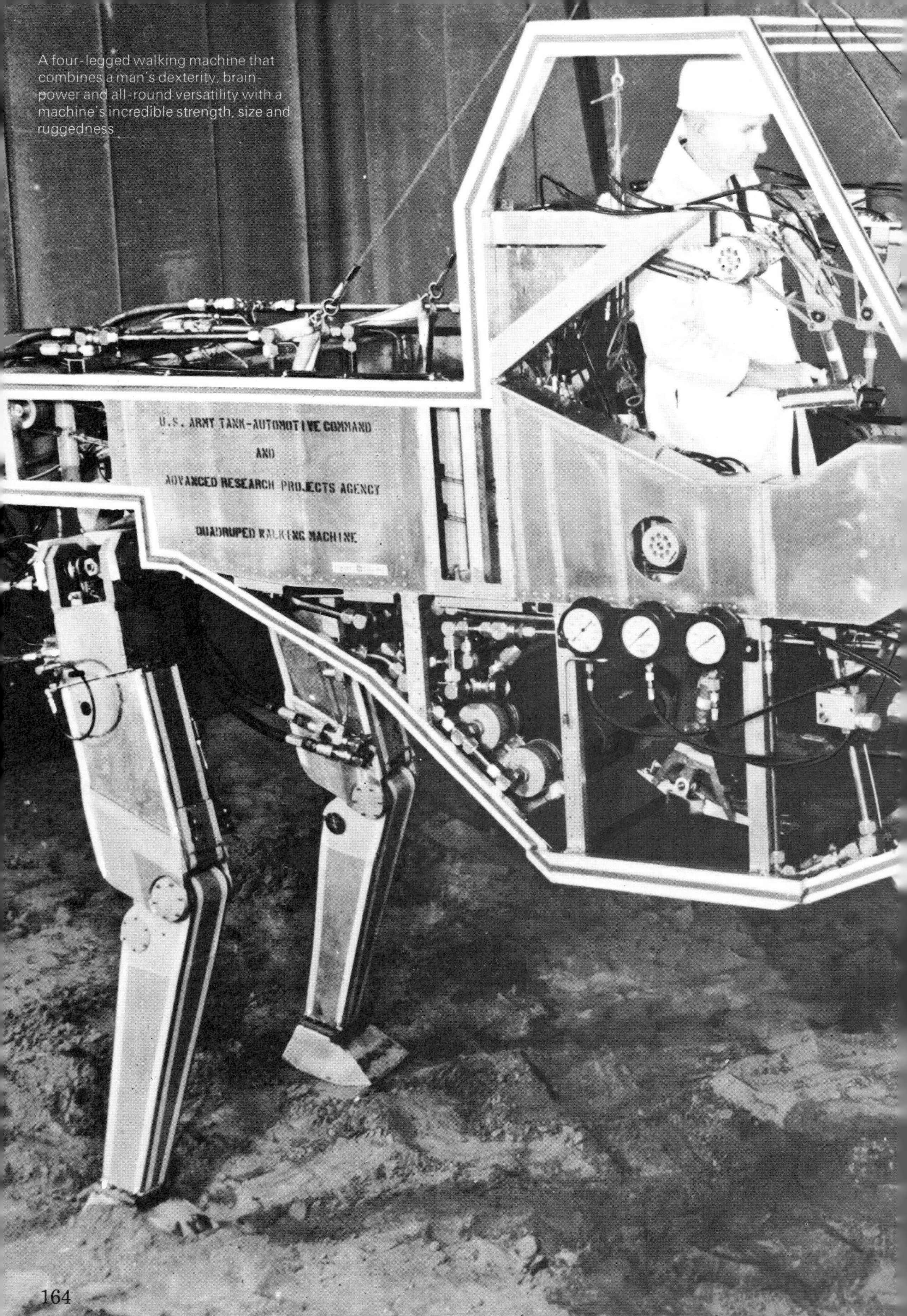

A four-legged walking machine that combines a man's dexterity, brain-power and all-round versatility with a machine's incredible strength, size and ruggedness

Machines like Men

Galaxy/Machines for Seeing/Listening by Laser/Space TV/Pictures in the Dark/Seeing through Solids/Jet Probe/Describing Colour/Machines that make Decisions/Computers that learn/Computers that decide/Fluidics-The Modern Pipes of Pan/AND/OR/NOR/NOT/Machines with Nerves/The Ultimate Power Source/

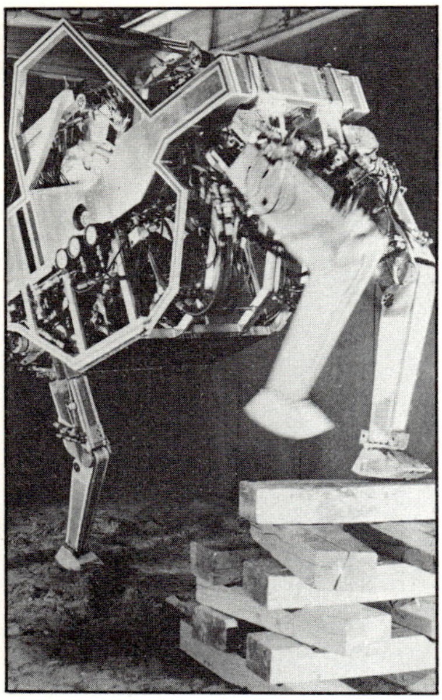

This machine climbs obstacles and balances on its two back legs like a metal prehistoric monster. It walks delicately along narrow paths at 5 mph using power from a petrol engine, hydraulic pumps, and accessories such as heat exchangers, filters, valves, and an accumulator. General Electric scientists who created the machine for the US Army believe that one day its grandchildren will be fording rivers and galloping up mountainsides

Recently the United States Army and the General Electric Company announced 'A Continuing Programme to Develop Advanced Man-Amplification Machines for Military Applications'. They meant that they had built a $1\frac{1}{2}$-ton prototype machine which, they claim, 'combines its operator's dexterity, brain-power and all-round versatility, with a machine's strength, size and ruggedness'. In plain terms, it is a device intended to help engineers to develop equipment that will improve the mobility and capability of the foot soldier in the most severe operational conditions.

By means of an advanced system of levers, control-linkages, and servo-mechanism, the walking machine mimics and amplifies the movements of its operator. The right front leg of the unit is controlled by the operator's right arm, its left front leg by his left arm and so on. In an initial series of tests, the prototype robot successfully walked across level ground, turned round, climbed obstacles, lifted a small military vehicle out of a mud hole, and hoisted a 500-lb. load on to a truck with one foot.

Whether or not one is tempted to laugh at the comic potential, or cry at the nightmare implications of this achievement, is neither here nor there. The fact is that the entire purpose of building machines is, and always has been, 'Man-amplification'. Nowadays, mankind is getting remarkably clever at reproducing and improving upon his own performance. The range of 'Machines like Men' widens month after month.

Galaxy

It is, nevertheless, surprising to learn that a machine has been developed to pursue the time-honoured human habit of star-gazing. It is called Galaxy, and has been developed at the Royal Observatory, Edinburgh, where astronomers use it every night when the sky is sufficiently clear to make observation possible.

Galaxy is a form of seeing-computer. It uses extremely sensitive photoelectric cells to scan photographs of the sky and record precisely the location, size, and intensity of each minute speck of light which indicates the presence of a star. These details are stored in a computer. Presented with a second photograph, the machine can then compare it with the mass of detail recorded from its first observations, and if any star has changed its location, size or intensity, the computer will instantly recognise and indicate this fact. Galaxy is being used in conjunction with the 16″ Schmidt telescope, and an Elliott 4130 computer system. Now a second machine has been ordered for the Royal Greenwich Observatory at Herstmonceux, for installation in 1971.

Although the Edinburgh system has been in operation for only a short time, it has already recorded and measured more stars than ever before. It is the first machine of its kind anywhere in the world to bring complete automation to the processes of optical astronomy. An indication of its accuracy is that the required specification demanded a precision of one micron in the position of the star-image, and a quarter of a micron in the size, and that the machine should find and measure 1000 star-images per hour

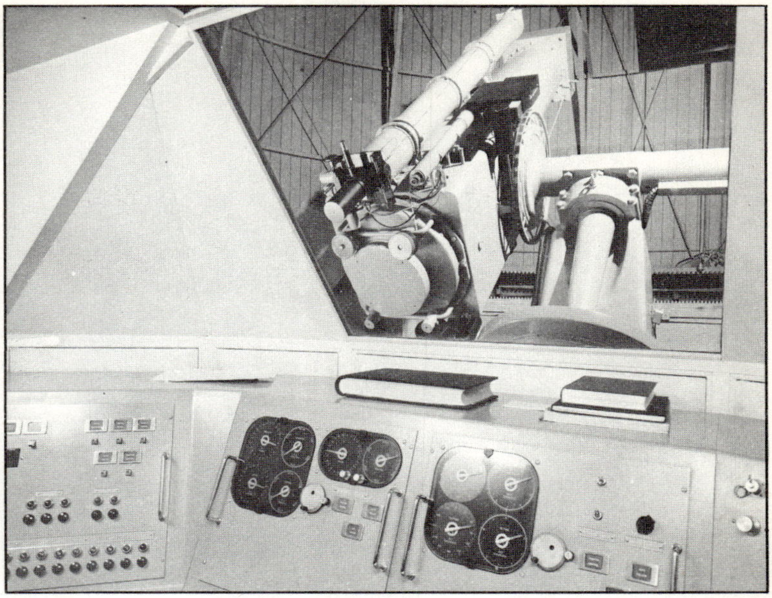

Above From photographs of the night sky like this, a new computer called Galaxy has in only a few months measured and recorded more stars than in the whole previous history of astronomy. Galaxy, developed at the Royal Observatory, Edinburgh, uses extremely sensitive photoelectric cells to record the location, size and intensity of each star on the photograph. The information is stored in a memory bank until a later photograph of the same area is fed to the computer. Then, if any star has moved or changed in size or intensity, the computer will spot it and give warning. Galaxy is due to be installed at the Royal Greenwich Observatory at Herstmonceux in 1971. Meanwhile doctors are becoming interested in it because it could help them study human blood cells and, perhaps, lung infections caused by microscopic particles of dust

Right Royal Observatory, Edinburgh, where Galaxy is used in conjunction with a 16-inch Schmidt telescope and an Elliott 4130 computer

entirely automatically. It has now been run for more than 2000 hours. There has been no major breakdown, and only a negligible amount of time has been lost through minor faults. The astronomers now face a major task in programming the computer to be able to cope with the tremendous potential output of this machine.

Galaxy, and its capability for observation, is now arousing medical interest. It can, for instance, study, record, and compare the kind of dust particles that cause silicosis in human lungs with other irritant substances. It can also be used in the micro-study of blood and in many other applications involving the comparison of micrographs or electron micrographs.

Machines for Seeing

Electron Microscope/Although the electron microscope is no longer new, the constant addition of devices to improve its performance, and the ever-expanding range of applications to which its immense power is being applied, mean that the impact which the instrument has yet to make can hardly be underestimated.

For instance, the *scanning* electron microscope, or SEM, is a relatively new tool that offers the scientist or engineer both high magnification and a remarkable depth of focus, a combination of characteristics not possible with optical or earlier transmission electron microscopes. It produces almost three-dimensional images of incredible quality, such as magnifying an ant's head. But, more valuable in the technological age, the SEM can examine such samples as electronic circuits for flaws and minute fractures by scanning them with an electron beam in a vacuum, and then detecting the electrons that are scattered from the surface under study. A picture of the sample is then created on a cathode-ray tube, in much the same way as a picture is formed on a television screen.

One of the most important research centres of the British Textile Industry is the Shirley Institute of the

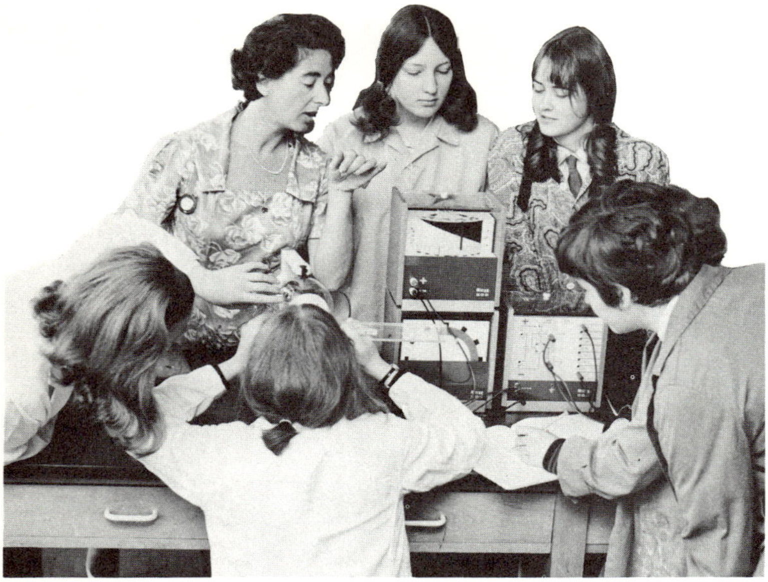

Today experiments are conducted in fifth-form science classes which 20 years ago were taught to second-year university students. To keep up with changing needs new and more sophisticated equipment is required in school laboratories

Cotton, Silk and Man-Made Fibres Research Association at Didsbury, Manchester. Speaking of the Institute's newly acquired electron microscope, Stanley C. Simmens, Head of the Microscopy Section, said, 'Beyond what the eye can see, even when aided by an ordinary optical microscope, there is a whole wealth of information that can be made use of to improve materials and to explain their characteristics and behaviour. Some of the electron microscope photographs taken in the short time since the equipment was installed have already revealed new information about the structure of man-made fibres.'

At Didsbury, the new electron microscope is being employed to provide much-needed information on the internal structures of fibres and also on the nature of fibre-surfaces. The way in which molecular chains are arranged into groups or bundles within the fibre is a key to its properties. The information which the electron microscope provides on the nature of these groupings, which are too small to be visible under even the most powerful optical microscope, will lead to an understanding of the changes in characteristics produced in fibres by chemical and mechanical processes, including texturing as well as other newly-developed treatments. The work is also relevant to the study of adhesives technology and dyeing, and since it is the surface of textiles that comes into contact with the skin, greater knowledge of fibre-surfaces will also have an important effect on the kind of garments that find their way into the shops in future.

It is perhaps easier to understand the power of the electron microscope and its importance to this type of research when you realise that a single thread from a 15 denier nylon stocking, when magnified, could have a width of about 30 feet.

Electron-probe Micro-analyser/Another new device closely associated with the development of the electron microscope and which also has staggering powers of magnification is the electron-probe micro-analyser. Amongst

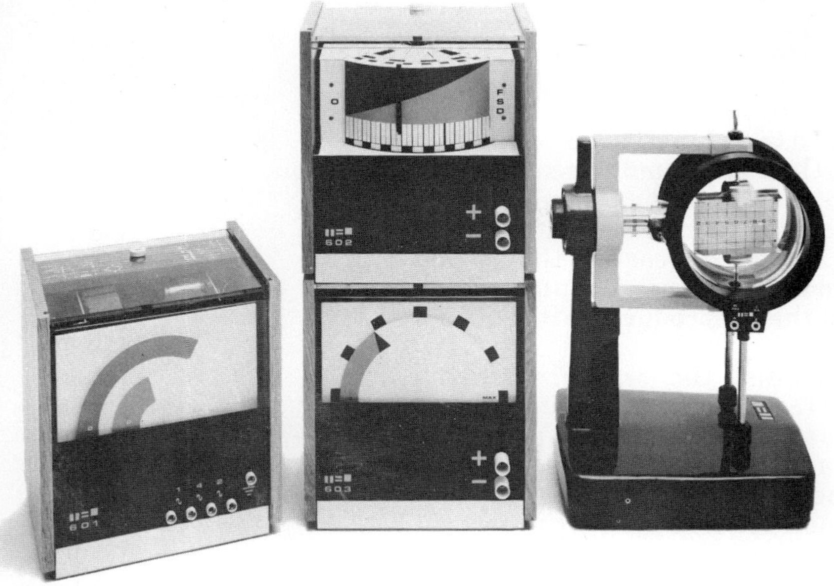

The hardware for a new approach to atomic physics in school. The high-tension module (left) supplies up to 500 volts. The dc kilovolt module (lower centre) supplies 5000 volts. The apparatus above it (upper centre) is a universal meter which could be used to measure the output of any of the supplies, or the current through the vacuum tube (right of picture) which is used to measure the ratio of the mass and charge of the electron. When the experiment is in progress a blue line appears across the screen in the vacuum tube. It can be moved up by applying voltages to the plates above and below the screen, and down by the magnetic field from the Helmholtz coils fitted either side of the tube. When these cancel out the relationship between electric and magnetic fields, others can be investigated. The modules have transparent tops so that the student can see the works inside

the newest instruments available to the research physicist, it can analyse the elemental composition of a material by bombarding it with a stream of electrons and examining the energy 'bounce off'.

Minute Mass

This function is comparable to another instrument employed in the investigation of the 'ultra-small'–the mass spectrometer. This uses electric or magnetic fields to sort ions, charged atoms and molecules according to their mass. In order to accomplish this, conventional mass spectrometers have up to now required electrodes with precise contours, machined to tolerances of less than one thousandth of an inch, which are hard to align.

Recently a unique new device which uses coarse wire-mesh similar to anti-mosquito screens for its electrodes has been announced in the United States. It is claimed that the new instrument can outperform the conventional mass spectrometers equipped with electrodes machined and aligned to the most exacting tolerances obtainable. It weighs less than 2 ounces and can analyse the composition of gases in concentrations as little as one thousand trillion times less than normal atmospheric pressure.

The small size and high sensitivity of the device make it ideal for applications on board spacecraft, but there are many other industrial and research potentials. It could, for example, serve as an air-pollution monitor to detect contamination. Medical laboratories might use it for blood-gas and respiratory analysis.

Listening by Laser

Holography, a technique for producing a three-dimensional image of a subject, uses split laser beams. But lasers can also be used to transmit as well as to reveal.

The scientists employed at the Lockheed Palo Alto Research Laboratory have just announced what they describe as 'a giant step forward in the communication capacity of a single channel'. They claim that by using the technique they have just perfected, over seventy television channels could be transmitted by means of a single laser beam using a microwave sub-carrier with a bandwidth of one billion cycles. To substantiate their claim, the research team concerned transmitted every available VHF television channel in their area simultaneously over a shaft of green light from a laser. At the other end high-quality television pictures were reproduced on a battery of normal television sets. For good measure, they added a computer-display link, and an FM music channel, which were also transmitted and then reproduced without loss of quality.

Like the electron microscope, the versatility of the laser beam in an aspect of technology still in its infancy makes it almost impossible to underestimate its potential, both in direct industrial applications and towards relieving pressure on the overcrowded lines of communication causing so much concern all over the world (see p. 98).

Space TV

The disaster which overtook the Apollo 12 colour-television camera after some 45 minutes of operation during the first transmission of colour-television pictures from the Moon provided an intensely dramatic spur to those technologists already engaged in the endless effort to improve colour-television cameras. If pointed at the Sun, lenses in a conventional television camera concentrate the light and heat into a tiny spot on the sensitive target of the television tube. The target's temperature soars as much as 2 million degrees per second and burns out almost instantly.

The Westinghouse Corporation, makers of the Apollo television cameras, have now developed a television camera tube that will not fail if pointed accidentally at the Sun. It also largely eliminates the problem of over-exposure, one of the serious hazards of television outside-broadcast techniques–particularly in space. The key to this development is the perfection of a heat-resistant target which is much less susceptible to burn-out, even when exposed to images one hundred thousand times brighter than normal operating levels. Conventional 'targets' consist of a layer of aluminium oxide, a thin film of aluminium and a layer of potassium chloride, so light and porous that 99 per cent of it is a near-vacuum. The aluminium oxide provides a supporting layer, the potassium chloride is the storage element, and the aluminium is the layer from which the actual television signal is produced.

In the new 'targets' a fine metal screen is used in place of the aluminium oxide support. The mesh acts as a heat absorbant and in addition, since it is a good conductor, also serves as a cooling mechanism to prevent the local build-up of excessive heat. This permits a much larger over-exposure of the target without ill-effects and with no reduction in picture-quality.

The heat-proof capability of the tube will prove of value in industrial applications where there is a risk of accidental over-exposure of a sensor. There can also be little

A television camera tube that largely eliminates the problem of over-exposure and 'burning out', the most serious hazards in television from space. The new tube has a unique heat-resistant target which cannot be damaged if the TV camera is accidentally pointed directly at the Sun. Such an accident occurred during the Moon walk of the Apollo 12 mission

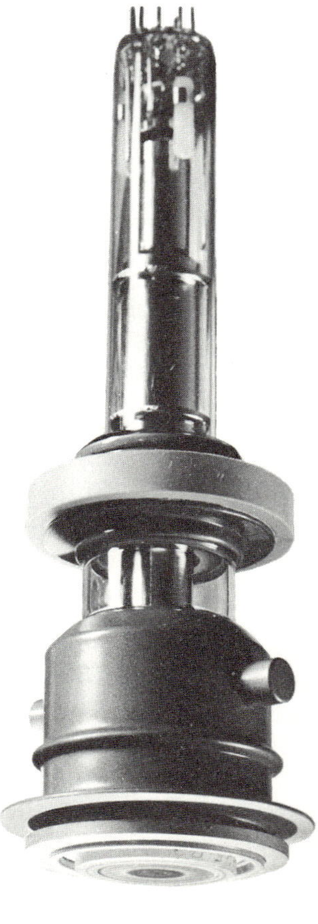

doubt of its significance in terms of military applications.

Pictures in the Dark

At the opposite extreme of the lighting problem for colour-television systems another significant advance has also been reported–this time from the Standard Telecommunication Laboratory at Harlow in Essex.

From our *Tomorrow's World* studio we have successfully transmitted television pictures using the light of only a single candle. In the case of the low-light television camera, the key component is an image-intensifier tube which is attached to a standard vidicon television camera. This system is so sensitive that it will produce reasonable television pictures even by starlight. A fast-acting, automatic gain-control circuit is also incorporated and, since the whole system is highly portable and can, if necessary, be powered by a battery, it has many potential uses. It could be used for unobtrusive night surveillance by police and security organisations, or as a navigational aid on land, sea, air or under the water. A television camera that can see in the dark could be of immense use to astronomers or to zoologists studying nocturnal animals in the wild. Work is also going on to use the technique in war-zones, to monitor front-line positions at night.

Seeing through Solids

Perhaps even more remarkable is a new photographic technique called Neutrography, which is capable of taking a picture of a burning candle behind a solid wall of lead. This results from an intensive three-year development programme in Pleasanton, California, and permits the photographic inspection of 'hardware' that could not even be effectively examined by X-rays.

Its true value lies in the fact that it offers industry a new method of detailed, non-destructive, internal inspection. The production-detail and quality-control of metal-clad internal structures can be easily checked in such 'problem' cases as dry-cell batteries, guns, and miniaturised mechanisms of almost every kind.

The technique is similar in some ways to that applied to X-ray photography. The object to be neutrographed is placed in a neutron beam produced by a small nuclear reactor. Neutrons pass through it and create an image on a sheet of film. The total neutron cross-section allows observations and measurement of internal details within metal containers without resorting to any type of destructive test or stripping-down.

The same technique can be employed to study the flow of liquids within storage tanks and fuel systems, and to examine the condition and distribution of explosives used in the latest types of missiles. If a living plant is fed a boron solution, which is opaque to the neutron beam, neutrography can even be used to record the feeding cycle about which, even to this day, comparatively little is known.

This new technique goes on where X-rays leave off–and it is because of this that it could have an important future in industry.

Above The television set is now an accepted teaching aid in almost every classroom where pupils watch educational broadcasts from BBC and ITA. But some schools are going further: they are installing television videotape recorders and even their own fully equipped electronic studios to produce programmes tailored to the needs of their own classes. This studio has been built at a school in Dover

Right Television cameras fitted with zoom lenses. Though they can be operated by amateur cameramen, they are sophisticated enough for professional use, too

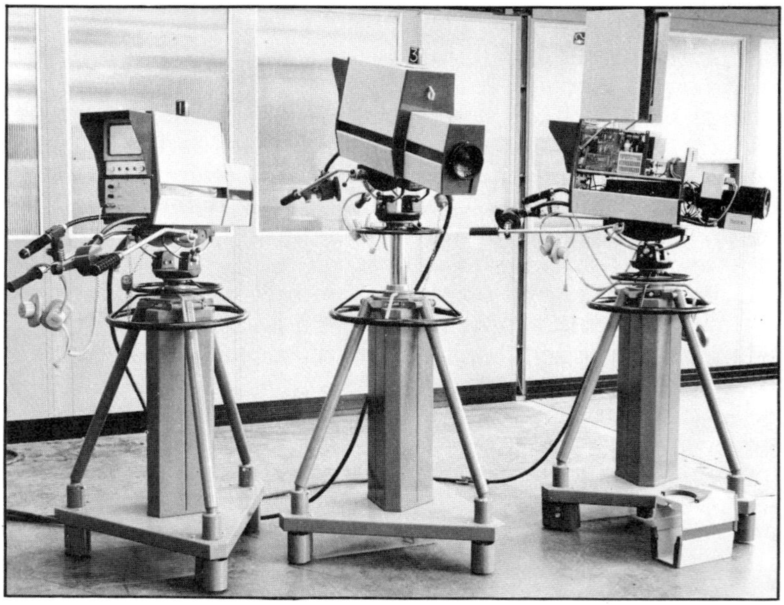

Jet Probe

A less complicated but no less remarkable approach to the problem of being able to 'see what's going on inside' has also recently been developed by the Solar Aviation Works in San Diego, California. Although originally developed for the Atlas missile, its application in industry, particularly the aircraft industry, promises to be of the highest importance. Amongst many other things, it provides designers and aero-engineers with something they have urgently needed, literally since their work began–an internal view of an engine running under full operating conditions.

Many designers believe that one way to build more powerful jet engines, without making them bigger and heavier, is to run the blades faster and, therefore, hotter–i.e. at critical temperatures. But to do this effectively means getting information from inside the engine. For the first time, this new development makes this possible.

It consists of three basic components. A form of quartz crystal has been developed capable of withstanding the intense heat generated near the turbine blades of a jet engine. This is used as a kind of observation window which is connected to an organised bundle of glass fibres encased in a flexible steel cable, which in turn is connected to a device known as a photo-voltic–a silicone cell which converts light-waves to electric impulses. In effect, the bundle of glass fibres acts as a flexible lens, transmitting light from the 'crystal window' to the silicone cell.

Given this chain of communication, the problem was to discover whether it was capable of the individual observation of each blade-tip which, under full operating power, would pass the quartz crystal at the rate of 20,000 blades per second. The system turned out to be so accurate that it could detect a single blade that was beginning to overheat.

Effectively, what this means is that a jet engine attached to a control system fed with information from the 'quartz' window can now be run at optimum efficiency automatically, so that its maximum power output can be secured at varying stages of flight without the risk of structural or mechanical failure within the turbines.

Some aero-engineers go so far as to believe that this development will lead to the introduction of a completely new generation of jet aero-engines as well as improving the operational efficiency of the existing power-plants.

In the 'Jet Probe', as it is called, the actual measurement of a turbine blade-tip temperature is achieved by photo-electric comparison of its colour with established 'norms'.

Describing Colour

This process of measurement by comparison–particularly of colour–is being increasingly used in a wide variety of applications. But the assessment of colour by the human eye is obviously a subjective judgement. Such a state of affairs is clearly unacceptable to the demands of scientists, so the precise definition of colour shades has become an imperative requirement.

One of the latest miniaturised video recorders for use at home or school. It records pictures and sound on magnetic tape

To meet these demands a 'Colour Atlas' has been created. It represents a major advance for all colour-using industries and technologies, especially in areas where it is vital to describe or reproduce a particular shade. Until now, reference documents have covered a range of between 200 and 2500 shades.

Eight years of work by specialists in the British Dyestuffs Industry lie behind the production of the Atlas which, although it contains a mere 1379 colour samples, includes a selection of neutral grey filters, the application of which provides further permutations, bringing the grand total of colours in the Atlas to an incredible 27,580. The use of the Atlas is complimentary to the Instrumental Match Prediction technique, developed by ICI a few years ago. For this a method was devised for employing a computer to calculate how much of the three primary colours (red, yellow, blue) would be needed to match any shade whose true colour-values had been scientifically measured.

Precise colour-match operations can now be completed within a matter of minutes from information received from anywhere in the world. Even so, the staggering number of precisely delineated colours thereby achieved, and capable of being matched, still falls far below the capability of the human eye, because a person with normal sight is capable of distinguishing more than a million colour shades.

Machines that Make Decisions

The development of automation is as significant to modern industry as the advent of steam-power in the last century. Techniques and apparatus capable of performing the necessary functions are being developed for every conceivable type of application–from driverless tractors to automatic sorting, weighing and packaging machines; from giant tool-manipulators such as 'Unimate' to computer-controlled mechanical draughtsmen; from a lawn-mower that will tend the suburban garden unaided to a hospital bed capable of keeping its automatic eye on the desperately sick patient.

The range of activity seems to cover almost every gainful task in which man employs himself. Machines can be given their orders, which they will continue to obey until their task is completed, in a variety of ways ranging from punched card or magnetic tape to on-line computer control; and from the ability to follow the contours of a template or pattern, to response to light, colour, temperature, or even the human voice.

The latest developments indicate the arrival of a new generation of machines which are not only capable of performing specific functions faster and more efficiently than mere humans, but also may altogether replace the necessity for human supervision, because they are able to make their own decisions. For example, a computer can be programmed to make the kind of decisions that allow it to play a game of chess. Perhaps most important is the computer's ability to produce an unprecedented amount of relevant information in an incredibly short space of time, information upon which human decisions can be

Computer systems so simple that even a child can use them. With the help of a slide projector and a tape recorder this small boy has a private geography lesson. If there is anything he doesn't understand, he can ask the computer a question by tapping the keys on the console

based. It is said that President Nixon's decision to invade Cambodia was based on information from the Pentagon's computer for strategic studies. He could, of course, have ignored the advice, but it is only one step away from the situation in which the computer will itself be capable of giving executive commands. Already one computer can command another computer, which in turn can command a machine or range of machines.

The advantages of the latter chain of command were recently indicated by Frank E. Spindler, an Executive Vice-President of the Westinghouse Corporation, when he announced a new machine-control system which provides each individual machine tool with its own computer in place of the now conventional, numerical-contouring control-technology.

'The system has two important benefits,' said Mr Spindler. 'It can be easily and economically adapted to any production system that now exists. It also means that a machine with the new control will not itself become obsolete, because it can be adapted to new control-schemes with an economical change of its programming. These benefits are of particular interest to manufacturers who are looking forward to that ultimate called the automatic factory–that is banks of machine-tools run by a central computer.'

What this means is that, by gradually installing machine-tools with their own computerised control-systems, a manufacturer will avoid the sudden cost and risks which would otherwise face him when he finally decides to convert to total factory-automation, since the computerised control-system on each machine would be easily adaptable to the ultimate control of a single central computer.

It is particularly significant that it is claimed that this 'local' computerised system will be competitively priced against existing numerical controls.

Computers that Learn

Another significant advance reported from the Research Organisation of Westinghouse indicates that this highly developed type of computerised control is by no means confined to the metal-cutting aspects of industry. Patents have been filed for a new type of adaptive control that will enable metal-rolling-mill computers to teach themselves how to handle different grades and sizes of metal. This new system promises to save the industry years of man-hours and thousands of tons of rolled metal.

Under the old trial-and-error method of programming a rolling-mill computer, skilled personnel had to be constantly on call to cope with the problems and calculations inherent in rolling a new piece of metal. It is claimed for the new system that, in time, the computer will acquire sufficient information to roll any further example of that type of metal completely automatically.

The first piece of metal is rolled on the basis of 'predicted programming'–that is, the computer is fed the product of human experience. Thereafter it observes and calculates the results achieved, corrects the original setting, and is ready for the next run. This process is continued until the entire operation is complete.

After only a few pieces of a given grade and size of metal, the computer is able automatically to adjust the controls of the rolling mill to the particular requirements and will automatically adopt the same settings whenever it meets the same problem again. Based on its own experience, the computer is able to complete an ever-widening variety of operations regardless of the grade and size of metal fed to it.

Two hot-strip mills in the United States have already proved the commercial viability of the system.

Below and right The 'Talking Page' is a teaching system to enable children of all intelligence levels to learn subjects from reading to algebra with a minimum of guidance and at their own pace. This electronic teacher is a metal box about the size of a portable record-player. A specially designed, illustrated book fits on top and a gramophone record slots easily inside. The book and the record are co-ordinated in such a way that the child can select passages of sound easily and precisely as he turns the pages of the book. A child who cannot read puts on the headphones and listens. A voice tells him to move a small lever at the side of the box until a red pointer lines up with a picture on the page. The voice takes over again and the lesson begins. The child can move back or forward over the lesson as he chooses. Trials have shown that the average five-year-old masters these simple movements in under ten minutes and before the end of one lesson is able to fit the book, slip the disc into its slot, and start the machine. The talking page is virtually indestructible. There are no exposed working parts and it is battery-operated

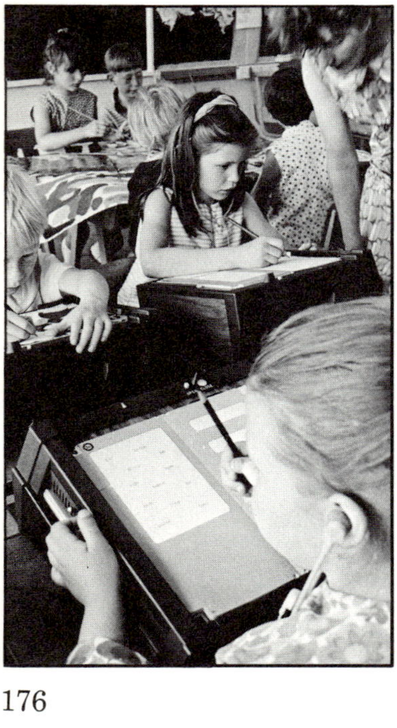

Computers that Decide

However, recent work at the Royal Aircraft Establishment at Farnborough is certainly of equal significance. It promises to give machines not only the ability to perform set tasks automatically, but also to decide the best way in which those tasks should be tackled.

In any type of metal-cutting machine-tool, the efficiency of the machining depends on several factors. Obviously the cutting edge must be correctly shaped for the operation and material, and given this, there is still the question of the rate of cut. This is in turn dependent on the rate of rotation, the depth of cut, and the feed-rate–the speed at which the component is advanced towards the cutting edge. A further factor is the type and supply of lubricant.

All these facts are closely interrelated so that the efficiency of any machining process is generally a compromise. For example, for maximum cutting efficiency the speed should theoretically be increased continuously as the diameter at the point of cut is decreased. But it is not possible to achieve this even with a human operator without exceptional instrumentation and controls. Tool wear is also a matter of economic importance: since the wear-rate of a cutting-tool increases sharply if the working temperature rises above a certain limit, the maximum efficiency will be obtained only if the temperature is kept just below that critical point.

The RAE team has been working on the control of the cutting speed of a lathe, using the working temperature at key points as the controlling signal. Studies of feed-rate and force on the tool have also been carried out, and since tool temperature is far more dependent on cutting speed than on feed-rate, control of the latter is being determined by the force acting on the tool. Based on these operational criteria, an automatic, electronic control-system has been perfected to govern the temperature of the tool and the workface which it is cutting, by carefully controlling the spindle speed.

Bottom left A typewriter that talks is teaching children to read faster than any schoolteacher could. Using it, the average 12-year-old can progress 11 months in only 6 hours' tuition. Bright 3-year-olds and backward adults have used it with equal success. When the keys are pressed the typewriter speaks the letters as well as printing them. Then the roles are reversed. For instance, when the machine says 'A', the child must press the right key for the letter or the typewriter will not work. Cost: approximately £16,500

Below Glassware to make classroom experiments simpler and cheaper. Five basic components are all you need for ten different chemistry experiments. Each piece press-fits together snugly, so doing away with old-fashioned cork joints and stoppers

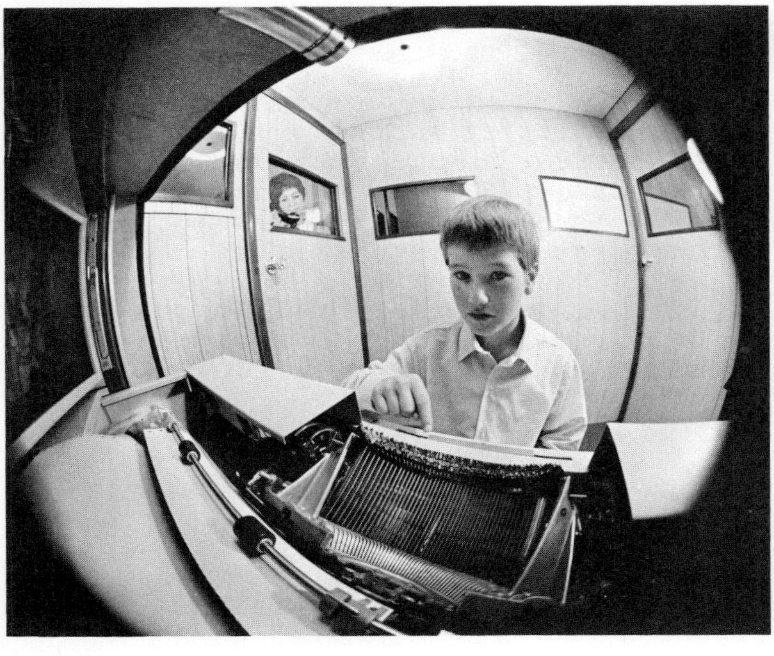

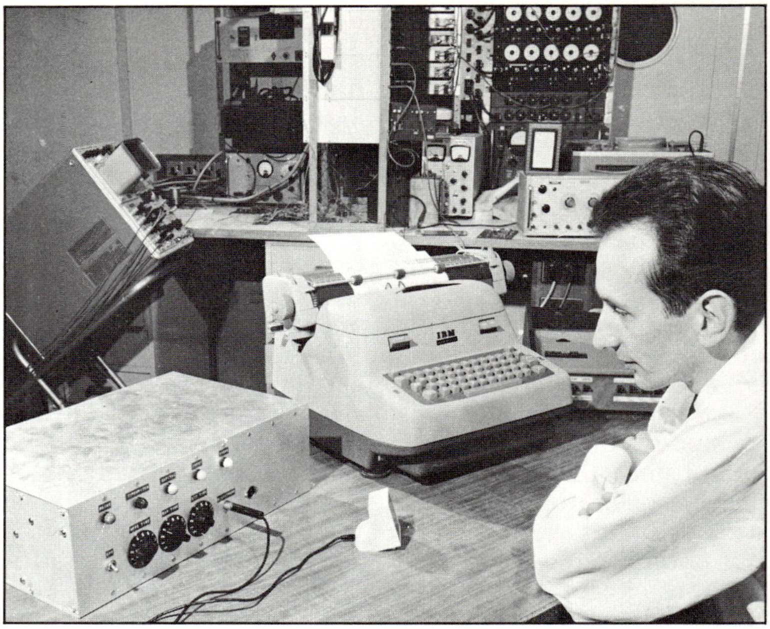

If this typewriter is ever perfected it could mean that severely disabled people will be able to conduct their correspondence unaided. The idea is that by using a verbal morse code the electric typewriter will register the appropriate letter. For example, to type 'A' you say 'di dah'. It has been found that after a few hours' training it is possible to operate the typewriter accurately at up to 20 words a minute. In addition the system can be harnessed to other electronic devices. Simple voice commands will make it turn on lights, TV sets—in fact, anything that can be operated by an electric switch

For increased production on rough machining, it pays to increase the feed-rate rather than the cutting speed. The limiting factor is the risk of breaking the tool. To avoid this hazard, strain gauges of a special type are mounted in the tool-holder to keep an automatic eye on the load imposed. The system simultaneously controls cutting speed by temperature and feed-rate by tool-force, thus ensuring that cutting operations from start to finish are performed efficiently.

The results so far have proved so encouraging that the system is to be applied to a large, numerically-controlled lathe in the near future.

There are many highly technical advantages and possible applications of this work, including among them the development of a 'machineability index' through the employment of this automatic machining-process on test samples.

Not the least important point made by the RAE's scientists is that all their considerations have been based on present-day machines which have not been designed with any automatic control in mind. In order to reap the maximum benefit from both speed and feed-control, a new generation of machine-tools should be specially designed for the system.

These are but three examples of highly complex and specialised work which is being undertaken in research centres all over the industrial world. For the layman, and particularly for those who tend to dismiss technology as being beyond the ken of normal mortals, they are difficult to understand. But what is important is that they bring closer the day when the vast majority–perhaps over 90 per cent–of the present-day labour force of industry becomes redundant. Whether or not the same measure of progress is being maintained towards the solution of the social and political problems that will inevitably follow is, at best, open to very serious question.

Fluidics–The Modern Pipes of Pan

If when painting a ceiling you have been maddened by that little trickle of paint which suddenly makes its way down the brush, along your fingers, over your wrist and up to your elbow, take comfort. You have demonstrated the basic principle behind a whole new technology which has grown up during the past ten years.

A less disagreeable experiment which also demonstrates the principle calls for nothing more elaborate than a dessertspoon and the kitchen tap. If you hold the spoon under a gentle flow of water two interesting phenomena can be observed. The vertical column of water falling from the tap can be adjusted until it is roughly cylindrical in shape. However, when this cylinder of falling water comes in contact with the spoon, it will tend to follow the contours of the spoon itself and will assume, and maintain, a new and different pattern of flow.

A slightly more complicated experiment produces an even more dramatic effect. If you blow air into a Y-shaped tube from the vertical 'leg' of the letter, the air will tend to come out of one or other of the angled segments. Then, if with your finger you close off the pipe through which the air is escaping, it will not only make its way out through the other extension of the Y, but will continue to do so after your finger is removed.

These phenomena have been known and observed for a great many years but it was not until 1960 that scientists began to apply them–the principles of fluidics–to control-mechanisms. As we have seen in so many other technological advances, the stimulus was space research. The object was to achieve control-mechanisms in rockets which would be reliable because of their simplicity, yet be capable of functioning in the most hostile conditions–intense heat, vibration, corrosion and so on. Rocket engineers in both the United States and Russia saw the possibility of fluidics controls, and a completely new technology was born.

The illustration overleaf shows four main types of pure fluidic devices. These are the fundamental elements of the whole technology.

The first diagram shows the 'Wall Attachment'. This is the Y-shaped tubing of our own experiment. However, instead of using a finger to interrupt and redirect the flow of the main column of air, a pulse control jet is incorporated in the system. When the control is applied, as in the diagram, from the right, the main flow is directed through the left segment of the Y. A momentary influence from the pulse control jet is sufficient to redirect the main column. A similar momentary pulse control jet injected in the system from the left would deflect the main flow through the right-hand segment of the Y, and again this direction would be maintained until a further interrupting influence is applied.

The Turbulence Amplifier shown in the second diagram makes it possible to transform a high-velocity jet along the line of laminar flow into a low-velocity, confused air-stream, which is deflected into the vent at the bottom of the section.

The Vortex Amplifier in the third diagram performs a

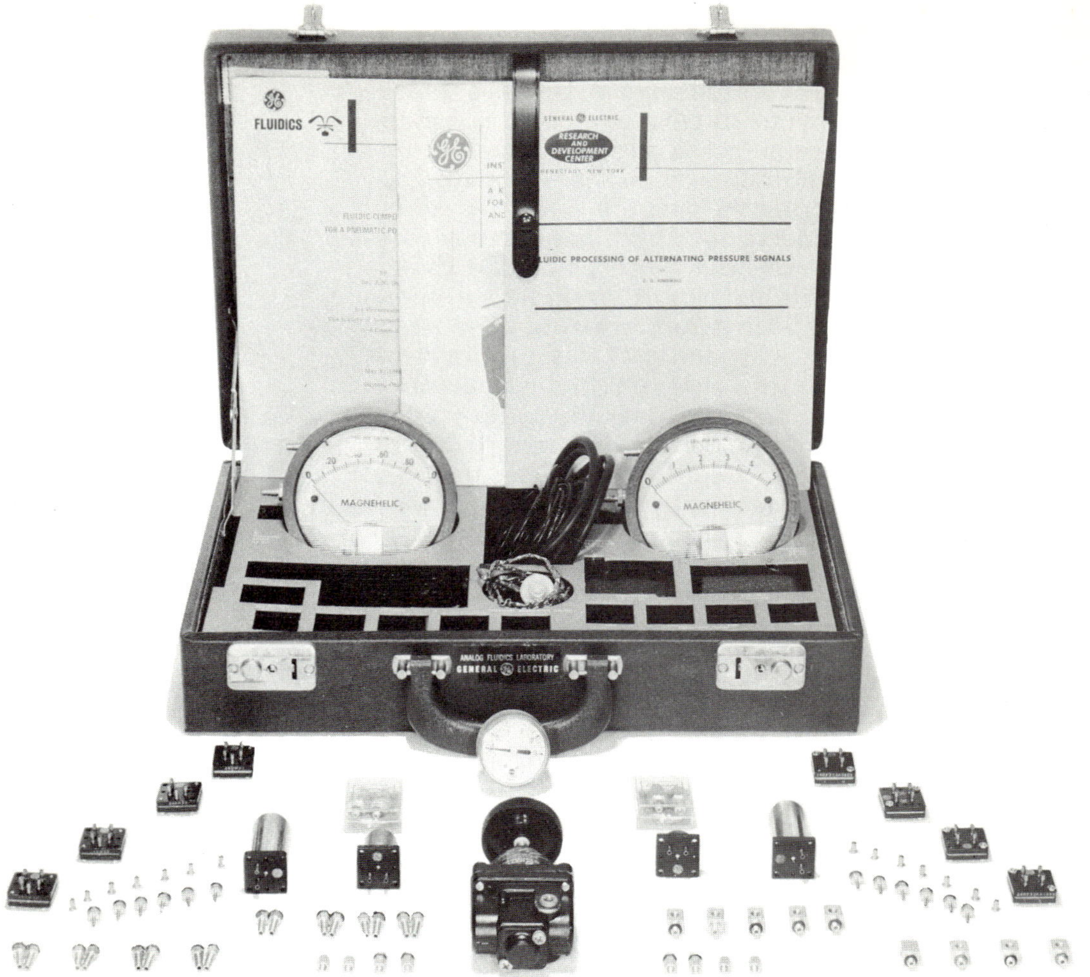
FLUIDICS
RESEARCH AND DEVELOPMENT CENTER
MAGNEHELIC
MAGNEHELIC
ANALOG FLUIDICS LABORATORY
GENERAL ELECTRIC

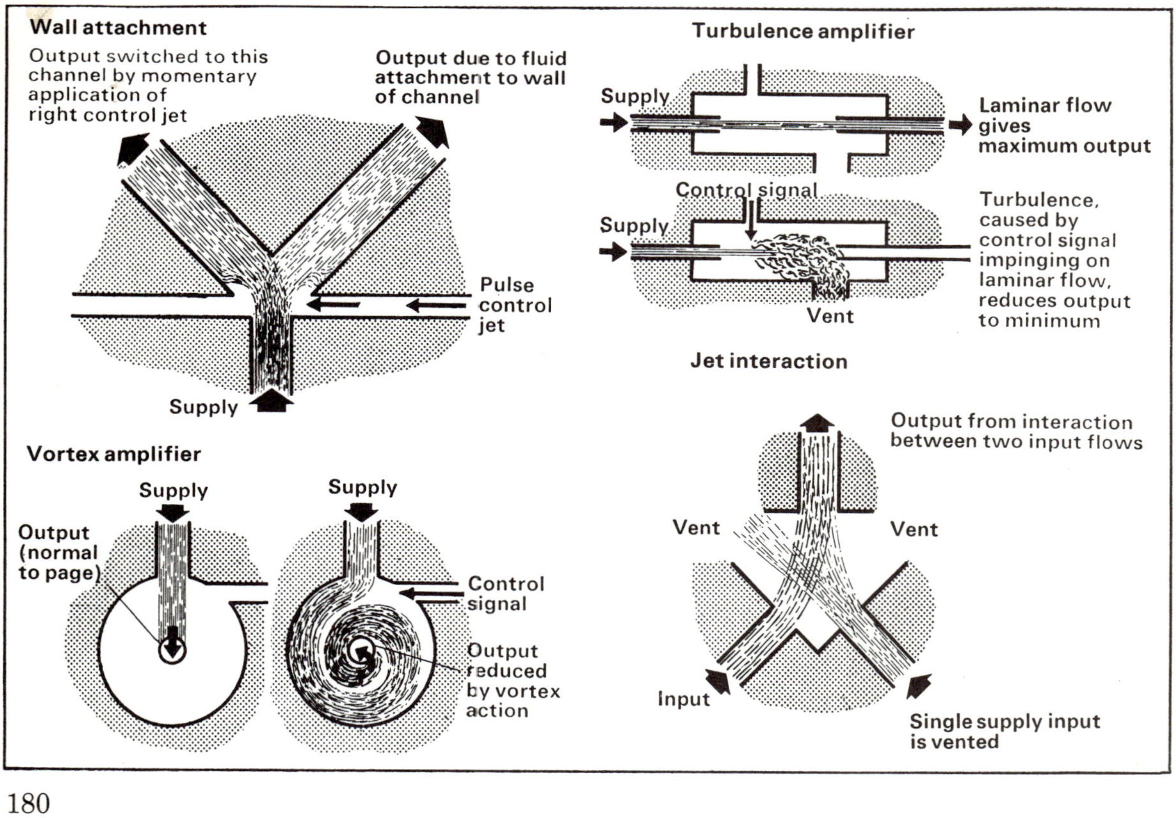
Wall attachment
Output switched to this channel by momentary application of right control jet
Output due to fluid attachment to wall of channel
Pulse control jet
Supply
Turbulence amplifier
Supply
Laminar flow gives maximum output
Control signal
Supply
Turbulence, caused by control signal impinging on laminar flow, reduces output to minimum
Vent
Jet interaction
Output from interaction between two input flows
Vent
Vent
Input
Single supply input is vented
Vortex amplifier
Supply
Output (normal to page)
Supply
Control signal
Output reduced by vortex action

similar function except that in this case a 'whirlpool' is induced as required.

The Jet Interaction diagram shows how a power jet can be used to drive a fluid across a gap and into a channel, to produce an output pressure. In a sense, it is the reverse of the effect of the first 'Y' diagram.

As the diagrams show, these instruments have no working parts: there are no valves, no springs, no gears, –literally nothing capable of wearing out. The diagrams also suggest the ease with which such components can be manufactured. In fact, most are made from channelled plastic blocks. The channels are moulded into two half sections which, mounted one on top of the other, complete the necessary pattern into which the fluid is to be fed.

Much has been written about fluidics in recent years, and a wealth of highly technical literature is now available. The rate of new publications shows no sign of slackening, as general professional interest in the subject continues undiminished. But industrial interest, as measured by the amount of hard cash invested in research, manufacture, marketing and applications, is not growing as fast as experts originally forecast.

The principle of fluidics can be applied, as the name suggests, to any fluid, though compressed air is very often the most convenient material, particularly for industrial control-systems. So far it has been exploited mostly in low-power applications in conjunction with compressed air. Control-devices can, however, be made large enough to be used in major industrial plants such as petrochemical refineries and sugar factories.

AND/OR/NOR/NOT

Between them, the various types of fluidic devices can provide a wide range of 'logic functions', which are capable of imparting to a machine the automatic ability to assess variables in a given situation and react accordingly (see p. 177). These logic functions themselves enjoy delightful names. There is, for instance, the 'AND function', in which all inputs are required before an output is possible–in other words, the device will not give a 'go' until all its requirements have been met.

In the 'OR function', on the other hand, any of the inputs will give an output. The potential ability to switch in an alternative supply is an application which immediately suggests itself. Then there is the 'NOR function' in which the application of any one of a variety of signals will stop the output; and finally in the 'NOT function' the input stops the output.

This may all sound very confusing and meaningless, but the automatic control of modern machinery requires this type of selective action, and the techniques of fluidics are capable of achieving it more simply than any comparable electronic devices.

In addition to obvious uses fluidic controls are currently earning their living controlling, for instance, automatic sewing-machines, drilling, brazing, and many other industrial applications. Thomas C. Keay (Dundee) Ltd uses a fluidics system which automates the sewing-up of sacks on a special machine. The roof linings of the

Left Fluidics is the contraction of the words 'fluid' and 'logic'. The system uses the interaction of streams of fluid to control instruments and information-processing equipment without the use of moving parts. It is especially useful where electronic systems cannot be used due to heat, nuclear, or electromagnetic radiations. The flow of liquid along a tube is simple to understand, but the way in which it can be harnessed to control highly complex mechanisms is much more difficult. Manufacturers of fluidic systems are producing these special kits for universities and technical colleges so that students can learn the fundamentals of this new technology with the latest hardware. The kits provide all the equipment needed to build complete fluidic control circuits, involving such functions as switching, sensing, logic and counting

Bottom The four fundamental elements (see text)

This partially built Plessey fluidic mounting assembly shows how interconnections are made between nozzles on the undersides of the plastic sub-bases. The elements are screwed to the top of the base units and the identification labels on their reverse sides can be clearly seen

Hillman 'Imp' are lined up by a fluid logics circuit which is part of the welding process. (Previously they did it by hand, which was difficult to do accurately and required the operator to move about the machine to ensure a correct line-up. The process is now fully automated.) At the Culter Paper Mills, Aberdeen, a similar system is used to detect the edge of a two-ton roll of paper. Correct alignment of the paper roll is essential before it enters the final cutting machine. At the Colgate-Palmolive Factory in Dublin, a solution has been found to the problem of filling detergent bottles which tend to foam under vacuum- or pressure-systems. Washing-up liquid is now packaged there by a fluid logic system.

A fluidic-controlled machine in Sussex has replaced four people who used to hand-test the resin-bonded mica used in electrical applications–a particularly exacting task. But perhaps one of the most important applications of fluidic control is in the machine-tool industry, and several leading firms are collaborating on the design and development of suitable devices.

The de-icing equipment, specially developed following the public outcry after the loss of three British trawlers in Polar waters in January 1968, is also activated by a fluidic control-system. A series of rubber pads is fitted to the bridge front and up the main spars and masts of trawlers. Thin pneumatic tubes are embedded in the rubber, and when ice begins to form the tubes are inflated, causing the pads to swell and break the sheets of ice before they have time to assume dangerous weights. A similar system has been employed on the leading edge of aircraft wings for many years.

A basic problem in fluidics is that the speed at which these devices and systems react is considerably slower than electronics, with which they are in many respects strikingly similar. Fluidics enjoy the speed of sound, electronics the speed of light. Even so, fluidic devices have been proved to be as fast or faster than mechanical or electro-mechanical equivalents, and therefore fast enough for many purposes. This is borne out by the list of industries in which the techniques are now in everyday use–wallpaper, precious metals, radioactive areas, aerosol-filling, glass-cutting, hypodermic-needle manufacture, printing machines, ceramics, tile- and brick-making, cosmetics, tobacco, food, plasterboard, leather-stamping, plastics, thread detection, carpet-making and clothing, the automotive industry, sawdust and sand-weighing, and biscuit manufacture. Most of these are in the United Kingdom but fluidics are being used in Canada, the United States, Australia, Israel, Poland and most West European countries. The Russians are using it in connection with rocketry.

In a recent official publication, the British Ministry of Technology summarised the potentially competitive situation: 'Most of the development of pure fluidic devices and systems has been carried out in the USA, whilst most moving-part elements have originated in Eastern Europe. UK interest has built up steadily since 1960, with most work in the pure fluidics field. Research organisations in this country have contributed significantly to the volume of knowledge available on all aspects of fluidics. But in Mintech's view, the UK needs to develop an indigenous, large-volume, manufacturing capability in fluidics components if its place and standing in the technology is to be assured. The alternative course will be that overseas-developed and produced components, and quite possibly systems, will swamp the market.'

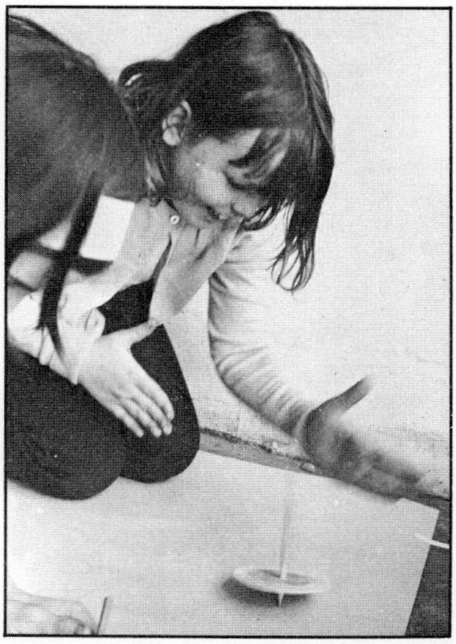

A new series of simple kits make it possible for children to explore the complex relationship between colours. Using spinning-tops and coloured gels as well as paints and brushes, youngsters between the ages of 9 and 15, the ones most often to suffer from the shortage of good art teachers, are able to discover for themselves the basic theories governing the uses of colour

Machines with Nerves

It takes remarkable men to make remarkable discoveries. Judged by any standards, Stanford R. Ovshinsky is a remarkable man. Born to poor immigrant Lithuanian Jewish parents in Akron, Ohio, he had no formal education beyond high school, and is a classic example of the self-taught scientist. Free of the inhibitions of a formalised scientific training, he has dared to probe into fields already discounted by the weight of accepted opinion, to defy conventional procedure and even accepted physical laws. He has continued the pursuit of his own theories in the face of active hostility.

Late in 1968 Ovshinksy published in the United States a paper entitled 'Reversible Electric Switching Phenomena in Disordered Structures'. It reported in detail his discovery of unusual electronic 'switching' effects in glassy semi-conductors. He sweepingly predicted that these materials and a great range of new devices they made possible would 'transform the electronics industry much as the transistor did 20 years ago'.

Immediately controversy flared in both academic and industrial circles. Scientists, many of whom were associated with the transistor industry, did their best to discredit Ovshinsky's devices. They said that his work was neither original, reliable, nor practical. One reason for the intense feeling was that much of it seemed to defy most of the accepted physical laws on which the semi-

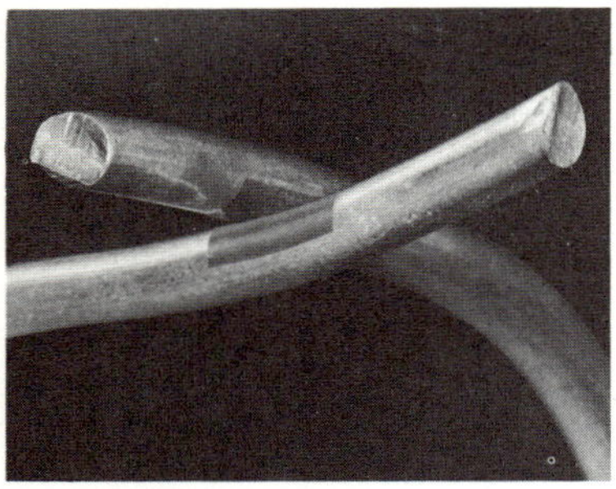

Two simple wires that could transform the electronics industry much as the transistor did twenty years ago. They are coated with a thin film of special glass that enables them to act as switches. Stanford R. Ovshinsky, who invented them, calls the system Ovonics

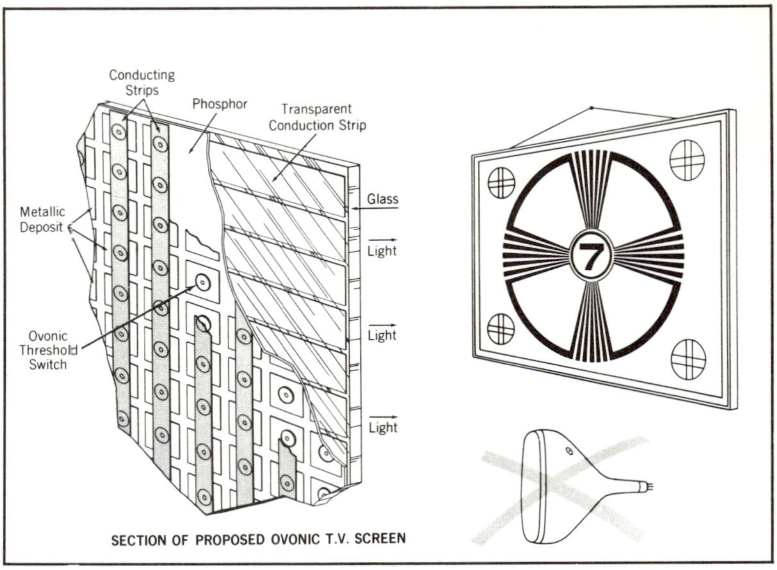

conductor industry had been built. The fantastic growth of solid-state electronics had been founded on the development of ultra-pure metallic crystals of germanium or silicon. These crystals have a geometric, orderly, atomic structure, into which minute and critical impurities are infused in order to promote the controlled passage of an electric current. Most of the new science of solid-state physics was based on the study of such orderly crystal structures, and an immense amount of research and money had been put into their development.

Imagine then the sense of outrage when it was realised that instead of working with pure crystals, Ovshinsky used a range of comparatively crude glasses, composed of many different elements with highly complex and 'untidy' structures. Glass is not truly crystalline–it is something between a liquid and a solid, and consists of an atomic structure made up of random, linked groupings mysteriously 'frozen' into position. Accepted solid-state theory held that such disordered structures would not be conducive to stable or useful semi-conductor effects. However, Ovshinsky had built and demonstrated devices, using his glassy materials, which displayed many of the same characteristics as transistors, as well as some unique properties of their own. Even more significant, these materials required nothing like the extreme care that went into the growing and processing of ultra-pure crystals, and therefore promised to be very much cheaper to produce.

Now Energy Conversion Devices Inc. of Troy, Michigan, the small company which Ovshinsky and his wife have founded, is about to introduce its first standard product–an entirely new type of computer memory. How this will fare commercially against the colossal vested interest which it challenges remains to be seen. But it is by no means without significance that research into amorphous materials has sharply increased and two large scientific meetings were held last year–one in New York and the other in Cambridge–to discuss glassy semi-conductors.

The components on both sides of the display panel are Ovonic Threshold Switches, one to control each electroluminescent lamp (centre of picture). In the past EL displays have not been commercially practical because of the lack of suitable electronic controls. The low potential cost of mass-produced Ovonic switches carried to its logical conclusion is expected to result in a thin-screen television set that will hang on the wall of your living-room like a picture frame (see diagram left)

The background to Ovshinsky's interest in these substances is as romantic as it is remarkable. Delving into the problems of automation, he became deeply interested in neuro-physiology–the study of the brain and nervous system. He became obsessed with the idea that a better understanding of how nerve cells store, switch, alter and transmit information might not only be valuable in understanding how the human brain works, but lead to more advanced electronic switching-devices and information-systems for purely technological applications. He refused to accept the theory that the computer could be compared to the human brain. He questioned the validity of a comparison between the fixed-junction transistors of the computer and the highly disordered, closely-packed molecular structures in the brain which produce more flexible and subtle effects more quickly than any transistor.

He concluded that a solid-state model, closer in structure to the brain, could be made of disordered inorganic materials. His determination led him to approach the Head of the Anatomy Department in Wayne State University Medical School and, because his ideas were so off-beat and intriguing, he was invited to test them in the School's laboratories. At about this time he married his second wife–a bright young Ph.D. in Biochemistry, who shared his interests in neuro-physiology. Together they set about the task of building amorphous devices that would simulate the switching actions of a nerve cell.

The conditions under which they worked have all the ingredients of a science-fiction romance. Furnaces in the kitchen, and crucibles of molten glass bubbling and hissing in the living-room, as the children did their homework and practised the violin.

In 1958 Ovshinsky succeeded in making a functional switching device based on a thin layer of amorphous oxide on tantalum metal. When voltage was applied to it the device would suddenly pass a current much as a sufficiently excited brain cell triggers off a nerve impulse. He went on to demonstrate how the same switch, with a different type of electrode, could act as a kind of memory device, becoming conductive under an electric pulse of one polarity and switching to non-conductivity at a pulse of the opposite polarity.

Whether or not this was a true imitation of brain cells became of secondary interest to Ovshinsky. He recognised that beyond his laboratory experiments lay a great unexplored range of amorphous materials in which profoundly important switching abilities might exist. Some limited work in this field had been done by a group of Russian physicists led by Boris T. Kolomiets, and by American researchers, but they had become discouraged when they found they could not achieve results more than comparable with crystal semi-conductors.

In 1960 Ovshinsky decided to drop everything and concentrate on electronics in amorphous materials. He managed to raise enough money to form a company, and set up business in an empty shop on Detroit's West Side. Diligent research narrowed the hunt down to a class of complex materials called chalcogenide glasses. After many experiments two basic types of switching action were found to occur in two different types of these materials. One he called a 'threshold switch'. It consisted of a film of glassy material between two electrodes. When an increasing voltage was applied across the material, at a critical level or 'threshold point' the material rapidly became a conductor. It continued to conduct as long as the current remained at or above the critical value. When the current dropped the material quickly reverted to a non-conductive state. The device could be used as an electronic switching-system for opening or closing a circuit completely and automatically.

The second type of switching action involved material of a different composition and constituted an entirely new kind of memory switch. In this the glassy material, when energised into a conductive state, remained in that condition even when the current was switched off. It returned to the non-conductive state only when it was given an energy pulse of a still higher current.

Precisely why these effects occurred is still not altogether understood. As Helmut Fritzsche, a young Göttingen-trained Professor at the University of Chicago, has put it: 'Without an ordered crystalline arrangement of atoms, the solid-state physicist finds himself in the same predicament as a psychiatrist without a couch.'

The reason for the colossal importance of switching devices of whatever kind is that they are at the heart of all telephone and telecommunications, from radio and television to microwave relay, and all modern automation-techniques and computer-operations, in which hundreds of thousands of tiny interconnected switches perform the task of programming, logic, computation, and processing. Ovshinsky's threshold switch might serve in many of this great range of switching applications, particularly in computers, where it would seem capable of performing the same function as transistors, but at a fraction of the cost

On the other hand, Ovshinsky's memory switch, with its reversible ability to retain conductivity without power,

This burner, originally designed as a laboratory tool for research into flame chemistry and physics, could be adapted for use as a space heater in the home. It will burn a wide variety of gas mixtures under closely controlled conditions

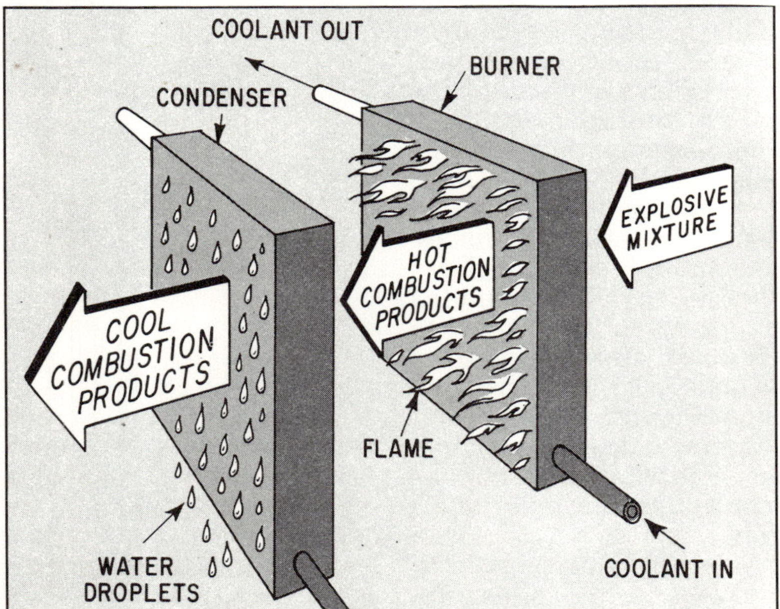

The same burner may be used to solve a problem in water-cooled nuclear reactors. It consists of a flat porous metal plate backed by cooling tubes embedded in a mass of sintered shot. A flammable gas mixture flows through the device and burns with an almost perfectly flat flame across the face of the burner. The potential application to nuclear reactors springs from the fact that in water-cooled plants, gamma rays decompose some of the water into hydrogen and oxygen. These gases are usually stored for about half an hour until any radioactivity that may be present falls to acceptable levels, at which time they are released into the atmosphere. The new burner, combined with a matching condenser, can safely burn the radioactive oxygen and hydrogen mixture to convert it immediately into harmless water and air has no counterpart in crystal technology. Computer memories have to rely on fixed-wire circuits or magnetic records, whose stored data is totally wiped out on each read-out and has to be renewed. The Ovonic switch–the name given to it by its inventor–promises faster, more flexible, high-density integrated memory-systems. Furthermore, the tellurium glass switches which he has built work equally well on direct current as on alternating current. Transistors can handle a.c. only with difficulty, and this ability suggests a possible new range of automatic controls. Another difficult area for the transistor is the handling of microwave energy. Ovshinsky's glasses, on the other hand, have extremely high switching speeds–in the range of one hundred trillionths of a second. This makes them potentially capable of filling the microwave gap.

Perhaps most important of all, they are unaffected by small impurities in manufacture, and unlike orderly semi-crystalline conductors, they are insensitive to radiation. This alone would open up to them a rapidly expanding range of potential applications. One is an electro-luminescent data-display panel for computers, which is activated by an array of thin-film threshold switches. Another is a computer mass-memory system, still in the laboratory stage, in which a pulsed laser beam swiftly puts rows of invisible dots on a thin film of the memory material in a binary coded pattern. The coded information is read out by a laser, and may be permanently stored or selectively erased in an instant by another laser pulse. Dots as small as half a micron in diameter have been achieved, and the target in view is to pack some 100 million dots–each one of them an information 'bit'–on to a square inch of film. This incredible density of storage could lead to a memory system of immense speed and capacity, unique in its versatility and many times bigger, better and faster than anything yet possible. A third application under development harnesses that type of memory system to a high-speed computer print-out. In this device a computer-directed laser beam 'writes' data characters, the printed word, or graphics on the glassy semi-conductor surface of a rotating drum, from which they are printed off in the same manner as existing electro-static machines. Because of the 'switchability' of the glass drum, the printed characters are permanent or instantly alterable as in other Ovonic memory devices.

Recently Ovshinsky and his team have found glassy materials which can be literally sculpted by a modulated laser beam. The raised characters which are created in the material make it possible for both gravure and litho plates to be made electronically under computer direction for normal printing applications. Finally, improvement of the electro-luminescent computer data-display panel may eventually achieve sufficient definition to make possible a mirror-thin TV screen which does not have a bulky cathode-ray tube behind it.

Controversy, argument and financial uncertainty still surround and confront Ovshinsky and his colleagues. But this story comes as a refreshing reminder that, even in these days when Man is intent upon replacing his own function in so many productive fields, it is the power of an individual brain and the strength of an individual character and personality which, as always in the past, provide the spearhead which rips through the frontiers of human knowledge.

The Ultimate Power Source

Ecologists, conservationists and others concerned by what they consider to be the excessive uses of power in contemporary civilisation draw little comfort from nuclear generators as an alternative to our depleting natural resources. They point out that the by-products of nuclear electricity-generating stations are amongst the most toxic materials known to man, that the disposal of radioactive waste material is already presenting a problem of the utmost gravity, and that the accidental discharge of active waste products is an ever-present threat.

Atomic energy electricity-generating plants use controlled nuclear fission as their power source. But there is an alternative–nuclear *fusion*. Whilst nuclear *fission* produces energy by splitting the nuclei of heavy atoms, nuclear *fusion* is exactly the opposite–it liberates energy by joining together the nuclei of light atoms. It uses fuel derived from a cheap and readily available source–water. Fission reactors, on the other hand, use uranium, which is scarce and expensive, or plutonium–a man-made element which is in any case derived from uranium. The waste product of the *fusion* process is an inert and harmless gas, helium.

Yet, as a practical source of power, fusion is still very much a scientific dream, because it is much more difficult to control than fission. Since 1962 the United Kingdom Atomic Energy Authority laboratories at Cullum have been tackling some of the immense problems–immense because the process is equivalent to creating a miniature sun on Earth. Even so it is likely that towards the end of the century, fusion will be tamed as an economic source of power.

Fusion is the process by which the Sun and the stars produce their fantastic and seemingly endless output of energy. The Sun is a huge nuclear fusion reactor, where the nuclei of hydrogen atoms are continuously colliding and combining with one another to form helium and freeing energy in the process which heats and lights the Earth as well as providing all the other forms of radiant energy which bombard the solar system. For five thousand million years, the Sun has been radiating four million tons of its substance every second.

Like most matter in space, the material of the Sun is in the form of plasma–superheated gas. Not least among the problems involved in creating plasma is finding a fool-proof way of storing it in a container. The next step is to keep the process going long enough for the energy released from the fusion to exceed that required to heat up the plasma in the first place–otherwise, of course, the result would be a completely non-productive process. Fortunately, since plasma consists of highly charged particles, it can be confined by magnetic fields. The mag-

CITY DISPLAY ORGANISATION

netic isolation of plasma at a temperature of up to 300 million degrees C, the design and installation of apparatus capable of this achievement, and the completion of an experimental programme to make practical applications possible from the theoretical studies, is the task confronting the Cullum research team.

Hydrogen, the lightest element known, is (in the form of deuterium, one of its more readily isolated isotopes) the most likely fuel for a fusion reactor. Deuterium can be produced from water easily and extremely cheaply. Theoretically a teaspoonful of liquid deuterium, costing a matter of shillings, could produce as much electricity as 10 tons of coal in a conventional thermal power station.

Deuterium gas, better known as heavy hydrogen, is converted into plasma by applying an electric current of several million amps. One of the effects of the electrical current itself is to attract the particles of plasma into a closely confined magnetic field which prevents it from coming into contact with the walls of the container. However, because the plasma is extremely restless an additional electro-magnetic field has to be provided by coils wound round the outside of the containing tube to act as an additional harness.

Valuable experience of plasma behaviour was gained at Cullum using an open-ended tube surrounded by an electro-magnetic harness they called the 'Eight Metre Theta Pinch Experiment'. This consisted of a quartz tube 8 metres long surrounded by a single-turn coil and mounted on a massive, reinforced concrete base. Now a closed-ended apparatus known as HBTX (High Beta Toroidal Experiment) is approaching completion at

Far left A model of the Priest Rapids Power Station, Grant County, Washington. Most hydroelectric plants depend on a water drop of at least 30 feet in order to generate enough energy to move the turbines efficiently. But now engineers have designed a straight flow turbine capable of working in a sea tide of only 10 feet

Above The new turbine wheel itself is allowed to spin freely in the tidal flow. A technique has been devised to seal the tips of the turbine blades so that the heavy equipment which generates the electricity can now be built into the side of the casing. It could prove to be a fundamental step in enabling vast power stations to be built on barrages across such areas as Morecambe Bay or the Wash

A tanker designed for the collection of waste corrosive products. It is constructed in lightweight glass-reinforced plastics and incorporates an unusual system for preselecting the load capacity. By moving a lever the driver can choose a 1500-, 1600- or 1700-gallon capacity and mercury probes within the plastic tank sense when it is full. An electro-pneumatic system of valves then cuts the air supply and the filler pumps stop automatically

Cullum. This will carry on the work done by the controversial ZETA installation (Zero Energy Thermonuclear Assembly) which relied on electric current passing through plasma both to heat it and to create the restricting magnetic field. If successful, the HBTX experiments may bring the era of nuclear fusion power into much more realistic perspective.

In HBTX the plasma circulates in an endless tube, shaped rather like an inflated car tyre, and since it will be contained within closed magnetic lines of force, the plasma can escape only by crossing the magnetic field. From this experimental rig, actual reactor conditions will be computed with a view to building more advanced experiments. If the rate at which the plasma crosses the magnetic field is low enough to allow fusion to occur, then a major step forward will have been achieved.

The tubular container of HBTX is made of quartz and surrounded by two conductors each supplied with 40,000 volts from a capacitor bank assembled on the floor above. The whole assembly is mounted on a reinforced concrete plinth 11 feet in diameter and 4 feet 6 inches high. Each of the four segments of the plinth weighs 2 tons. This massive assembly is necessary because an impulse force, or shock, of 400 lb. per square inch is exerted on the plinth when HBTX is operating. To protect the concrete and ensure a clean surface it is encased in a GRP which helps to insulate the plinth from the tremendous electrical charge of the conductors.

Needless to say, Cullum is not the only research centre where attempts are being made to harness the enormous energy of fusion reaction. It is no exaggeration to say that upon their success depends the likelihood of a continued source of power sufficient to meet the demands of the ever-expanding catalogue of machines which will influence the lives of generations to come.

But there is an inescapable price to pay. Water used in thermal generating plants to produce steam may be condensed and used again. Water treated in fusion reactors is irrevocably destroyed and lost for ever. So water, as well as food, could well be in short supply in the world of tomorrow.

Analysis from orbit—the first ever multi-spectral photographs taken during Apollo 9's Earth orbit mission. All four show the same area of California seen through filters that reveal different facts about the area. In every photograph the string of white clumps running top to bottom left is clouds
1 & 2 Red band and near infra-red band: These filters reveal different plants growing in the vegetation mass (top left and mid right), as well as different soil deposits running through the area
3 Infra-red colour: This filter reveals plant disease, if present, and surface water characteristics
4 Green band: This filter is predominantly useful for revealing underwater sea-bed formation (see p. 160)

From comparison and analysis of multi-spectral pictures the infinitesimal developments in natural features or plants due to climatic changes or biological influences like growth or disease can be spotted some time before such changes could be observed at ground level. It is said that the US has already been able to predict a Chinese wheat-harvest disaster months before the Chinese themselves knew about it

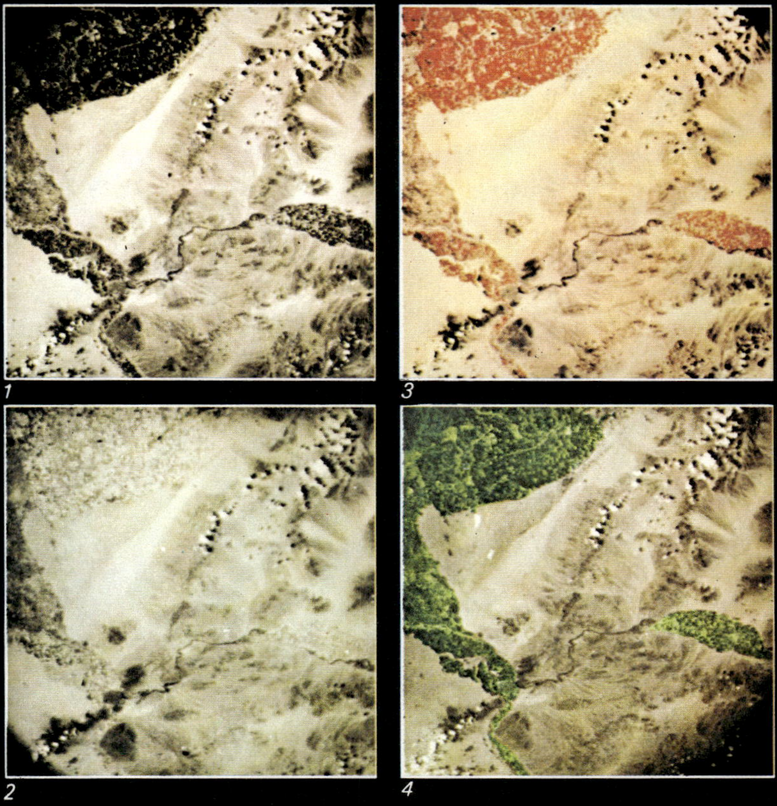

Right They call this equipment the 'speech microscope'. It was built to study speech quality on telephone circuits. Using it, researchers have discovered that about half the duration of some speech sounds can be removed without the listener noticing the difference. This work is aimed at producing economic telephone equipment designed to handle speech on a narrow band-width—so that more conversations can be squeezed down the same telephone circuit simultaneously. A single word is converted to a visual graph on an oscilloscope. Researchers can study what happens when various portions of the speech sound are removed. Sometimes the word might alter—e.g. from 'sticks' to 'dick'. But many words remain unchanged even when a considerable proportion of their total content is removed electronically

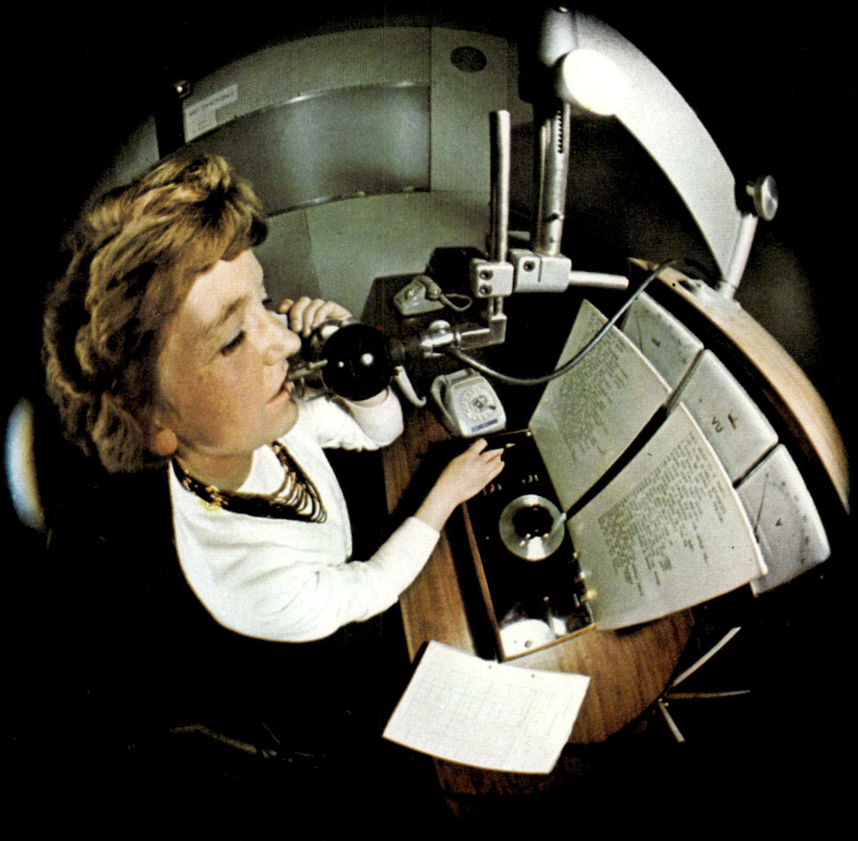

Right All three programmes on the screens are being transmitted simultaneously over the single shaft of green light from an argon laser. The signals are picked up in the usual way but instead of being fed directly to the sets, they are piped to a microwave subcarrier as FM. This microwave signal then modulates the intensity of the green laser light by using a wideband 'optical modulator'. The beam is then focused on to a photo-detector which 'de-modulates' the signal and converts it back into the original carrier waves. These are fed to the sets through the aerial sockets. Researchers at Lockheed believe that this system could carry more than 70 television channels and produce better-quality pictures than possible with conventional methods. Eventually laser light could be used to relay messages between satellites in space.

Far right Light that bends around corners: A laser beam is fed along a dielectric fibre waveguide. It is possible to use laser 'cables' for most kinds of communications

Bottom right A split laser beam carves a colourful pattern of light to create a 3-D picture called a 'hologram'

BIG JACK

Above The molten metal within this crucible is suspended in mid-air, held there by a powerful magnetic field. Scientists developed this system to overcome problems of contamination when manufacturing ultra-pure metals for semi-conductors. Previously, molten metal which bubbled against the sides of the crucible risked picking up molecules of the material from which the crucible was made

The newest material on Earth. A glass spherule found among Apollo 11 lunar rocks, enlarged 12,600 times. Its shape indicates that the spherule underwent melting at some point in its history

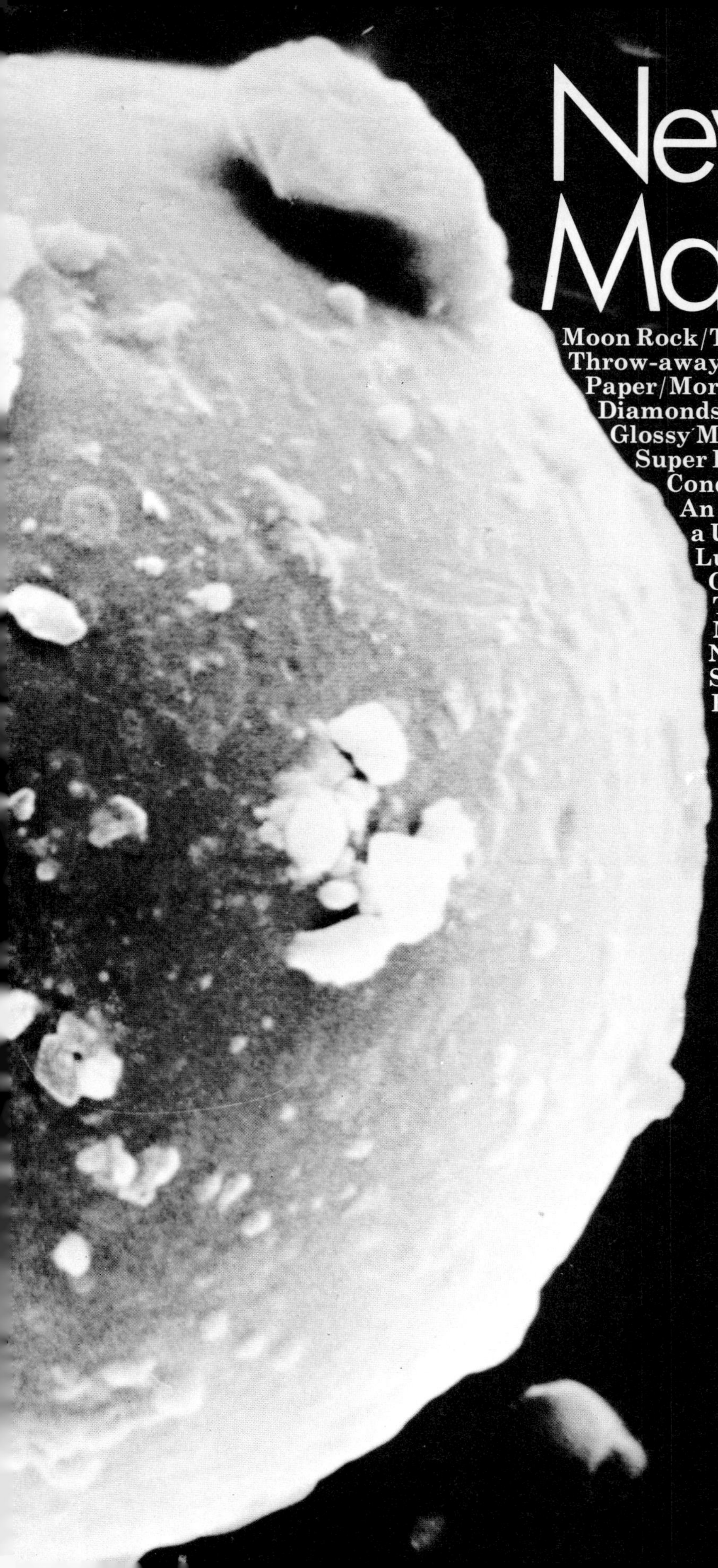

New Materials

Moon Rock/The Mixed Blessing/ Throw-away Milk Bottles/Plastic Paper/More in Less Space/ Diamonds/Stiff and Strong/ Glossy Metal/Fireproof Petrol/ Super Power from Super Conductors/Killer Fibre/ An Invention in Search of a Use/New Lamps for Old/ Lubrication Without Grease/The Exploding Tape Measure/Chemical Machining/Old Wine in New Bottles/Man-made Seaweed/The Trail Blazers/

Mankind, it seems, is perched on the brink of an enormous explosion of knowledge. Far-reaching discoveries are made in almost every field of scientific research month in and month out. We are learning something important every day. And all this is having a profound effect, not least on education. Nowadays children have to specialise at a desperately early age in order to take in the mass of information that they need to assimilate to qualify themselves for a profession–particularly one involved in any way with science and technology.

All these inventions and discoveries act as a stimulus to scientists, engineers and technologists the world over, who are constantly seeking new ways of tackling their newly-discovered problems, as well as new ways of approaching the old ones.

Despite the dire warnings of certain ecologists, who threaten imminent exhaustion of the world supply of natural resources, we do not seem to be producing very many fundamentally new materials. Rather, we are busy finding new uses for the old ones. When we do produce a new material–carbon fibres for instance–the excitement engendered by optimistic and premature publicity can often diminish the true stature of the material concerned.

The invention of new machines is even more prolific than the publication of discoveries. The demands of contemporary technology are so complex that a fashion seems to be developing for 'custom-built' machinery to cope with the specific demands of every aspect of every industry. And the purpose of all these machines is to improve the output and efficiency, and therefore the profitability, of the individual productive operator–to relieve him of the unpleasant, repetitive chores of his work, and enable him to complete his task faster and more thoroughly. In this context, certainly the most important advances have been made in what is known as systems engineering–if you like, automation. Much of this comes as the 'spin-off' of military and space research–the micro-electronic industry in its entirety is, for instance, a direct result of both.

America's vast expenditure on space and military research is now reaping very substantial dividends in the commercial applications of an extremely wide range of products and techniques so developed. And even in this extremely sophisticated field, there are still very new approaches being developed to some of the 'systems' problems. A striking example is in fluidic control–one of the very important areas of technological expansion this decade–a technique which can achieve more cheaply, more simply and in many cases more efficiently what up to now has been done by complex systems of electronics. As will be seen, this is one of the comparatively rare fields in which new technology is achieving simplification, as well as miniaturisation and the overall improvement of efficiency.

Moon Rock

Probably never before has so much interest and excitement been aroused by any material, nor so much effort and money been expended upon obtaining it, than for those few pounds of Moon rock brought back by the astronauts of Apollo 11. Everyone had their own ideas of what it might actually be–theories ranging from uranium to solid diamonds.

The truth was perhaps even more interesting. The Moon rock contained many separate materials, none of those so far isolated being particularly mysterious, and by studying them the scientists are building up new theories, now backed by at least some measure of concrete evidence, concerning the mysteries of the origin of the solar system.

For instance, analysis of cosmic-ray tracks in rocks studied at the Research and Development Center of the University of California at Berkeley indicate that the dry surface of the lunar Sea of Tranquillity has undergone extensive churning by natural forces. That research has also provided new evidence about low-energy cosmic rays which originate in the Sun. These pronouncements, made in the first few days of a study which has involved 142 teams of investigators at various research centres throughout the Western world, in themselves disputed many long-cherished theories.

The Berkeley team went on to report the discovery of pure iron particles not of meteoric origin among their lunar material. This supported other evidence that the Moon's history is very different from that of the Earth, where natural iron always occurs in an oxidised form.

The researchers found that many of the lunar rocks were riddled with cosmic-ray tracks, and they suspect that these tracks have been accumulating in the lunar surface since the Moon's creation. They are the result of the impact of charged particles that bombard the Earth and Moon ceaselessly from all directions in outer space. Low-energy cosmic rays are spewed out by the Sun, while high-energy cosmic rays are generated by sources as yet unknown in space.

By studying the tracks themselves, scientists hope to be able to determine whether the chemical composition and intensity of cosmic rays have changed over billions of years. From crystals that at one time composed the top few thousandths of an inch of the lunar surface, they identified tracks created by the nuclei of iron atoms (probably from the Sun) with energies lower than had ever been observed before. The findings of this research are expected to reveal new clues not only to the evolution of the Moon, but also to that of the Sun as well.

The distribution of tracks on individual rocks suggests that they must have been turned over at least once in the past. If the rock remained in just one position, track density would be high only at its top surface. In fact the top and bottom surfaces contain many tracks, while few have been observed in the interiors. So the surface of the Sea of Tranquillity must have been churned up in the past in a manner that belies its name. A similar study was applied to the 4-inch-deep core sample taken by the Apollo 11 astronauts. This, too, suggested that the loose topsoil had by no means remained undisturbed over the millennia.

Plastics in the auto industry. A complete lamp-cluster for an American car, moulded in polyester/chopped glass fibre pre-mix developed by BP Plastics. This simple unit replaces a large number of metal components and is relatively cheap to produce

One of the most important tasks confronting geologists studying the Moon rock is that of dating. Nearly all lunar minerals and glasses, like those on Earth and in meteorites, contain trace-amounts of uranium. The most common uranium isotope is uranium 238, the atoms of which undergo a breaking-up process, splitting into two fragments of roughly equal size. As the fission fragments move apart in the mineral glass, they leave a trail of damage a few atoms in diameter and about 400 millionths of an inch long. The research teams have learnt how to etch out the fission tracks to the point where they become visible under an optical microscope. For a given concentration of uranium, the older the rock, the more tracks are to be found.

Fission track-dating on one lunar glass fragment yielded a maximum age of 35 million years–making it three to four billion years younger than most of the rocks found close to it. This 'young' age suggests that the sample was formed when a meteorite crashed into the Moon less than 35 million years ago, melting old surface material which then solidified into a new rock.

X-ray refraction analysis has been used to study the crystals contained in Moon rock samples. Each sample is subjected to an intense beam of X-rays, which are deflected by the atoms in the crystal in characteristic directions, and are recorded on photographic film. In this way the unique 'fingerprint' of each crystal is exposed. Surprisingly the crystals are generally found to be of good quality, without major structural defects; although some of the larger crystals show signs of deformation, there are few that show evidence of meteoritic collision.

Another technique, the microprobe, which is a development of the electron microscope, is used to determine the chemical composition of microscopic structures within the rock samples. Studies such as these have revealed the presence of pure iron. This could not have been deposited on the Moon by meteorites because iron in meteorites always occurs as an iron-nickel alloy.

Finally, mass spectrometry has revealed that lunar dust contains an abundance of gases, predominantly hydrogen and helium. It is believed that these gases reached the Moon's surface by means of the 'solar wind'– charged particles that constantly stream out from the Sun. These studies may provide more detailed information on the composition of the solar wind itself.

Thus it will be seen that the analysis of the Moon rock samples is providing evidence not only about the origin of the Moon, but also new knowledge relevant to the Sun and to other energy-sources in space.

From the layman's point of view perhaps the most dramatic and, indeed, physically beautiful discovery amongst the Moon rock samples was the glass beads, about which there had been a great deal of speculation, and which provided probably the most exciting moment in the *Tomorrow's World* studio when we unwrapped those specimens which had been allocated to British research teams. The illustration (p.197) shows one of these glassy spherules, which was found by scientists at the General Electric (USA) Research and Development Center in Schenectady, New York. The spherule was photographed, enlarged 12,600 times, with a scanning electron microscope. Its globular shape indicates that it underwent prior melting. It may have been created when a meteorite crashed into the Moon, melting lunar material and splashing it long distances.

Of course, one must bear in mind the comparatively minute quantity of samples retrieved by the Apollo 11 and 12 crews. It is as if a crew of astronauts from outer space were to arrive in the middle of the Australian desert, snatch a few pounds from a few square yards, and use this as a basis for a report on the geological structure of the entire Earth. It was the great tragedy of the Apollo 13 mission that further Moon rock samples were not brought back to the impatient scientists waiting to study them.

The Mixed Blessing

The 'plastics revolution' is still in its early days. In a paper outlining the growth of plastics-usage, Bryan Knight of BP Chemicals recalls the Eskimo proverb that tells us to 'heed the portents of dawn, lest the Sun find you on melting ice'. In his view, this simple message is significant for those in the plastics and motor-manufacturing industries with respect to the coming of the all-plastic mass-production automobile. Recent forecasts suggest the rising of this particular sun as early as 1972, but informed opinion points to 1980 as a more realistic date. Certainly the increasing tempo of research and development in materials, techniques, and new applications gives cause for an optimistic view. The advantages of the use of plastics for car bodywork are self-evident, if only in terms of the material's resistance to rattles and rust. The growth from 4 lb. of plastics (mainly phenolics) per car in 1939 to the present figure of 40 lb. (UK total 32,000 tons) is modest in the extreme compared with the predicted jump to 256 lb. by 1980 (400,000 tons in the UK, $3\frac{1}{2}$ million tons throughout the world).

The glass-fibre body has been used in limited-run production models for many years. The basic disadvantage here is that it is still virtually a 'cottage industry'. The glass-fibre matting has to be laid by hand over a hand-made mould, then daubed with the bonding resinous paint, again by hand, before the next layer of matting is laid and the process repeated. The limitation in terms of genuine mass-production makes the technique totally unacceptable in that form. The process must be not only mechanised but automated. Experimental work is being energetically pursued all over the world. To mention only a few examples, all-plastic cars have already been produced in ones and twos; in the United Kingdom, both Chrysler UK and Rolls-Royce are conducting extensive research into plastic car-bodies; the Pressed Steel part of British Leyland is planning cars with bodies made from structural urethane foam and rotationally-moulded plastics, as well as pressure or vacuum-injecting techniques; ICI has rebuilt a Rover 2000 body in polypropylene; Honda is all set to mass-produce large body panels; following her deal with Fiat, Russia is making a huge investment in vehicle plastics technology; and interestingly enough Citroen has confidently predicted mass-produced all-plastic car bodies for 1972.

In addition to a new freedom in the ability to mould intricate shapes and integral fittings, plastics also offer the car designer durability, strength, safety, self-colour, fewer parts and easier repairs.

For some years, the 'missing link' has been the development of a strong plastic material that can withstand high temperatures, and the parallel development of manufacturing techniques, such as blow-moulding and injection-moulding, to produce very large structures at high speed. It is true that the application of glass-fibre reinforced polyesters could be made more efficient by the automation of the 'laying-up' process or the use of matched-metal dye-moulding. This is technically feasible, but leaves the material relatively expensive in competition with sheet steel. The most promising starters now seem to be what are known as ABS plastics and dough-moulding compounds (DMCs), as well as structural foams. Polyester DMCs have been produced for many years, but their main use has been in electrical components. They consist basically of polyester resin, chopped glass fibre, filler pigments, and a catalyst. Specially formulated resins in combination with 'non-shrink' additives have overcome previous disadvantages.

Polyester dough-moulding compounds have, as the name implies, a dough-like consistency. Frequently the material is supplied in rope or billet form, and by cutting suitable lengths of the extruded dough correct quantities can be processed with little difficulty. Application of heat and pressure causes the material to flow and cure, and moulding is generally carried out on compression-moulding presses, although transfer-moulding can be used, and recently injection-moulding has been successful.

From the point of view of safety, the ABS materials have a strong claim since they have high impact-resistance and high energy-absorption. They also mould beautifully into complex curved forms. But they are not satisfactory at temperatures much below −20°C because they tend to become brittle. They can also distort over long periods, and ABS cannot be finished to the high standards now expected by the car-buying public.

The dough-moulding compounds have many special advantages. They possess the required structural strength and retain it after many years' service. They can be easily processed, even using existing machinery designed for sheet-metal work. They can be internally coloured, or subjected to standard paint-stoving operations. They withstand extremes of temperature, to 140°C and higher, and they can be made highly resistant to shrinkage, a common drawback with most other polymers and one which has to be overcome to achieve a high standard of finishing.

The Union Carbide Company of the United States is working on a glass reinforced sheet for fabrication on metal-forming presses, using ABS, PVC, and polypropylene as the materials.

Other interesting developments include the possibilities of acrylics, with their high degree of transparency, and the introduction of the new 'super-plastic' metals such as Prestal.

An important factor in the development of plastic materials to replace metal in cars is the problem of dwindling natural resources. DMCs can be used for component parts which up to now have been made of zinc, aluminium or steel.

Throw-away Milk Bottles

At midday on 19 May 1970 at the Hatfield Peveril Dairy in Essex, Sir Basil Engholm, then Permanent Secretary to the Ministry of Agriculture, Fisheries and Food, pressed a button to start the first production-line running. As a result, many Essex housewives are now receiving their milk in quart-sized non-returnable plastic bottles.

The new bottle is conventionally shaped but ribbed for strength. It is blow-moulded in a blend of high- and low-density polyethylenes developed by BP and ICI. The weight saving alone will be appreciated by every milkman delivering the new bottle. Empty it weighs less than one ounce, and has a heat-sealed aluminium-foil cap. One surprising advantage is that the quart size fits into the same shelf-area as a traditional pint bottle in glass, thus doubling the quantity of milk which can be stored in most refrigerators.

The use of plastics for milk bottles would, of course, eliminate both the noise problem and the danger of broken glass, although it could pose its own problems of disposal. The bottles are also blow-moulded within the dairy itself, making for further reduction in distribution costs. Two machines each produce 1500 bottles an hour. They are filled and sealed by a specially adapted machine, and at no time is there human contact with the milk, so the bottles themselves are virtually 100 per cent sterilised at the time of filling.

Plastic Paper

Another striking plastics innovation is Finoplas, a plastics 'paper'. This material has uses in the food, electronics, leather, cosmetics and other industries as a protective covering, or packaging, and is making dramatic inroads where, formerly, conventional, light wrapping-papers were used. It is a high-density polyethylene developed in Britain by BP Plastics and manufactured by a company at Louth, Lincolnshire.

This plastic has a cost-advantage over wood-pulp, the

Miles of conversation. There are 7 million telephone connections in Britain today; by 1973 there will be 10 million. All over the country, particularly in commercial centres, the underground cable-ducts are filling up. Part of the answer to this problem may be provided by a new, ultra-thin polythene insulation for telephone cables which makes them 25 per cent thinner, so creating extra room for more lines in existing ducts

conventional material which it replaces, which, because of world demand, is becoming increasingly more expensive. Estimates suggest that Britain could be saved up to £2 million a year on paper and pulp imports in this way.

It has many other practical advantages. It is lighter than tissue paper, with greater tear resistance. It is more transparent. A certain amount of 'give' enables awkward-shaped objects to be wrapped easily. It seals in flavour and odour. It is resistant to water, chemicals and heat. It takes printing well, and is available in ready-cut sheets easily dispensed from a pull-out pack.

But the possibility is that one day newspapers and magazines will be printed on this material. Ordinary newsprint and other paper-litter blown into the sea virtually dissolves in a comparatively short time. Potential damage from a single copy of a newspaper printed on insoluble plastic and blown off a picnic beach

There are 8000 separate wires in this telephone cable, each one perfectly insulated by a coating of polythene thinner than tissue paper. Apart from space saving, another advantage is that water is less likely to seep in and disrupt communications. The cable is packed so tightly there is no room for moisture

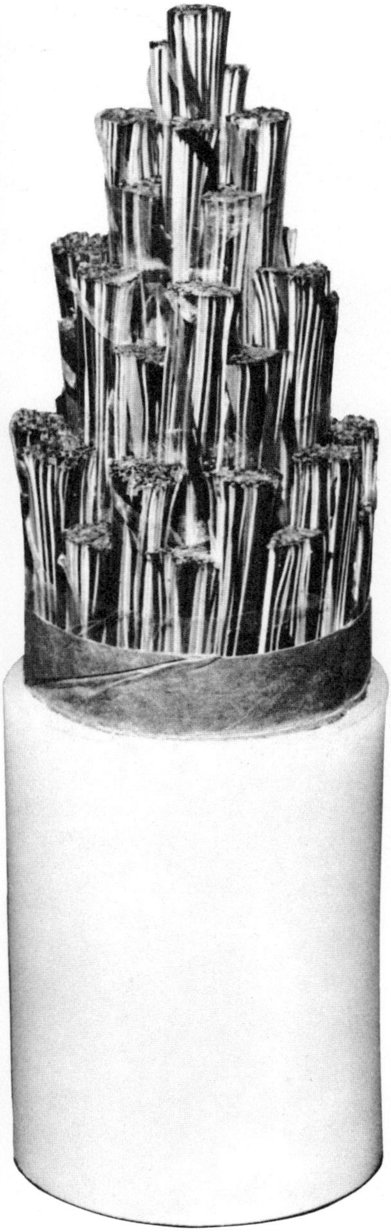

is really quite appalling. By clogging the water intake, it could stop the engines of a fair-sized ship. It could suffocate fish and sea-birds by the hundred–quite apart from the unsightly mess of its spread-out pages returning on tide after tide to the shoreline.

More in Less Space

However, it is the very qualities of indestructibility which make 'plastic' paper simultaneously a menace and the potential solution to a problem of mounting severity. One of the difficulties in coping with the rising demand for telephones is that the underground cable duct-ways, particularly in business areas, are filling up. There simply is not room for the installation of the additional lines for which subscribers are waiting with mounting impatience. There are more than 7 million telephone connections in Britain today; by 1973 the number will be nearer 10 million. A very real contribution to *this* problem of putting a quart into a pint pot is made by the British Insulated Callender's Cable Company whose research division has perfected a very thin cellular polythene insulation for copper conductors.

This new product, Hyperden, increases what is called the pair-density by at least 25 per cent, so that in an existing duct-system the number of lines which can be installed is increased by a quarter. The high cost and public inconvenience of extending the network of telephone ducts is thus avoided. BICC believe that Hyperden cables could supersede all others for subscriber-telephone networks, because they will help to hold down the cost of telephone services by enabling more efficient use to be made of existing duct space, and through the reduction in the number of necessary joints by giving a more reliable service. Large quantities of the new cable are being exported to Hong Kong where there is an explosive growth-rate confronting the telephone service.

Diamonds

The synthetic manufacture of highly durable substances is by no means confined to the plastics industry. Since diamonds, nature's hardest and most glamorous material, were proved in 1797 to be a form of carbon, Man has attempted to make them artificially. Man-made industrial diamonds have been on the market since 1958, and the Speciality Materials department of the General Electric Corporation at Worthington, Ohio, following up their pioneer research work, has become one of the world's major producers of industrial diamond abrasives.

Now the same team has succeeded in the laboratory-creation of gem-quality diamonds. They are produced by subjecting graphite, a form of carbon, to extreme pressures and temperatures in special apparatus. Very high-quality diamonds, some of them weighing more than a carat, have been produced white and clear as well as in a variety of colours including hues of blue and yellow. Unfortunately, the process of 'growing' large diamond-crystals in the laboratory is long and complicated, and these carat-sized gems are currently many times more costly than natural stones dug from the ground. As

Dr Arthur M. Bueche, the Vice-President for Research and Development, honestly points out: 'We simply don't know whether it will ever become possible, in the future, for synthesised gem diamonds to compete economically in the market'.

However, the production of industrial diamond abrasives which are used for grinding, polishing, sawing and other industrial applications, has proved a resounding financial success, even though the total output from the General Electric plant has been no more than 15 tons since 1955. Such limited production is not surprising, since the action of a molten-metal catalyst and the simultaneous application of pressures of nearly 1 million lb. per square inch, with temperatures above 2500°F, are required to make a tiny diamond the size of a grain of sugar. Even so, the score card for synthetic diamonds now includes carat-sized industrial crystals, semi-conducting diamonds, a process for growing diamond crystals without the use of a metal catalyst, and the first hexagonal diamond made by static-pressure metals.

Left A goal that has tantalised scientists for nearly two centuries, the creation of gem diamonds in the laboratory, has at last been achieved by American scientists. This one was made from graphite, the soft black substance used in 'lead' pencils. It is three-quarters of a carat in weight. Although the diamond has undergone slight polishing, it has not been cut and retains the shape in which it was 'grown' in a special apparatus for subjecting graphite (the pile of black powder) to extreme pressures and temperatures. But this stone costs many times more to produce than similar ones dug from the ground and it is not yet known whether synthesised diamonds can ever compete economically in the gem market

Above This diamond was grown in a continuous process extending over a period of weeks. The process starts with a tiny seed crystal of man-made diamond, about as large as the full stop at the end of this sentence. The seed crystal, along with a metal catalyst and man-made diamond powder, is subjected to high pressures and temperatures in a special press. As the temperature climbs the metal catalyst melts and the diamond powder dissolves. Meanwhile, the end of a tube containing the seed crystal is kept cool enough so that the seed will not melt. By carefully controlling the heat and pressure, the carbon atoms from the diamond powder can be made to migrate through the molten metal catalyst and deposit themselves upon the seed crystal, forming a large diamond

Stiff and Strong

It was the extraordinary natural qualities of carbon which attracted the interest of aircraft engineers faced with the need to develop a material for aircraft structures which would offer a better stiffness for a given mass than aluminium alloys. They came up with carbon fibres. Carbon is the sixth lightest element. What is known as the inter-atomic bond (the force which holds one atom to another) is stronger in carbon than any other substance. However, the specific properties of a material increase as the lining-up of its molecules approaches perfection, i.e. the crystalline state. In the case of carbon, the crystal is diamond, which does indeed have a specific stiffness over twice that of the present production carbon fibre, but it would of course be hopelessly brittle even if it could be produced as a structural material. The form in which carbon can be manufactured which is nearest to perfection is, therefore, as a fibre with an oriented or molecularly organised structure. In spite of the enormously high specific strength of carbon fibres, it is

Above The aft and outer structure of an 'Oberon' class submarine of the British Navy, moulded in polyester resin and glass fibre. Some experts believe that plastics will play an increasing part in shipbuilding. As long ago as 1967 Vosper Thornycroft Shipyards constructed the first glass-reinforced plastic-hull section, which was then put to severe tests by the Naval Construction Research Establishment. The result is that the Navy has now ordered the world's largest plastic-hulled ship—a mine-hunter that will look similar to the familiar coastal mine-sweepers. It will be constructed of a solid glass-reinforced laminate. Decks, bulkheads and other relatively flat components will be moulded separately and subsequently bonded on to the hull-structure. The hull itself will be laid up in a steel female mould constructed of bolted sections. When it is complete, these will be removed and the ship will be launched in a normal manner

Right The first glass-reinforced plastics structural section being shipped away for tests. The year: 1967. Today the world's largest plastic ship is under construction

the degree of stiffness which they exhibit which makes this new material so exciting to the designer. This stiffness results from what is called the 'specific tensile modulus' and in carbon fibres this exceeds that of all existing structural materials. However, the rigidity is not always a virtue: it is often associated with brittleness and a relatively small breaking strain.

Carbon fibres alone are of little value in spite of their outstanding mechanical behaviour. If the material is to be of value as a practical load-bearing structure, it must have an additional substance to hold the fibres together and to transfer the load from one to another. So we have the concept of a reinforced composite body, of precisely the same structural form as the combination between glass fibre and resin (GRP) in car bodies. The fibres must also be aligned in the required direction to allow the desired properties to be obtained, and as with glass it has been found in practice that resins behave admirably as the partner material.

Carbon fibre reinforced laminates are probably the most theoretically highly qualified material ever devised, but there remains one important consideration: whereas glass costs much the same as plastics, carbon fibres, although much more effective, will, for the foreseeable future, be considerably more expensive.

One of the most dramatic and successful applications of carbon fibres has been by Rolls-Royce in the manufacture of compressor blades for jet aircraft engines. But its use is not confined to such elaborate and expensive products. High-performance canoes in GRP are now commonplace, and a kayak has been built in which interior stiffening-ribs of carbon fibre laminate were introduced. Such canoes are now used by the British Slalom Team. A similar technique enabled the bodyweight of the 1969 Le Mans-winning car to be reduced from some 153 lb. to 93 lb., of which only 3 lb. were carbon fibre.

However, there are a great many problems concerning the practical application of carbon fibre techniques yet to be solved. We announced their discovery in *Tomorrow's World* on 21 February 1968, but a good deal of premature optimism grew up around the story. There was a tendency to predict that carbon fibres were going to become the ultimate lightweight replacement for metal, but it may well be that the true commercial application will be found in fields infinitely less glamorous than those for which they were originally developed. In many old-fashioned machines an enormous amount of power is absorbed merely by the movement of massive component parts. A classic example is the massive frame of a weaving loom; a carbon fibre, composite frame has already been designed and there is the prospect of greatly increased weaving speeds as a result. It may be the unexciting applications such as this which will eventually earn for this most glamorous of new materials its true place in industry.

Glassy Metal

But whatever the fortunes of carbon fibres may be there is little doubt that fibres are now the favoured form for adding strength to structures. From the Battelle Institute's Frankfurt factory comes yet another fibrous material–metal in glass.

This is made by putting thin metal rods into a glass tube half a metre long with an inside diameter of roughly 2 mm. The tube and its contents are heated and stretched. The metal melts and the glass becomes a coating on its surface. It is then possible, by still further stretching, to reduce the diameter so that eventually a 2-mm metal wire becomes a glass-coated filament of 2 micrometres diameter ($\frac{1}{500}$ mm).

The process can be used even with brittle metals and compounds, and a wide variety of metals, alloys, silicides, borides, oxides and mixed oxides have been used. Certain steels with a high content of brittle components such as molybdenum have been turned into fine wire with a tensile strength about double that of carbon fibres.

Production so far has been limited to the laboratory but has proved not unduly expensive–100 grammes of microwire about 100 km long would cost £1000. The Battelle scientists, however, believe that commercial production will cut these costs considerably.

These fibres have been incorporated into a number of oxide and plastic materials to show how they could be used in practical applications. Like carbon fibres they appear to be particularly suitable for cylindrical and tubular products, and for components subjected to high bending stresses, such as turbine blades.

Tough Bubbles

Another important new material employing glass as a basic ingredient has been developed to make extremely heavy research submarines ride lighter in the water. It is being tested by the Lockheed Missiles and Space Company for the Naval Research and Development Center in the United States.

Called syntactic foam, this buoyancy material is composed of millions of tiny glass bubbles ranging in size from thousandths down to hundreds of thousandths of an inch in diameter. The glass spheres are bonded together by an epoxy resin plastic material.

Because of the purpose for which it has been designed, the syntactic foam must be capable of withstanding tremendous pressures. Depths of 20,000 feet are envisaged for this type of research submarine, and the resultant pressure is nearly 10,000 lb. per square inch. In order to be able to withstand these enormous stresses, the research vessels themselves have to be constructed from very strong, and therefore very heavy, materials. They are much heavier for their size than the larger naval submarines, and they also lack the natural buoyancy of conventional subs. When a submarine is submerged, it should be neutrally buoyant so that it tends neither to sink nor surface, but to maintain its desired depth, naturally. It is to meet this requirement that buoyancy materials such as syntactic foam are necessary to achieve stability at exceptional depth.

Syntactic foam gives approximately 28 lb. positive buoyancy per cubic foot and scientists are now working on buoyancy materials that will give up to 34 lb. But the problem is not only in achieving a highly buoyant material, but in developing one that is both buoyant and strong enough to withstand pressure even greater than that achievable by present syntactic foam. It is an interesting thought that this new material will almost certainly be rendered obsolete when another new material is developed, i.e. one capable of providing the necessary strength for the pressure hull of submarines without paying the colossal weight penalty at present incurred: perhaps carbon or other fibre reinforced plastic.

Fireproof Petrol

The manufacture of minute glass bubbles for syntactic foam is reminiscent of the techniques of microencapsulation. This technique enables droplets or granules of material to be encased within individual skins of suitable plastic material. The resultant advantages are by no means confined to those of handling: microencapsulated petrol, for example, could be thrown in sponge-like blocks from aircraft without further containerisation, and without risk of fire.

One evening in the *Tomorrow's World* studio this was demonstrated by holding a naked flame to a quantity of encapsulated petrol. Only with some difficulty was the material ignited. But then, with a nice little petrol fire going, a further demonstration was made of the potential of microencapsulation to fire-fighting equipment. An extinguishant, also microencapsulated, was sprinkled on to the fire, which immediately went out.

However, the necessity for the microencapsulation of extinguishant material for fire-fighting may well be avoided by the development by ICI of a new fire-fighting powder. It has several times more power than any existing dry-powder extinguishant, and therefore knocks down flames more rapidly. It is effective against fires of burning liquids and gases, and can be applied from hand extinguishers, tenders and automatic systems.

It has been known that potassium salts, particularly potassium oxalate, are more effective as fire-extinguishing agents than sodium bicarbonate, at present widely used. While most materials, when passed through a flame, remain relatively unchanged in size, particles of potassium oxalate explode–or more accurately decrepitate to produce minute particles of potassium carbonate. The problem is that potassium oxalate is poisonous and therefore cannot possibly be used. Further experiments, however, by reacting urea with potassium bicarbonate, have produced a new substance, and the white, non-toxic solid which resulted was capable of being made into a free-running powder. The high efficiency of this material is caused by the breaking-up of relatively large particles into extremely small ones, each of which has active surfaces; in addition the chemical reaction generated by these particles results in the production of fire-inhibiting gases.

A pilot production-plant has been built with a manufacturing capacity of several hundred tons a year, and an evaluation of the new material is being made by fire authorities here and in Europe and the USA.

Super Power from Super Conductors

An atomic-research establishment may need as much electric power for its magnets alone as a town of half a million people. Yet this colossal expenditure of power could almost all be saved by the use of what are known as 'Super Conductors'.

A super conductor is a material which, when cooled within a few degrees of absolute zero, will lose all its electrical resistance. In theory, once an electrical current is introduced into a ring of super conductive material, it will circulate indefinitely without further power. This phenomenon was demonstrated as long ago as 1911, but when the theory was put into practice, it was found that the magnetic fields generated by the circuit adversely affected the super-conducting properties of the material used, and interest in the idea died.

It revived, however, in 1960 when it was discovered that niobium-tin would retain its super conductivity despite the hazard of magnetic interference. Since then intensive research has been devoted towards the development of other super-conductive alloys involving such substances as caluminium and germanium as well as niobium.

Incidentally, it is interesting to note that niobium-tin is one of the metals employed by the Battelle Institute researchers in their work concerned with the manufacture of glass-encased, fine-wire filaments.

The laboratory search for super-conducting materials has involved combinations of all the 103 elements that make up matter. Accurately measured quantities are bonded together by fusing in an ultra-high-temperature argon research-furnace capable of the very rapid production of temperatures approaching 4000°C. The resultant pellet of fused material is then super-cooled in liquid helium, itself cooled in an outer container of liquid nitrogen. Temperatures of almost −250°C, only a few degrees above absolute zero, are obtained. An electric current is passed through the super-cooled pellet, and accurate measurement of the current lost determines the efficiency of the new material as a super conductor.

But liquid helium is extremely expensive and liquid

hydrogen, too, has obvious disadvantages in terms of industrial application. The search has therefore been for a material enjoying the properties of a super conductor at higher temperatures such as are capable of being produced by more convenient cryogenic (or super-cooled) gases, and fairly recently super conductors capable of practical operation in both liquid hydrogen and liquid nitrogen have been evolved.

In the United States the electric-utility industry has been spurred, by the soaring demand for electric power in urban areas, to develop extra-high-voltage power cables, using super conductors for practical underground operation at extremely low temperatures. An exploratory project, developed by the Edison Electric Institute and the Tennessee Valley Authority, is now in its second year. One of the incidental problems to emerge is that of electrical insulation for these underground cryogenic cable systems. Conventional underground power-cables use oil-impregnated paper for insulation, and this was soon found to be unsuitable for the very low temperatures involved. However, synthetic papers impregnated with cryogenic (low-temperature) liquids have been shown to retain excellent electrical qualities for flexible cryogenic cable systems. According to experimental evidence the electrical energy consumed by this proposed system would be 1/10th that now consumed by the insulation of conventional underground cables. The experiments also show that such a system could be more heavily loaded electrically than the insulation of conventional underground cables.

The experimental data indicates that synthetic paper insulation could be applied successfully to a 500,000-volt cryogenic cable system. The flexible cable assembly would be housed inside a hollow pipe, through which liquid nitrogen would be pumped. Liquid nitrogen is comparatively cheap, has a temperature of minus 320°F, and is inert. It could be used over and over again by continuously cooling and recirculating it. Pressure in the pipe would be less than that in present-day gas-transmission pipelines.

The operation of the proposed cryogenic cable system at the temperature of liquid nitrogen would therefore appear to be entirely practical. It would permit underground transmissions at loads several times larger than currently possible by conventional means, and such a relatively high cryogenic temperature would simplify the problem of constructing an efficient refrigeration system in order to keep the nitrogen in liquid form and also to minimise the problems of underground thermal insulation.

In practical terms, this preliminary research indicates that it may be possible to develop flexible, extra-high-voltage cryogenic cables which will be able to transmit power in blocks of 3 to 5 million kilowatts and more by the end of the next decade. Further research is required to determine the financial feasibility of such a scheme.

Meanwhile, in the United Kingdom the Plessey Group's Pre-formation Division has recently extended its operations into this field. Their first activity was to attempt to overcome the undesirable brittleness of the niobium-tin super-conductor alloy. Instead of employing the glass filament technique developed at the Battelle Institute, they have decided to mount the material as a thin film on a base of specially developed tape. The research team is hopeful that the development of this tape may revolutionise high-power magnet technology. For instance, a solenoid constructed of this material and measuring only $4\frac{1}{2}$ inches long by $4\frac{1}{2}$ inches in diameter, using almost no power-consumption, has produced a magnetic field so strong that it would have required an ordinary electro-magnet using a power supply of $1\frac{1}{2}$ megawatts and 500 gallons per minute of de-ionised cooling water. The niobium-tin tape will withstand the effect of magnetic fields nearly twice as strong as those possible without a decrease in the efficiency of other super conductors.

A possible application of extremely important commercial interest is the use of super conductors to build very large, high-current, low-voltage dc machines. Super conductors would reduce the size of an 8000-hp electric motor by 75 per cent, its weight by nine-tenths, and its cost by a third. Other applications include magnetic separators for metallic ores, in communications engineering, and energy storage. Further research in this country is now well in progress with the support of the National Research Development Corporation.

Killer Fibre

The fibrous material which is left after sugar has been extracted from sugar-cane is called bagasse. In every sugar-cane-producing country, thousands of tons of it are burned, allowed to rot, buried or dumped.

Yet it is possible to use bagasse as raw material for fibre-board production, and it is capable of tremendous development to provide an extremely low-cost substitute for other cellulose fibres in chip-board and hard-board; for galvanised roofing, soft woods in paper, cardboard, and viscose rayon production; and for fillers and extenders applicable to the plastics industry.

The problem has been that, in common with other dusty or fibrous materials, stored bagasse has disastrous effects on those working with it. It contains some 5–6 per cent of residual sugar, which provides food for certain micro-organisms that thrive in the hot, humid environment of the cane-growing areas. These micro-organisms attack the respiratory systems of people who are exposed to them. The condition is called bagassosis. When bagasse is stored for any length of time, fungi and bacteria grow rapidly and it becomes an extremely lethal material.

An Englishman, Charles Wright, has been concerned with fibre-board and resin-impregnated board production for many years. He had himself suffered from bagassosis during experimental work with the material. He decided to launch an all-out personal attack on the health hazards, and deliberately exposed himself to the disease by inhaling the organisms which were suspected to be its cause.

Tests conducted at the Brompton Hospital in London under the supervision of Professor J. Pepys, a leading authority on chest diseases, revealed a positive reaction in Mr Wright to the micro-organisms that had developed in stored bagasse from Trinidad and Louisiana. These organisms were recognised as being in the same group which causes the common agricultural respiratory complaint known as 'farmer's lung'.

While many fungicides and bacteriocides were known to be capable of destroying these organisms, they were all poisonous, and therefore unsuitable for use on humans. Eventually bagasse was treated experimentally with organic acids that, in small quantities, are non-poisonous. It was found that the addition of only 1 per cent of an organic substance called propionic acid to semi-dried bagasse effectively stabilised it. The result was that deterioration from fungi and bacteria during storage was prevented and so was the rise in temperature which led to the development of the harmful micro-organisms.

Mr Wright went on to develop techniques for the addition of a weak organic acid, propionic acid, to bagasse, and machinery for applying atomised propionic acid by spraying the material. These developments have been made the subject of world-wide patent applications.

All this means that for the first time an inexpensive, versatile building material is now available in many underdeveloped areas of the world which, unable to produce conventional building materials locally, have desperate housing problems and in many cases severe unemployment. Now that use of the vast amounts of bagasse has been made medically and commercially feasible, many of the serious problems which afflict sugar-cane-growing countries could become significantly diminished.

The stabilisation of bagasse also means that it can now be used in place of wood-pulp to make strong paper, and a bleached form is already being used for newsprint in Mexico. As its composition is similar to that of spruce, bagasse could also be used for the production of rayon. It also has an enormous potential in the plastics industry. Fibres are playing an increasingly important part in the development of new plastic materials, and because of the simple organic fibrous structure of bagasse, it is particularly suitable as a low-cost composite material in polyester laminates and moulding compounds, phenolic resins, and rigid foams.

The success of this work has also stimulated research into treatment with propionic acid of cork and other dusty materials that are dangerous to the workers who handle them.

But Mr Wright's original interest in bagasse was his determination to make fibre-board out of this waste material, and he has achieved remarkable success. Building board has been made from bagasse which had been treated with propionic acid twelve months previously. Although the bales had been stored in the open in Northern Germany for over a year, the bagasse showed no signs of deterioration. This low-cost board has a tremendous potential as a versatile building material in sugar-producing areas; it is twice as strong as the British Standards specification. The 'outside' of the board is covered with a layer of polyester resin, glass fibre and pebble chippings. The back is also faced with polyester resin and the sides are finished with phenolic resin. The result is an attractive and practical material.

An Invention in Search of a Use

Another exciting new material, in which *Tomorrow's World* became closely involved, was metal foam. It all started when a group of scientists were experimenting with plastic foam–the sort of material used for synthetic sponge, stuffing for cushions, draught-stopping tape, and so on. One of them suggested that they should try to coat the plastic foam with metal. The idea sounds simple enough, but in fact it took them $2\frac{1}{2}$ years to solve the technical problems involved.

They prepared the foam by washing it in detergent to remove every particle of dirt and grease, since any impurities would result in flaws in the finished metal-plating. In the early experiments they used an ordinary washing-machine and spin-dryer, and unscientific as it may seem, it is still far the best and cheapest way. The foam is then stretched and pinned on to a rigid frame, which is electrically wired up, and dipped into a bath where it is electroplated in the ordinary way.

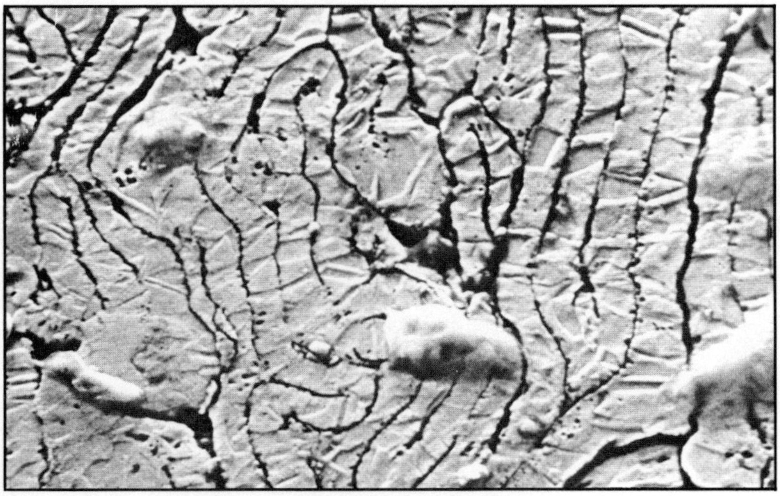

Left One of the latest man-made materials. Metal foam magnified under the microscope 1330 times. This sample is made from Nickel Chrome

Bottom left A new process that creates feather-light foam metal of extraordinary strength and rigidity has just been developed by British researchers. To make it, they use plastic foams as a base upon which to deposit the metal; then the plastic is burnt away leaving behind a structure that is 95 per cent air. The metal foam has a wide range of properties. It is an excellent flame arrester. Its electrical and thermal conductivities are low. It can be used as a filter. It can be used to separate oil from water. It can be made in almost any shape. And if you are rich enough, you can get it in silver and gold. Now the researchers who developed it are looking for an equally wide range of industrial applications for their new material

The technique which took all the time to develop, and which is still an industrial secret, is the method of preparing the plastic foam so that it not only carries the necessary electric current, but also accepts the coating of metal deposited on it in the plating bath. Suffice it to say that this can now be done with foam of any density, and with a wide variety of metals from nickel to gold and silver, and the rate of plating can be controlled so accurately that precisely the required thickness of metal is achieved every time.

At the end of the coating process, which takes as long as two days, the combination of metal and foam is burned in an oven at carefully controlled temperatures to remove the plastic without damaging the metal.

It was at this point that we were first let into the story. Having successfully produced an apparently exciting new material, as an exercise in fundamental research, the Chairman of the Dunlop Company, Sir Reay Geddes, was not unnaturally anxious that his experimental team should find an application for it. Meanwhile, blissfully unaware of this work proceeding in Birmingham, another team, at the BP Research Laboratories 100 miles to the south, was grappling with a problem which has haunted combustion engineers for years–the search for the universal burner. The need for such a device has recently been highlighted by the introduction of North Sea gas into the domestic mains.

As is well known, conversion equipment has to be installed on every appliance before the switch can be made from town to natural gas and the expense and

A new flame spray-gun for spraying on metal coatings. Although hand-operated, its output and the quality of the finished result is practically comparable to expensive heavy-duty equipment that is usually used in this process. The spraying speed has been achieved by a new gas-head design which gives high thermal efficiency as well as a greater spray. The result is sprayed-metal coatings at a lower cost

inconvenience which are involved have become a major headache. The invention of a successful universal burner would obviate this and many similar problems. In theory, it is a device capable of producing a satisfactory flame without having to be adjusted, from a wide range of fuels–natural and town gas, paraffin vapour, naphtha, methane, propane, and so on. The problem is centred on what is known as flame speed. The struggle to provide a compensation for the different speeds accounted for the BP researchers' experimental work in search of the universal burner that would make it possible.

The solution which they have achieved is a device known as the matrix burner. In essence, it consists of a metal honeycomb mounted in a frame. The necessary air for combustion is allowed to pass up through the bores of the tubes contained in the burner. Gas is fed into the 'free' spaces between the tubes, and air and gas combine at the upper surface of the honeycomb to complete the correct mixture in the combustion zone. The manufacture of even experimental burners of this type proved both difficult and time-consuming, but a practical design was eventually achieved. However, as soon as the BP burner research team saw the results of the Dunlop metal foam experiments, they realised at once that here was precisely the material they had been looking for.

The potential commercial applications for metal foam range far beyond the requirements of the gas industry. Retimet, as it has now been called, has shown itself to be an excellent material as a flame-arrester, to stop a gas flame from spreading, while allowing free passage of the gas itself. (The earliest classic example is the wire gauze contained in the Davy lamp–the original safety lamp for miners. The application is, in a way, that of the matrix burner in reverse.)

Flame-arresters are important in oil- and gas-supply operations, particularly in dealing with partly-emptied vessels. They are also necessary for the safeguarding of electrical equipment in mines, dusty industrial atmospheres, and, of course, in the petroleum industry. The metal foam also has possible applications to safety doors in mine ventilation-shafts and 'roads', in dangerous atmospheres sometimes developed in the holds of ships, and in association with the increasing use of hydrogen in industry and atomic-energy plants. Because of its extremely open structure–over 90 per cent of its volume occupied by air–it is very light in weight, but its uniform three-dimensional structure gives it remarkable rigidity and strength. Metal foam is therefore an extremely versatile material suitable for use in all branches of engineering–mechanical, chemical, aviation, medical, marine. It is a good thermal insulator, has low electrical conductivity, and is an excellent absorber of reflected sound. With the ever-increasing problems of noise, these acoustic properties may well prove particularly valuable. And, of course, the material is attractive by purely aesthetic standards, and will excite the imaginations of decorators, interior designers and artists.

In fact, the degree of interest which was aroused by the announcement of its development in *Tomorrow's World* is reflected by the fact that within seven days we received, and passed on to the researchers, no less than 135 enquiries from people and organisations professionally interested in the industrial potentials.

New Lamps for Old

As has been apparent throughout this chapter, what we call new materials are invariably a new combination of old ingredients, or the application of new treatments in existing manufacturing processes. However, a considerable technological effort is at present being devoted to the improvement of materials which have been in everyday use for many years. The target has been as much to minimise their undesirable qualities as to improve their desirable ones.

A good example of this is found in the latest treatment of coated fabrics. Any farmer will tell you that, though plastic sheeting is all very well in its way, it cannot 'breathe' like old-fashioned canvas. Anyone who has used plastic covers to protect a boat laid up for the winter would readily agree. Yet the advantages of the material need no elaboration here. Now J. van Tilberg of Alkmaar, in Holland, have perfected a new method for imparting 'breatheability' into fabric covered with such materials as vinyl, rubber, or polyurethane. First of all, the material is cooled to a temperature at which the coating fabric becomes brittle. The temperature depends to a large degree on the nature and amount of the plastic coating used–for vinyl-impregnated fabrics it lies somewhere between −20° and −100°C, which is achieved by immersing the material in liquid nitrogen.

In its brittle state, the material is then passed between a pair of rollers, the coating is disintegrated, or crazed into a pattern of minute cracks, although the supporting fabric prevents it from completely falling apart. The micro-cracks make the fabric porous and therefore capable of 'breathing'. The material is then reheated until the micro-cracks begin to fuse together again. Clearly if too high a temperature is used and the fusion becomes complete, the entire object of the operation is defeated, but given the correct heat, porosity is maintained, as well as the basic characteristics of protection, flexibility and durability.

Lubrication Without Grease

Since Man first made machines, he has had his lubrication problems. When two working surfaces are in contact, there must be a film of lubricant to separate them and reduce friction losses, prevent excessive wear and overheating, and generally improve the efficiency and smooth running of the machine. This is by no means as easy to achieve as it sounds, and in order to overcome some of the problems, 'dry-bearing materials' have been developed to operate in the complete absence of oil or grease lubrication, and to suffer negligible wear even under the most adverse conditions. They are used when the bearing must operate at temperatures too high for conventional oils and greases to withstand (100–150°C). In other applications it is the freedom from the expense of

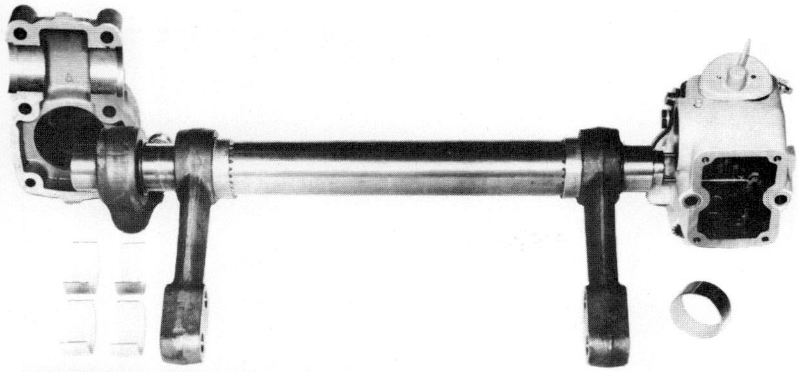

On the farm the latest hydraulic lifting mechanisms for tillage and other agricultural equipment make use of lubrication without oil

having to provide an oil-based lubricant which may be the attraction, and indeed in some cases circumstances may make conventional lubrication totally impossible.

The majority of these dry-bearing materials are based on a polymer called Polytetrafluorethylene. This has exceptionally good temperature-resistance and chemical stability, and under certain conditions has such a low coefficient of friction that it can reduce the effect of one surface bearing on another to that of ice on ice. Its resistance to wear, however, is poor.

Recently it has been reported that this failing can be overcome, so that it can be used as the basis for materials which have as much as 5000 times the wear-resistance of the original polymer. The most successful dry-bearing material based on PTFE is a steel-backed material with a porous-bronze lining impregnated with a mixture of PTFE and lead. This material has been developed by the Glacier Metal Company and is known as Glacier DU. Its extraordinary resistance to wear is demonstrated by the fact that it can carry a load of 80 lb. to the square inch for the equivalent of 1600 miles of rubbing, while suffering less than 2000ths of an inch of wear.

Although no dry-bearing has the ability to operate under such severe load and speed conditions as a bearing with full-pressure oil lubrication, DU is now operating successfully under a wider range of conditions than any

Take porous bronze, impregnated with a mixture of polytetrafluorethylene and ordinary lead, bond it to a steel backing and you have Glacier DU, the most effective dry-bearing yet invented. Dry-bearings that do not wear out have been developed to operate at low friction without the need for oil or grease lubrication. They are used in the docking probes of command modules to ensure a smooth connection with the lunar module. Nearer home, the incredible loads on the new Western Avenue extension in West London ride on dry-bearings lodged between the underside of the elevated roadway and the top of its concrete supporting pillars

Dry-bearings are now being used in the motor industry, too. The king-pin assembly of the Daimler Fleetliner bus incorporates plastic-lined dry-bearings

other bearing material. In the textile industry the elimination of oil avoids the danger of staining the yarn or cloth, and in agriculture the farmer can forget all about lubrication maintenance. It will work successfully immersed in water, and in oven conveyors temperatures up to 250°C have presented no problem.

These remarkable qualities make the material attractive to the automotive engineer. It is important in the design and manufacture of the all-plastic car as it has already proved its capability as a dry lubricant for king-pin and suspension-bushes, as well as a variety of linkages, in many of which conventional grease-lubrication can be applied in addition. It has also been used in aircraft jet engines such as the Rolls-Royce Avon and the new RB Two-Eleven.

Perhaps the most dramatic application is the use of DU in the Apollo space programme. On its return from the Moon, the lunar module makes contact with the command module via a docking probe which opens out, umbrella-like, from the mother spacecraft's nose. A DU bush forms the collar of the umbrella and more of them work elsewhere in the mechanism.

In total contrast, another interesting application is the use of DU as a structural bearing in civil-engineering applications, such as the pivot bearings in bridges and fly-overs. The very low friction coefficient obtained, despite the enormous loads encountered, is an important factor in the design of these structures.

Further recent work with DU now makes it possible to use the substance in direct relationship to plastics, instead of steel. The recent development of high-speed moulding on automatic powder-presses has reduced the cost to levels even acceptable to the extremely cost-conscious motor industry. Intricate parts can be moulded to finished size without any final machining operation.

Nylon bearings are being used in an increasing number of lightly-loaded applications, sometimes with a graphite or molybdenum sulphide lubricant added. They are very cheap, but their wear-resistance is comparatively poor compared with the PTFE-based compounds.

Another important plastics material, acetal resin, is also showing remarkable response to small amounts of the lubricant. Again, the material is steel-backed. A combination of the good mechanical properties of the acetal resin and the firm bond to the steel gives the material a surprising degree of load-carrying capacity and fatigue strength. Bearings such as this have been run experimentally as the 'big ends' of internal combustion engines and proved themselves capable of surviving severe stresses. Exciting as it is, that work is still in the experimental stage. Meanwhile, the remarkable powder which can make a dry surface as slippery as ice is playing an important role in the smooth running of the housewife's spin-drier, and all the other high-speed machinery now making its appearance around the house.

The Exploding Tape Measure

An item in *Tomorrow's World* which none of us will forget was our demonstration of Jet Axe. This is an

Jet Axe—Instead of drilling holes for dynamite geologists have a plough that plants explosives

explosive method for cutting neat circular holes quickly, simply and safely in almost any material–including quarter-inch sheet steel. Our demonstration involved some extremely impressive explosions just outside the door of our studio at the Television Centre. Immediately after the demonstration, the police were included amongst 'interested parties'!

Now ICI have produced a tape which, in a similar manner, produces a highly controlled and directional explosion.

In their constant search for new mineral resources, particularly oil, field geologists have to build up a three-dimensional map of the underground surfaces and structures. In the past this has been done by setting off seismic charges of dynamite and studying the echoes of the explosions; by drilling out a sample core; or by a combination of both methods. The new explosive, Geoflex, offers a far more convenient system for subterranean 'sounding'. It is in the form of cord which can be made up into almost any length from a few feet to half a mile or more. It is in fact a multi-core cable containing the explosive PETN, covered by two layers of fibres and a tough plastic sheath, which cushions the explosive against external shock. It is exceptionally resistant to accidental detonation from rough handling, and if set fire to, it will burn vigorously but will not explode. It has the added advantage of being able to survive long periods of storage in the most extreme conditions without deterioration.

The explosive tape is laid by ploughing it into the ground a few feet below the surface. Geophones (extremely sensitive microphones) are placed on the ground in a precise pattern, and when the tape is detonated they measure and record the returning echo-signals. These traces are then fed into a computer which produces a far better geological map than has been available before. The entire operation is not only quicker, but it can be operated in areas that have always been considered impossible for seismic mapping–desert areas, for instance, where the subterranean rock to be studied may be covered by a layer of sand over half a mile thick.

Soundings to depths of up to 20,000 feet can be obtained in this way because the shock wave generated by the explosive cord is extremely controlled and stable, and the blast is not wasted in the air, but is directed downwards.

The ploughs are the only special equipment needed by the exploring seismologist and geophyscist. They are designed for use with the type of standard tractors that would probably already be in the possession of prospecting teams. A single plough can replace three or four exploration drills, and the combination of a plough and tractor with two or three operators can outpace the observation team. The plough can often lay sufficient explosive tape in a few hours to keep the observation team busy for an entire day.

The system can be put into operation in swamp or jungle as well as desert. It has also already proved of great value in submarine operations. It has been used in the North Sea, and in the Persian Gulf a high-speed hovercraft has been specially fitted out for this type of prospecting in the 30-foot-deep shallows. We recently flew over the newly-developed oil-prospecting areas on the northern coastline of Alaska. Looking down at that incredible waste of snow, ice, and terrifying mountain ranges, it was remarkable to think that men were conducting a meticulous scientific survey of one of the most inhospitable places in the world. Their work is now to be assisted by explosive tape. Instead of a plough they will use a circular saw to cut the necessary slots in the ice and permafrost, and they will have to design a new plastic core for the Geoflex itself, because the existing material is too brittle in Alaska's −50°C. There can be little doubt that work on the many applications of cryogenics, involving extremely low temperatures, will provide a ready solution to this problem.

The new explosive cable feeds into the head of the specially designed plough during 'planting'

Chemical Machining

Etching is the process in which a chemical is used to dissolve away material in a controlled pattern and so produce a three-dimensional design on a surface. The traditional method used by engineers to achieve much the same objective is called 'machining'–unwanted material is cut or ground away, often while the object being worked is rotated on a lathe.

Now a new technique, chemical machining, is gaining increasing importance, particularly in the aircraft industry. Metals are 'machined' by immersing them in an acid or alkali solution to etch specific shapes and sizes into the finished article. Precise patterns can be achieved by masking areas of the component, so protecting them from the corrosive action of the etching fluid. In this way it is possible to form shapes in metal which would not be practical with conventional machining, forging, or casting techniques. Other important advantages are that no machining stresses are produced in the component, parts with very thin sections can be produced without distorting the whole, and sometimes components can be chemically machined on both sides at once.

Rolls-Royce have recently opened a large chemical-machining plant for the manufacture of components for the RB Two-Eleven engine. Other components are being made for the Dart, Spey and Conway jet engines. The actual process of etching follows the traditional method precisely. After meticulous cleaning the surface is roughened so that the masking material, a rubber solution, will obtain a good 'grip'. A template, or pattern, reproducing the shape of the areas where the masking material is to be left on, is placed in position and traced out on the rubber with a sharp-pointed scriber, before the unwanted rubber masking is simply cut away by hand. The areas masked are larger than the required finished areas because the chemical undercuts the mask as etching proceeds. The prepared component is then dipped into a tank containing the etching solution, where it remains until the appropriate amount of material is eaten away. The temperature of the etching solution has also to be carefully controlled since it can cause large variations in the etching rate. Needless to say, if varying depths of etch have to be produced in any particular component, the masking material can be removed section by section, so that the process can be repeated and continued as necessary.

Work of extreme delicacy can be undertaken in this way, but despite the old-fashioned origins of the technique, the most exacting production controls required by components for the aircraft industry can be maintained on an entirely economic basis.

Old Wine in New Bottles

The advent of plastics in the toy-making industry has brought into every home a range of models, games and toys which a few years ago could have existed only in the fairyland of childish imagination. Now the techniques of moulding in polyester promise to bring into the price-range of every home reproductions of the great sculptures of the world–in a plastic coated with metal powder which is virtually impossible to distinguish from any conventional casting–until you pick it up.

But art students are not the only people interested in polyester casts. An Irish surgeon is making casts of the complicated heart arterial system to help his research into coronary thrombosis. Instead of a plaster mould, he uses an actual heart into which he injects polyester resin. The resin is forced gently into the coronary arteries, following in every minute detail the course pursued by circulating blood, and after the resin has hardened the tissue of the heart structure is dissolved away. A complete and durable polyester cast of the entire arterial

Rolls-Royce at Hucknell, which produces vital parts for aircraft engines, is making use of one of the newest forms of industrial processing–chemical machining. And not a lathe in sight!

system remains. By using resins of different colours, separate casts can be made of the arteries or veins alone, or of blood vessels and chambers together. Such models, taken from the hearts of people who have donated their bodies for medical research, have proved of immense value both as a teaching aid and for research.

Many similar examples are found in the shipbuilding industry, now busy modernising itself in order to reap the advantages offered by new technological developments. Vospers Shipyard at Southampton has enjoyed a reputation for advanced and adventurous work for many years particularly in the field of high-performance boats. They are now building a new mine-hunter for the Royal Navy which promises to be one of the most interesting hulls ever constructed in glass fibre (GRP).

The problem is to construct a ship from material which exerts no magnetic influence capable of detonating the latest type of magnetic mine. Hence the choice of GRP. It is more costly than steel but not more expensive than laminated wood, the only other material that could be used in this type of vessel.

As a first step towards the evaluation of the new material in operational conditions, a prototype vessel is being built to the existing mine-hunter design. The intention is to provide a really efficient hull for a future generation of mine-countermeasure ships. So far, glass-fibre technology in the Royal Navy, and elsewhere, has been confined to use in small craft, sailing boats, and certain exterior components such as casings and fins for *Oberon* class submarines. The new ship, with an overall length of 153 feet, will be much larger than any previous vessel constructed of glass-reinforced plastic, and will cost somewhere between £1½ and £2 million.

What governs the physical properties of a GRP fabrication are the resins and type of glass fibres used. The new ship will utilise glass in the form of woven fibres which provide a good strength-to-weight ratio and good fire-resistance. In addition to its non-magnetic qualities, a GRP ship is free from corrosion or attack from marine organisms, a factor which has made the material so attractive to yachtsmen.

Incidentally, the importance of magnetic influence in this type of craft is such that it will even be impossible to carry canned beer on board.

Early in the research programme which lies behind this development, it was found that, although a sandwich-type construction had many attractions and could develop adequate strength under static loading, it was unable to cope with the shock loadings which such a vessel would be expected to undergo in a seaway. But a single-skin structure has been found to be entirely satisfactory and will be used in the mine-hunter: a transverse framing system is proposed with a deep vertical keel, as on a yacht, and it will be interesting to see whether or not carbon fibre-reinforced laminates will eventually be adopted for this requirement.

Any shipyard building vessels of this type must be equipped with special facilities. Large covered slipways are required so that the ship can take shape in a well-

Complicated shapes achieved by the use of chemical-machining techniques

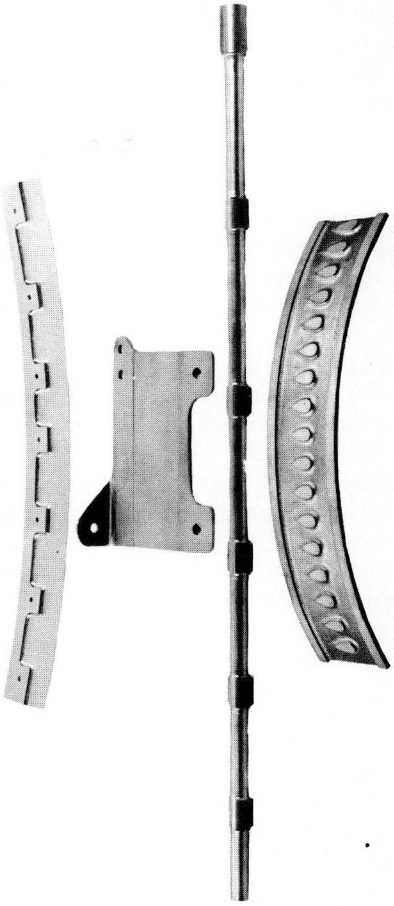

ventilated, temperature-controlled environment. The streamlined hull has to be 'laid up' (the layers of fibre and resin bonded one on top of the other) in a special steel mould which has eventually to be removed before the ship is launched. More special moulds for flat panels are needed for the fabrication of decks and bulkheads. All these components will be made on site. Special mechanical equipment will be required to dispense and saturate the glass reinforcement with resin as required. An interesting offshoot of the development will therefore be that, for the first time, the entire hull and superstructure of the vessel will be constructed from the most elementary raw materials near the slipway.

Man-made Seaweed

An even more surprising example of the application of plastic techniques to marine problems is the manufacture of artificial seaweed. Natural seaweed often plays an important part in the creation of sandbanks because, by reversing the scouring action of the tidal or river currents and by changing the wave-pattern and removing some of the energy from waves, small particles of waterborne material like sand are deposited among and around the tufts.

So polypropylene 'artificial seaweed' has been developed to protect underwater pipelines and similar structures. It is already being offered by ICI to drilling companies in the United Kingdom. It works by harnessing the action of currents in exactly the same way as natural seaweed, and trials have already shown that it is capable of building up substantial sandbanks. It can also be used to protect and secure the legs of drilling rigs themselves.

The system can be employed in depths of up to 20 fathoms, always assuming that the sea-bed has sufficient sand in the vicinity to make the build-up possible. Estimated costs so far suggest about £30,000 per mile of pipeline protected. The original purpose of this work, which was started five years ago, was the prevention of coastal erosion: a small-scale trial at Bournemouth a year later, followed by two much bigger trials off the East Coast, proved extremely promising. In a matter of months, 5-foot-high banks of sand had built up, and subsequent surveys have confirmed that the situation there has become stable.

The artificial seaweed itself is made of tufts of polypropylene tape tethered to supporting tape or plastic ropes. One end is held on to the sea-bed, and as polypropylene is lighter than water the free ends float upwards in much the same way as the fronds of natural seaweed.

Banks of sand around a pipeline prevent stresses and strains on the pipeline itself and also reduce the possibility of damage by anchors. Presumably the effect of the implantation of such major artificial 'growths' on local marine life will be similarly assessed and evaluated in due course.

The Trail Blazers

Radio isotopes are produced by bombarding common substances with nuclear energy. This process of irradiation changes the atomic weight of the substance, and it then emits characteristic radiations which are easily detected and measured, although its chemical composition may remain unchanged. Radio isotopes are produced in research reactors like the ones at Harwell, production reactors at nuclear power stations, or cyclotrons.

It is estimated that British industry saves well over £12 million a year by using radioactive 'tracer' isotopes in a wide variety of manufacturing processes and instrumentation. Many thousands of lives are now saved annually by using radio isotopes to detect and treat illness and disease. But exposure to radiation sources, except under carefully controlled conditions for diagnosis and treatment, is immensely dangerous. It can result in death or serious injury in a matter of seconds. So scientists have had to learn a great deal about the amount of radiation which a body can withstand without damage. Very effective safeguards in the form of shielding, instrumentation, remote-control techniques, protective clothing and decontamination have been developed. The principal centre for this work in the United Kingdom is the Radio Centre at Amersham in Buckinghamshire–the world's leading producer and distributor of radio isotopes.

When they are first removed from the reactor or cyclotron in which they have been bombarded, few radio isotopes are ready for use immediately. They invariably require further refinement, and are sometimes subsequently built up into complex compounds. The Amersham Centre, which is part of the United Kingdom Atomic Energy Authority, is run very much as a business. It dispatches 75,000 consignments of radio isotopes to order in a year and earns more than £1½ million per annum in foreign currency.

Since most of the chemicals used in radio-isotope production are in themselves dangerous, processing is carried out in screened and airtight compartments equipped with remote-control apparatus or rubber gloves sealed into access ports in order to ensure that the operators are protected. Various radioactive materials emit different kinds of radiation, each presenting its particular hazard. For instance, alpha rays can penetrate only a fraction of a millimetre into human tissue; beta rays may penetrate several millimetres; and gamma rays require very thick shields to reduce their intensity to a safe level. Injury can be sustained by 'internal radiation' by breathing or swallowing radioactive particles or absorbing them through the skin; or by 'external radiation'–exposure to a radioactive source some distance from the victim. Lead and concrete shielding is used to protect personnel from gamma-activity, whilst thin barriers made from plastics or metal are adequate protection from beta rays. In addition alpha-emitting substances are processed in sealed boxes to prevent the escape of radioactive dust.

At Amersham the radio-isotopes are refined in laboratories and then transferred to the special Radioactive Sources department where, if required, they are blended with other substances and finally encapsulated in metal containers. Until the final encapsulation has been completed, the materials are handled on special 'hot benches' of solid concrete on which stand 'process-boxes' made of our old friend GRP.

A special, remote-controlled 'railway' has been built to convey materials from one process-box to the next, and gamma-ray protection is afforded by concrete walls several feet thick and by a lead brick wall at points of special exposure risk.

GRP was chosen for the construction of the process-boxes for a number of reasons, not least because it is imperative to guard many isotopes from contamination by metallic impurities. Other arguments in its favour were that it withstands temperatures of up to 100°C and resists acids and solvents. A special grade of resin was specified and developed for this application. Meticulous care was also required to ensure that the containers would be totally impervious. Another advantage is that GRP does not tend to collect dust as a result of static attraction and it is easy to clean and can be moulded into smooth curves that avoid dust traps. And the boxes are reasonably light in weight and can be moved easily when they have to be decontaminated.

An additional advantage resulting from the fact that GRP is an easily worked material is that fittings of perspex and glass windows, air and electrical connections, glove ports and other fixtures are not difficult to mould into the main structure.

The contribution made by these apparently simple qualities towards the expansion and advancement of radio-isotope technology is very real. Ever more complex and powerful compounds for ever-widening fields of application can be expected during the next decade. Already there are very few aspects of scientific research, and indeed everyday life, on which radiation technology does not already have its influence.

Work already completed at Amersham tells its own story. Two years before the manned landing in the Sea of Tranquillity, foils of polonium were supplied for the Surveyor V vehicle which soft-landed on the Moon, and 'scattered' alpha-particles on to its surface so that instruments on board the spacecraft could measure the radioactivity of the rocks, and transmit the results back to Earth. From these measurements it was possible to deduce information about the Moon's surface.

During the next few years similar advanced activation analysis systems developed in this Buckinghamshire town will probably be deposited on Mars in preparation for the manned expeditions anticipated for the 1980s and 1990s.

In medicine, an experimental method for treating certain types of leukaemia is being developed, which allows the blood to circulate for long periods through an external plastic tube fitted to the patient's wrist. The tube contains a strontium-90 source which irradiates and destroys specific blood cells without harming the rest of the body.

In our recent *Tomorrow's World* programme from Edinburgh, we demonstrated mobile X-ray machines which used gamma-ray sources for 'on the spot' analysis of rock samples by geologists. This method means that ores no longer have to be sent to the laboratory before their composition can be determined, and Amersham-produced isotopes are the key ingredient to this time-saving technique.

In agriculture, polonium beryllium neutron sources are being used to measure the moisture contents of soils with extreme accuracy.

Finally, Amersham is now supplying radium and americium sources to the United States where a domestic fire-alarm system is being developed which will be 'triggered' when smoke enters one of two balanced ion chambers.

Left The rarest gems in the world—man-made diamonds up to one carat in weight. They were created a few months ago by scientists in the United States, who can turn out high-quality stones in a variety of colours from white to various hues of blue and yellow. Although the diamonds have undergone slight polishing, they have not been cut and they retain the shape in which they were 'grown'. But it is still much cheaper to dig your diamonds out of the ground. Estimates suggest that this handful cost approximately 1 million dollars to manufacture

Bottom left A novel method of casting sculpture, using polyester resins instead of conventional materials. It is cheap, too. This life-size head will cost less than 10 shillings to make. It can be given a realistic metallic finish by simply adding metal particles to the plastic mix. The result is tough and durable

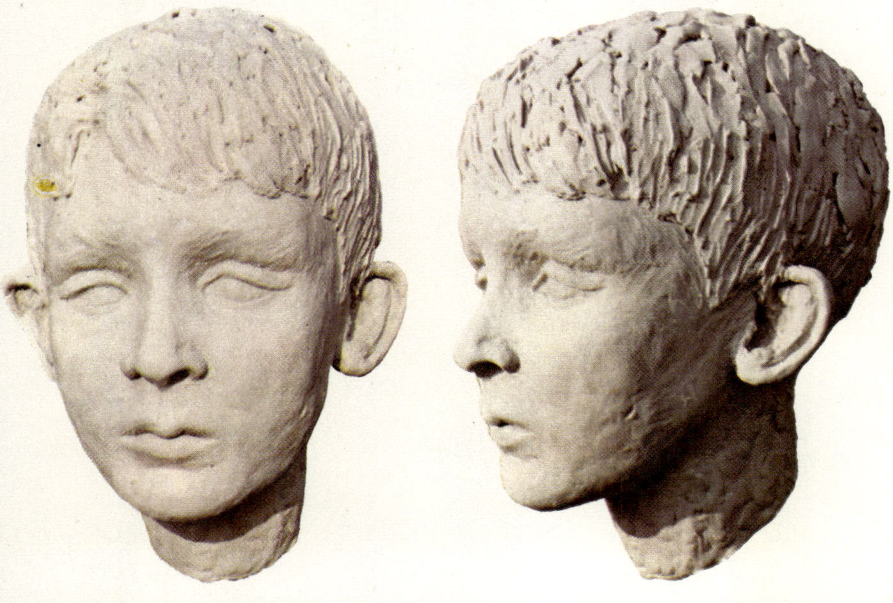

How to Eat Petrol and Thrive

Hunger / The Search for Protein / Meatless Meats / Take 2000 Gallons of Milk /

Meatballs, hamburgers, and a juicy meat loaf—made from the latest man-made, artificial minced meat. Ingredients include spun soya protein, dried egg-whites, vegetable protein, artificial flavours and colour, and vitamins. It is cheaper than fresh meat but just as nourishing.

As every housewife knows, the basic difficulty about feeding people is the never-ending demand of the human body for energy-giving sustenance in relatively small quantities at regular and comparatively short intervals. This calls for constant supply–and there's the rub.

In many ways Nature herself is apparently opposed to such an arrangement. Crops, no matter what the climate, grow more or less seasonally. There is a time for sowing and a time for reaping–and the gap between the two can be a time of hunger unless appropriate measures are taken. In addition, the questing nature of Man has always tended to put him into hostile environments. Ever since he climbed on to a floating log in order to cross a river to see what was on the other side, Man has been obliged to take with him iron rations for the journey, at least since he became prudent enough to realise that the supply of roots and berries on the far shore could not necessarily be guaranteed.

These, of course, are fundamentals which do not take into account economic factors. It could be argued that it was these basics of supply and demand which led Man to realise the possibilities of buying, storing and selling food–almost certainly the origin of all commerce. Today the food industry is the largest in the world. For most families food is the major item of the household budget. In the civilised world we spend a quarter of our earnings on food–sometimes much more.

Long ago Man discovered that his food would keep longer if it were dried and kept in a cool place. Grains in particular, and even meat, fish, and certain fruits could be kept for long periods if they were cut into thin strips and dried in the sun or over an open fire. Another early discovery was that rubbing salt into food, or soaking it in salt water, helped to preserve it as well as improve its flavour. Subsequently ways were developed for cooking and packing food in sealed containers so that a wider variety could be stored or carried on journeys. And, of

Pre-cooked, pre-diced artificial chicken was used to make these tasty-looking dishes. Already food like this is making its appearance in American restaurants and, according to reports, customers like it. The makers say that in recipes 1 lb. of artificial chicken is equivalent to 3 lb. of fresh meat after cooking. It can be mixed with poultry left-overs and one big advantage for commercial caterers is that it may be kept hot for hours without losing its fresh, rich appearance. As with artificial beef and ham, it is made from a base of spun soya protein which is actually 'knitted' on special looms to produce a chewy texture similar to fresh meat

course, the Eskimoes have been refrigerating their meat from time immemorial. What present-day food technology is doing is to improve all these preserving processes; experiment with totally new techniques, such as irradiation; develop ever more efficient methods of packaging; and finally invent completely new foods, in an attempt to make palatable substances capable of meeting the constant energy demands of the human body.

The modern food industry is an extraordinarily complex chain of events that begins before seed is sown, fish spawned, or animals born. It involves every discipline of science, every natural law, every process and device of modern technology. When you buy your food, you are paying not only for its production but for collection, preparation, transport, distribution, packaging, advertising, research, capital investment, wholesale and retail distribution, storage and delivery. Bearing in mind the colossal weight of overheads all this represents, it seems a remarkable achievement that food prices are as low as they are, so low in fact that it is often uneconomic to attempt to produce such uncomplicated garden crops as potatoes, and even tomatoes and soft fruit, on an amateur basis. As we straighten our aching backs after a couple of hours' old-fashioned toil in the vegetable garden, we tell ourselves that it is worth it because of the flavour. But even this comforting thought is becoming open to serious question, and there can be little doubt that in the very near future the amateur vegetable garden will only be supportable as a 'fun medium', to be tended and cropped purely for pleasure, as the flower garden always has been.

Yet all these considerations are little more than social niceties in the face of the real threat which now confronts mankind–hunger. The family is outgrowing the larder–and not only the larder but also the farms, the fisheries, and the factories which supply it. The bulk of technological effort is currently devoted to improving the situation of the 'haves', rather than the 'have nots'. The production of protein from unconventional sources is a case in point. This technology has made remarkable strides in the United States during recent years, but the product is being added to the already abundantly-stocked shelves of American supermarkets rather than finding its way into countries where hunger is a way of life.

Hunger

From his earliest origins, when he has wanted to celebrate something, Man has had a feast. The tradition persists in every known culture, but an occasional feast does little good for a hungry child. No amount of food and special treatment will ever restore to full health, and develop to complete manhood, the child deformed by malnutrition. There are millions like him. Since birth he has suffered from undernourishment, and starvation has done him permanent injury in a subtle, and until recently unsuspected, way.

A Mexican paediatrician, Dr Joaquin Carvioto, is a specialist in the protein-deficiency disease, Kwashiorkor. Ten years ago he noticed that even when his most severely affected child had recovered, he seemed dull-witted and less intelligent than average. These observations were borne out by scientific assessment: further testing in mountain villages in Mexico and Guatemala confirmed that poorly-fed infants were consistently less advanced than those that were adequately fed.

To explain the phenomenon Carvioto considered several possibilities. Underfeeding left children vulnerable to a variety of diseases, while conditions amongst the poorest families invited attacks by parasites and the effect of exposure. Genetics might also have explained why undernourished children failed to measure up in intelligence. Then a far more ominous truth was discovered. The human brain accomplishes 80 per cent of its growth in the first 3 years of life, even though the rest of the body grows to only 20 per cent of adult size in the same period. Before birth, from approximately the fourth week after conception, the brain starts to develop, and critical to the production of its new cells is the supply of amino-acids. These aminos are supplied by the mother, who derives many of them from the protein in her own diet. After birth the infant gets them from its mother's milk, and then from its regular diet. But if at any point in the sequence the supply of amino-acids is deficient, the brain does not grow. Under the stress of inadequate protein, so often a significant factor of poor diet, instead of expanding with live and functioning brain cells, the nerve tissues may be permanently stunted.

As a result of Carvioto's work, and similar research elsewhere, it has now been established that the bulk of cells in the human brain are grown in the first 6 months of life. Some post-mortems have revealed that undernourished children lack up to 20 per cent of the brain cells normal for their age. The implications are appalling. Malnutrition can mean the mental degradation of whole societies, and rehabilitation may never make up for inadequate feeding during infancy.

On the other hand, one more happy possibility has emerged from similar studies at Washington's Children's Hospital. Researchers report that mongoloid children treated with certain amino-acids walk and talk earlier than expected, which may mean that ways can be found to correct nutritional brain-damage chemically. Dr Bernard Rimland of San Diego's Child Behavior Research Institute goes further in suggesting that nutritional improvements might someday help to raise the average IQ from 100 to 130. But again, Dr Rimland has been working with the 'haves'. With the nightmare spectre of hunger haunting so many millions, it is difficult to accept as a serious contribution towards the advancement of food technology the work at Michigan State University devoted to the development of cherry-flavoured butter, and chocolate cheese.

Temperature-sensitive tags which guarantee that food is sufficiently cooked before serving. At present they are designed for use in school kitchens, factory canteens and hospitals, where large quantities of food have to be prepared, but one day they could be adapted for the home. The tags are attached to the container in which the food is cooking. If temperature and cooking time are correct, the purple tag turns to green. Failure to change colour indicates that the food is not fit to be served

The Search for Protein

Scientists working in food technology tend to object to the application of such adjectives as 'new', 'synthetic' and 'artificial' to protein. Protein is protein–whether it be seaweed, or beef, or the soya bean.

The search for new sources of edible protein, its development and application, is almost certainly the most important positive contribution which technology can make towards the elimination of malnutrition. As will be seen, an important sector of this search is the reappraisal of what in the past were often considered as poisonous or waste products.

Fish Protein Concentrate/What its designers claim to be the world's first commercial plant to produce high-purity protein from the sea is about to go into production at Canso, Nova Scotia. The $5 million extraction plant, producing 30 tons of pure protein supplement per day, requires an input of 200 tons of fresh fish. The fish will be purchased under contract from independent trawlers and seiners as well as from a nearby modern fish-processing plant. Species of fish not normally used for food, as well as fish waste from the processing plant, provide the raw material.

The process, which employs isopropyl alcohol-solvent, has been developed after many years' research in Halifax by the Canadian Fisheries Research Board. At first small quantities were produced under pilot plant conditions. Now it is a practical and commercial production technique. The new plant will extract protein from fish, and fish waste such as skate wings, to produce a bland, near-tasteless powder containing up to 95 per cent of high-grade animal protein. This can be stored

Canadian researchers built this equipment to turn any fish in the sea into a high-quality protein powder that can be added to all kinds of foods. The machine uses alcohol to extract oil, fats, excess water and smell from the kind of fish that are often caught but cannot be sold in the shops. When the process is complete, there remains a tasteless, odourless powder that has more nutritional value than milk protein

indefinitely and can be added easily to standard foods and beverages to increase their nutritive value without affecting their taste.

The cost is significantly low, but it is particularly interesting that the sales potential of the new plant seems to have been assessed against a local market only. The Canso principals have drawn attention to the fact that, according to recent Government surveys, there are 1 million undernourished people in Canada and 10 million in the United States. Theoretically, with protein from the sea at only 35 cents per lb., a person could receive his full requirements of annual protein for only two-thirds of a cent per day, or less than $3 a year.

There are other fish protein extraction plants in operation elsewhere in the world, but they start with fish-meal as their raw material. The Canso technique does away with the interim process. More development work is now being conducted on the production of fish protein flour with special characteristics, for example, one which will swell in water to resemble the original fish muscle. This 'texturing' is of major importance in expanding the acceptability of such 'new' foods, and is the subject of major studies elsewhere.

Clean food. This ordinary-looking cage has been called one of the most significant developments in food preservation in recent years. It has been designed to irradiate fish fillets with radioactive Cobalt-60, which kills 99 per cent of all bacteria normally found on fish. The advantage of this technique over other pasteurising methods is that the fish remains cold and raw. But the surviving 1 per cent of bacteria are still capable of reproduction, so the fish must be kept refrigerated. Even so they can be stored three times longer. Now researchers are to use the machine to irradiate bacon, grain products and potatoes. They could all be in the shops within 5 years

Protein from Carbohydrates/In Britain, the National Research Development Corporation is backing a development programme aimed at producing protein from carbohydrates, with an investment of up to £500,000 spread over three years. A team of scientists and engineers has been working on the project for the last $4\frac{1}{2}$ years at the Rank, Hovis, McDougall Research Centre at High Wycombe. They are attempting to create a new high-quality protein for use in protein-enriched foods and as the basis for entirely new food products. They have decided to call it A3/5.

Within five years it is hoped that it will be possible to produce A3/5 very cheaply anywhere in the world. It can be made from potatoes, yams, cassava, sugar or any carbohydrate crop.

Even in Britain, where in theory there is more than enough protein, malnutrition exists, because the old and the poor often do not eat enough meat, fish, eggs and cheese to meet their daily protein requirements. To overcome this, suggests NRDC, A3/5 could be used as a supplement in foods like bread, breakfast cereals and biscuits.

It is claimed that A3/5 can be made to taste and look like normal meats, such as chicken, veal and beef, as well as other foods. Of equal, if not greater, importance is the question of cost, and although not yet in a position to state a price to the consumer, Dr Ezra Bard of NRDC regards the development of A3/5 as one of the most significant advances–'possibly of this century'.

Professor Arnold Spicer, who has led the research team, says, 'It will affect the life of millions upon millions of people'.

A3/5 is produced by a fermentation process similar to brewing, where micro-organisms are grown in tanks under carefully controlled conditions. The process is

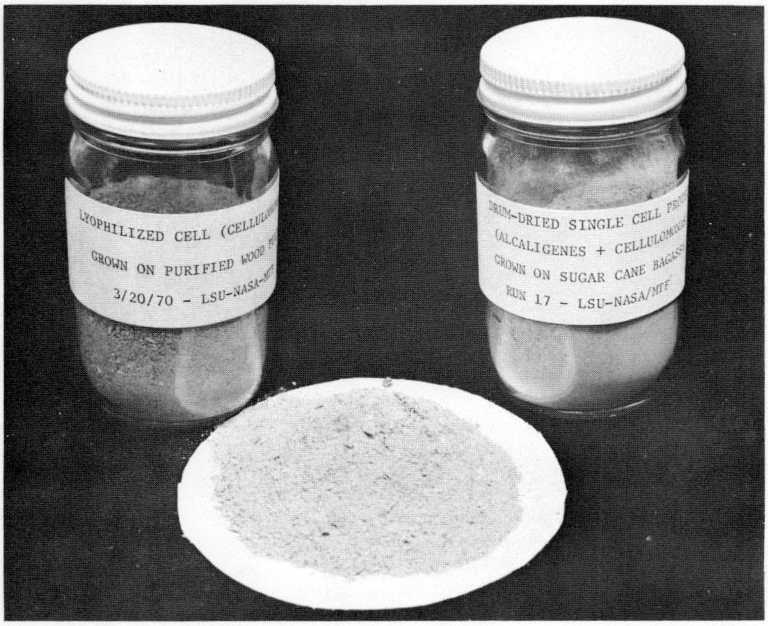

A coarse brown flour that could be the basis of a large portion of the world's food supply in 25 years' time, according to the American scientists who have produced it. It is made by using sugar-cane bagasse, the waste cellulose left over when all the sugar has been extracted. And it could be the cheapest food in the world, too

based on continuous production–a constant feed-in of fresh nutrients and a constant removal of grown culture. After further refinement the product can be dried to granular or powder form, or processed further in a moist state into simulated meats.

But A3/5 protein still has a long way to go. Feeding trials on laboratory animals have shown that it can sustain normal growth and good health; long-term nutritional feeding trials will now be conducted with farm animals. In about three years, assuming that all the experiments are completed successfully, it will be evaluated for human consumption.

Protein from Cellulose/Bagasse, the fibrous residue from sugar-cane processing plants, has until recently been regarded as a messy, and positively dangerous, waste, (see pp. 213-4). But one day *you* could be eating it. Recently published work at Louisiana State University, under the sponsorship of the American Sugar-cane League, now promises to turn it into a useful foodstuff. The world production of bagasse during 1966–7 was about 51 million tons, and most of it was used as fuel for sugar-mill boilers.

The fermentation of cellulose is certainly not a new idea, but it is a concept which has so far failed to find much application–at least in a peace-time economy. The process that has been developed at LSU treats cellulose material such as bagasse, straw, grasses, etc., with a mild alkali, which amongst other things oxidises the material and makes it suitable for further biochemical treatment. Meanwhile another department at the University discovered that a bacterium could attack native cellulose substances. It had rapid growth-rates and simple nutrient demands and it was the culture of this organism in controlled conditions that turned out to be the key to the production of protein from sugar-cane waste.

An economic analysis was prepared for a proposed plant, using sugar-cane bagasse as its raw material, to

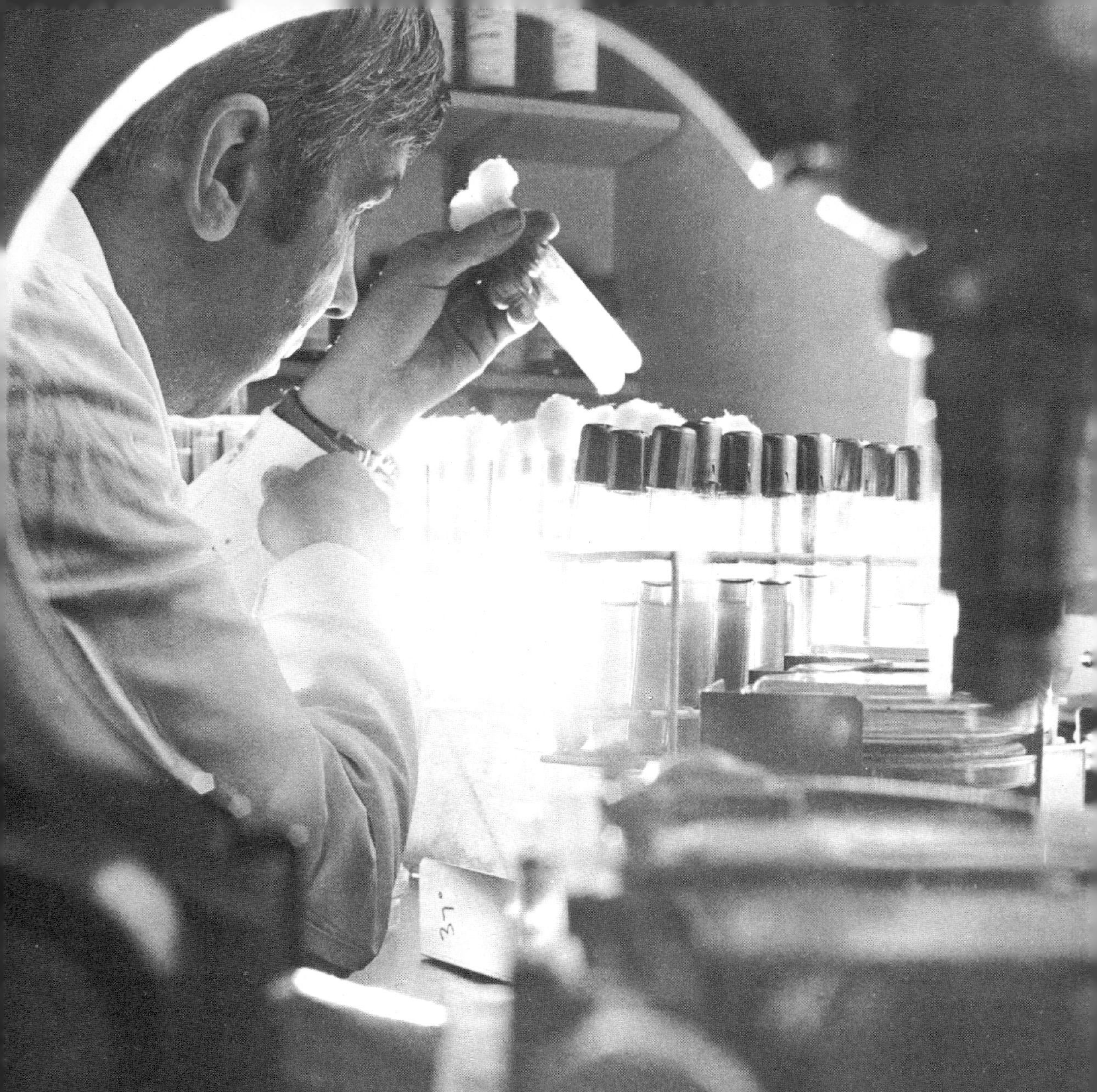

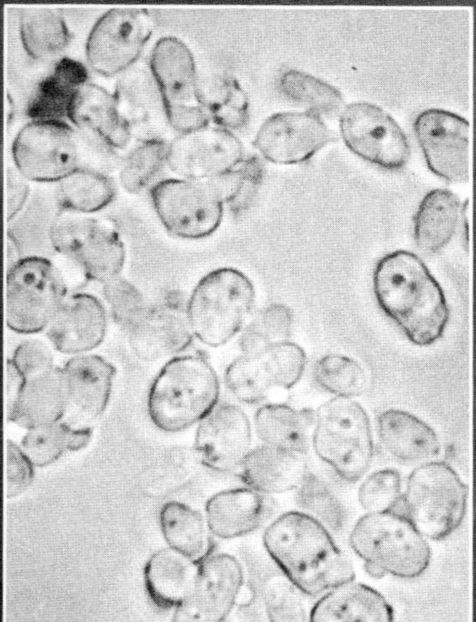

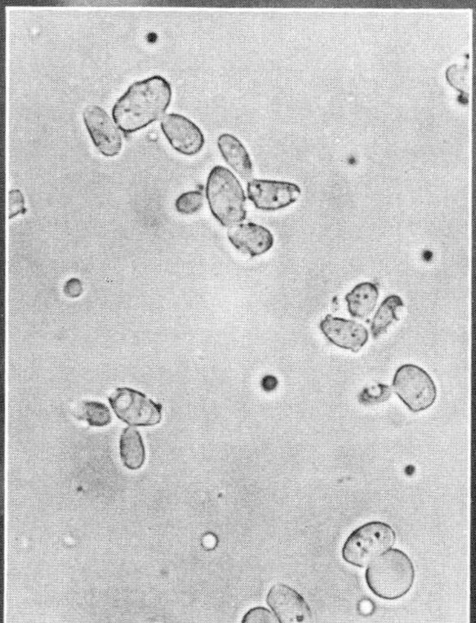

Background An analysis check of yeast samples grown on the by-products of an oil refinery. From this unlikely source scientists are producing food that could help solve the world food shortage

Left Growing yeast. From these large fat cells scientists can create food. The yeast is growing on hydrocarbons which show up here as small black and white droplets

Far left Proteins from petroleum. A micro photograph that shows yeast growing on hydrocarbons at BP's Grangemouth Refinery in Scotland

produce 10,000 tons of dry cell product in a working year. It was assumed that the culture could be continuously maintained at a cell-density of 10 grammes per litre, which would double itself every 2 hours. This study suggested a finished price of 6·7 cents per pound of dry product, which compared favourably with comparable substances produced from hydrocarbons and was considerably cheaper than protein from other more conventional sources.

The product was found to have a good amino-acid pattern, to contain over 50 per cent of crude protein, and to be an acceptable food for rats when fed in amounts of up to 75 per cent of their total diet. But there is a long way to go before it can be served up for your dinner. Already it has been suggested that processing facilities could be made available at the sugar mills where bagasse is first produced, and since much of the world's sugar-cane industry is centred upon economically-backward areas, the result of this work could benefit some of those who do not get enough protein to eat.

Protein from Petroleum/Like so many major advances in science and technology, the unlikely possibility of obtaining protein from petroleum was discovered by accident. In 1957 the BP Laboratory at Lavèra, near Marseilles, was asked to investigate the de-waxing of kerosenes and gas-oils by using microbes to break up the waxes. Waxes tend to be a nuisance in petroleum products because they freeze in cold weather, with disastrous effects on pipes and filters. It did not take long for the microbiological research team at Lavèra to find suitable 'animals' that were fastidious enough to eat the unwanted wax and leave the rest of the gas-oil uncontaminated. At this point the research team headed by Robert Champagnat realised that, while performing their job on removing the wax, these microbes were actually producing protein.

In order to be able to understand how it was possible for such an unlikely substance as crude oil to be capable of supplying vital human food, it is necessary to understand a little more about the family of complex organic compounds generally described as proteins. Like hydrocarbons, proteins have certain things in common with each other: they all contain carbon, hydrogen, oxygen and nitrogen–and most contain sulphur. Normal paraffins are hydrocarbons with a straight-chain molecular structure refined to a high degree of purity. It was therefore perhaps more logical than may at first appear that two possible roots for the manufacture of protein from petroleum products became apparent to the BP team, and have been under development since 1963. At Lavèra gas-oil is used as the 'feed stock' for the bacteriological activity, while at Grangemouth in Scotland pure, normal paraffins are used. Both processes are complimentary since together they extend the range of hydrocarbons available for protein production.

In simple terms the object is to grow yeast. What is unusual about the BP processes is that yeast is grown in strictly controlled conditions on a material that is not generally regarded as a biological food. The more conventional materials on which yeasts are grown are normally only seasonal and are by no means universally available. But gas-oil and paraffin are available all the year round in most countries of the world. A further advantage is that the petroleum-based method for yeast production is some thousands of times more rapid than conventional techniques and the protein it produces can be used for feeding humans as well as animals.

Gas-oil, as used at Lavèra, is the first 'cut' taken direct from the refinery's atmospheric distillation column–the first treatment to which the newly-arrived crude oil is subjected. The initial process in the protein plant is one of continuous fermentation. At this stage yeast micro-organisms, together with nutrients provided by mineral salts, oxygen from the air, and nitrogen from ammonia, are introduced continuously under controlled conditions. The micro-organisms feed on the wax content of the gas-oil, multiplying rapidly to form a yeast cream which is then fed into the harvesting stage of the plant.

This consists primarily of a process in which the wet-yeast cells are spun out at high speed from the gas-oil which is then returned to the refinery tanks for further processing into conventional petroleum products.

In the third, or purification, stage the wet-yeast cells are put through a specially developed solvent process

British Petroleum processes for the production of protein

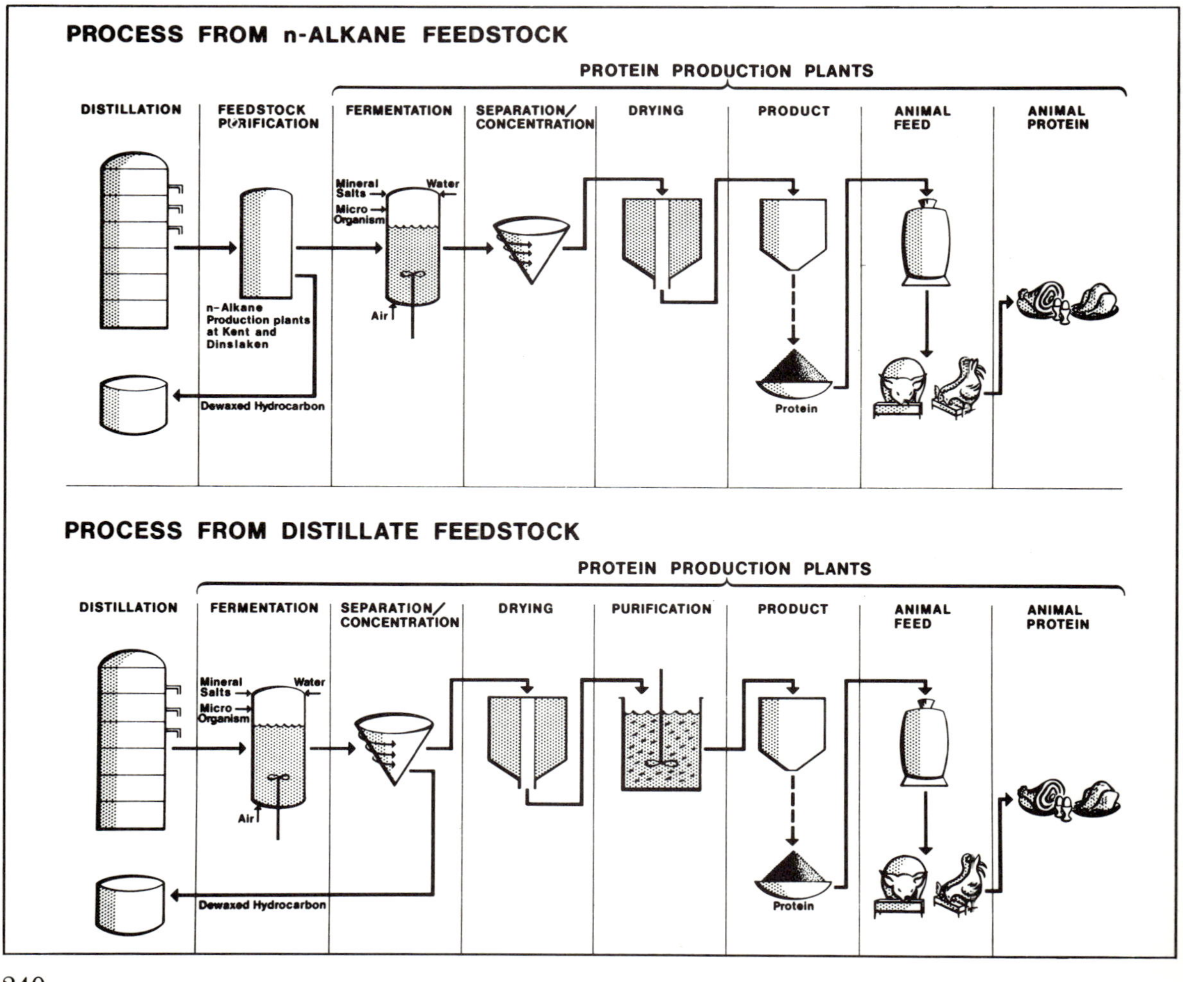

to remove any remaining traces of oil. When the solvent is then removed the final product appears in the form of a yellowish, odourless powder which can be added directly to animal-feeding stuffs.

Paraffins of the type used at Grangemouth are obtained from petroleum distillate by the use of molecular sieve techniques. Such plant is in existence at BP refineries on the Isle of Grain, and at Dinslaken in Germany. The cost of this feed stock is one of the major items in the overall cost of the resultant protein, and the successful development of the essential refining technology is an integral part of this method.

As with the gas-oil process, the paraffin treatment is one of continuous fermentation. Because of the high purity of the feed stock, it is wholly assimilated in the fermentation process and consequently the subsequent separation of the protein concentrate is much simplified. In particular, it does not involve a solvent extraction as in the case of gas-oil. Because of this, the product contains the natural fats (lipids) of the bacteria, to approximately 8–10 per cent of total weight.

Apart from this, there is virtually no difference in food value between the paraffin and gas-oil products. One ton of normal paraffin produces approximately one ton of protein concentrate, and there is no hydrocarbon remaining after the fermentation. The yield of protein from the gas-oil is only 10 per cent of the total fed to the fermentor, but the remainder is, of course, highly valuable unmetabolised gas-oil. For this reason, any future protein plant using gas-oil as a food stock would have to be an integral part of a refinery.

Throughout the development of both these techniques constant and intensive animal-feeding trials have been conducted. Two research centres in Holland were commissioned for this purpose–the Central Institute for Nutrition and Food Research at Zeist, which carried out toxicity tests, and the Institute of Agricultural Research into Biochemical Products at Wageningen, which was commissioned to undertake tests to assess the value of the concentrate as an economic and practical source of protein food for farm animals. Their conclusion is that there should be immediate application of these foods on the farm, especially as an ingredient in pig and poultry diets. There is no reason to doubt its nutritional suitability for cattle, sheep or other farm animals, but the research points out that many of these are capable of using cheaper and even simpler sources of food.

The concluding paragraph of the report, published by BP, is of special interest: 'Animals' eating-habits do not vary from country to country. The opposite is true of humans. BP's view therefore is that a more immediate effect on the world's protein shortage will be made by giving proteins to humans indirectly via animals, which will have the added advantage that it will release the more conventional proteins currently being fed to animals, such as oil seeds and fish meal. This method is admittedly less efficient than that of using protein directly for human nutrition, but on the other hand no matter how good a product may be from the point of view of the

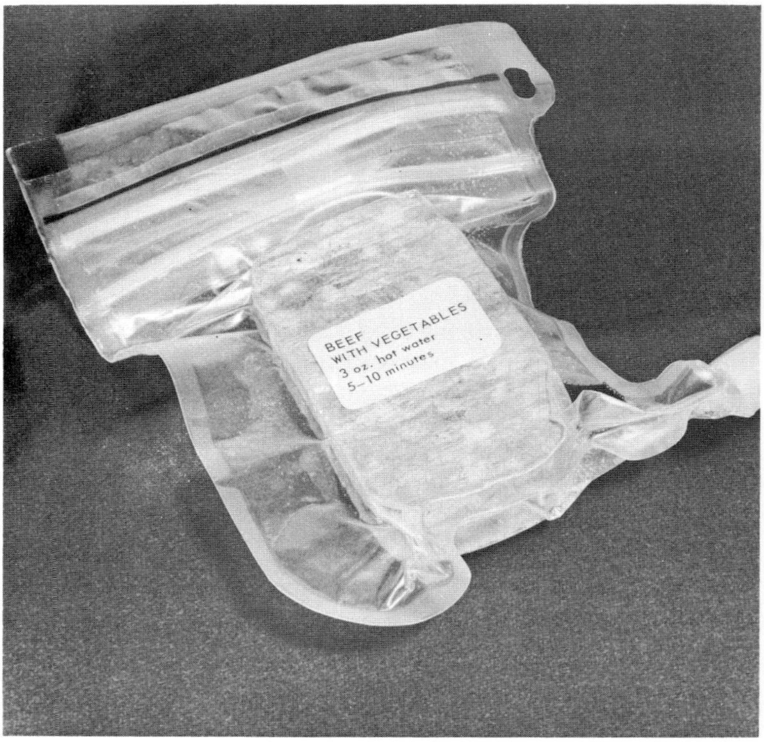

Before and After—The most famous food of the decade, beef and vegetables taken to the Moon by the crew of Apollo 11. By the addition of a little hot water a hard block of dehydrated stew is turned into a soft, mushy main course, similar to toddlers' food

biochemist and the toxicologist, it is useless if people will not accept it. The effective biological value of food which is not eaten is exactly zero.'

Meatless Meats

There is nothing new about the addition of protein concentrates to a wide variety of food products. Until recently most of these additives were manufactured from such commonplace sources as soya beans, cotton seed, peanut, sunflower and other oil-seed crops. But these additives are becoming foods in their own right as scientists work out ways to turn them into substances that have 'bite', 'chewiness', 'texture', and other forms of 'customer appeal'. As a leading authority in this field, A. D. Odell, who directs the Isolated Protein Programme of the General Mills Corporation of the United States, puts it in a recent paper: 'The primary objective of texturisation in protein food is the reproduction of the bite, chewiness, mouth feel and performance of a much prized segment of the human food spectrum. All cultures are conservative eaters, in the sense of loyalty to traditional foods. Change in dietary habits is brought about only very slowly in either the rich or poor consumer. If, in response to the urgent need to extend existing or to tap new sources of protein, engineering and food technology do no better than create amorphous powders or flours which are bland or unpalatable, or unfamiliar or unrewarding to eat, they may fail to create a useful food product. No matter how nutritious the food may be, no matter how brilliant the technology which created it, the food may be ignored by the affluent and shunned by the needy.'

Effectively, this is precisely the conclusion of the BP report on proteins from petroleum, but that seems to be

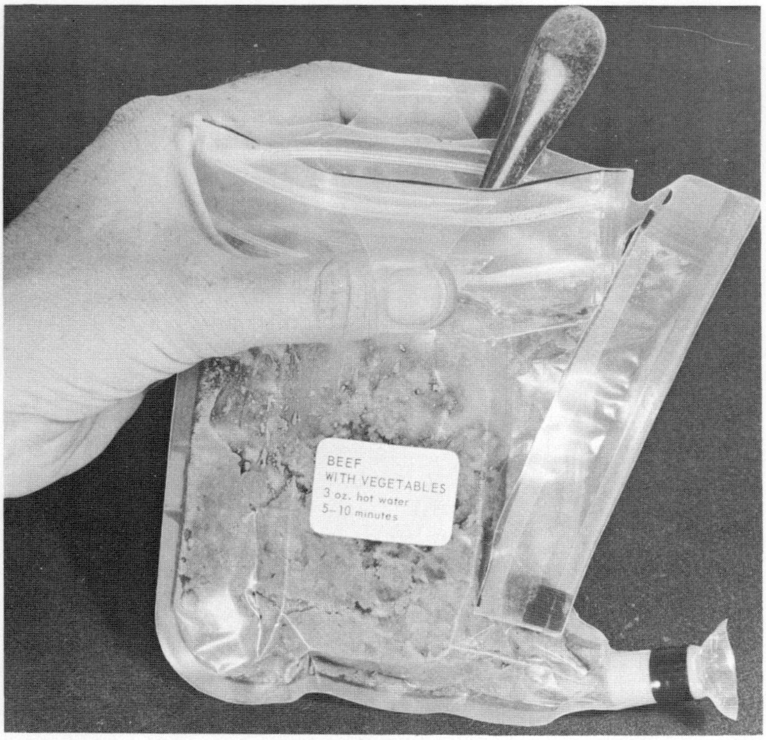

the extent of agreement between British and American researchers. Odell goes on to argue that ways must be found to convert unappetising powders into appetising items that resemble familiar foods, are pre-cooked, and are easy to prepare or can be eaten straight out of the packet. If such formulated foods have long shelf-lives, high nutritional value, and above all are attractive to eat, they should find ready acceptance in the market-place. The American claim is that all these objectives can be achieved with ease by textured protein foods based on oil seed. Certainly it is more efficient to feed humans with these new foods than simply to use them to fatten up cattle–which, in their turn, end up on the table. In fact the case for textured protein foods is that they get the nutrition on to the consumer's plate five to twenty times more efficiently than via the animal route. They can also serve as the vehicle for the bacteria, yeast, mould and fish protein powdered concentrates already mentioned.

Many firms, particularly in North America, are involved in this new technology. The objective is to produce products similar to meats for the immediate use of the American housewife, for food processors, and what are described as 'Institutional Feeders'. It is claimed that there is a real need for such products to *supplement,* not to *supplant,* traditional animal meats. In such products texture-producing fibrils, spun from 90 per cent pure protein, seldom constitute more than 40 per cent of the dryweight. The remainder consists of vegetable or animal fats, carbohydrates, oils, and other ingredients whose function is merely to improve colour, flavour and consistency.

Because the manufacturers of these new foods are very sensitive about words like artificial, synthetic, imitation,

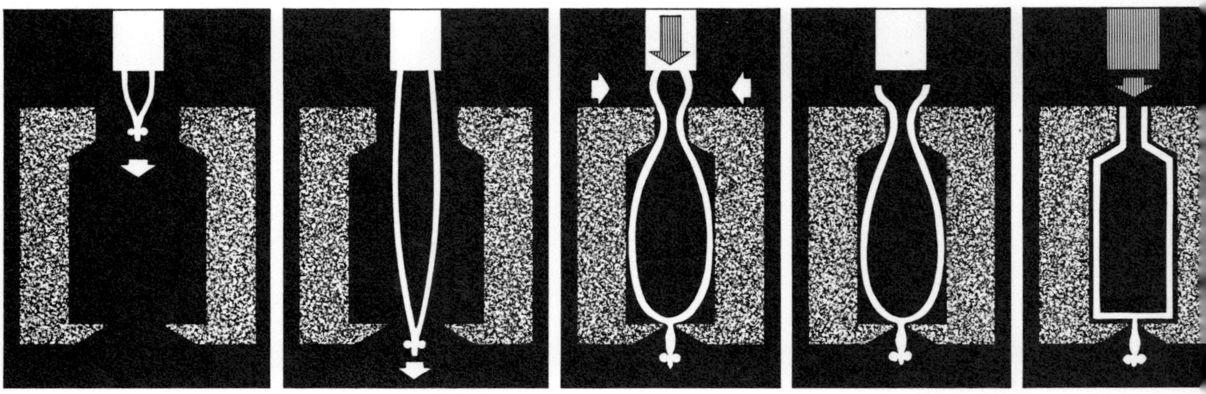

Plastic throw-away milk-bottles could replace glass within a few years. Development work is going on in several countries to produce the ideal lightweight, plastic bottle. Already one firm in France has designed a simple machine that actually makes the bottle, fills and seals it within a few seconds.

From left to right
1 Sealing bottom of an open parison tube
2 Draw-down of parison tube
3 Closing of mould and pre-blowing
4 Neck-pinching and cutting of parison tube
5 Blowing
6 Filling
7 Metering foam-removal
8 Neck-sealing
9 Deflashing
10 Opening of mould and ejection

and so on, they prefer to call their product an 'analogue'. An analogue of bacon is already becoming popular in the United States. A recent report stated: 'Approval and acceptance in this country [the United States] seems certain to simplify introduction in less developed countries, where it can and should be manufactured locally from indigenous crops.'

It is calculated that the cost of these 'analogue' foods will be between one-third and half that of their cooked natural counterparts. Their nutritional value has been demonstrated by elaborate studies undertaken by various manufacturers and research groups. These have included human consumption by convict volunteers who were put into hospital, fed on a diet consisting of up to 90 per cent protein concentrates, and subjected to the most exhaustive metabolic studies. Groups of students were also used, first as in-patients and then as out-patients with, it must be admitted, slightly less successful results in terms of voluntary acceptance. A programme of study has also been completed on post-Kwashiorkor children in Guatemala.

Odell is by no means alone in his belief that textured protein foods should provide viable local industry based on materials that have until now been regarded as uneatable waste. The overall optimism of the American attitude is summarised in the concluding paragraph of another recent paper: 'As the ultimate in convenience and stability, capable of meeting all ethnic, religious and geographic requirements, such products, when flavoured and textured to meet local preferences, can provide for a few pennies per day the bridge across the protein gap.'

Take 2000 Gallons of Milk

There is scarcely a single ingredient of the daily menu anywhere in the world which has not been exposed to the long cold look of the food technologist. New ovens for bakeries, which travel vertically rather than horizontally, have been developed in this country. A special conveyor-belt that bombards oysters with microwaves so that they open their shells automatically has been developed by a Massachusetts fishery. An automatic centrifugal sheller for cashew nuts, which goes on to roast them in the heat of their own burning shells, has been granted a manufacturing licence by the NRDC. A Wiltshire catering firm now uses compartmentalised plastic trays in which the

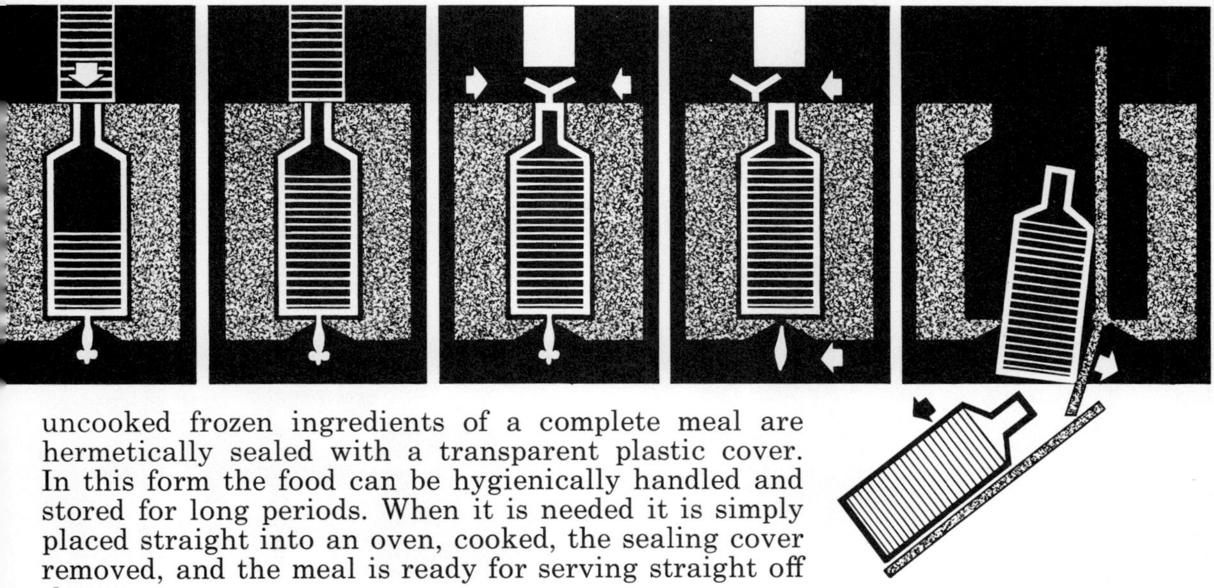

uncooked frozen ingredients of a complete meal are hermetically sealed with a transparent plastic cover. In this form the food can be hygienically handled and stored for long periods. When it is needed it is simply placed straight into an oven, cooked, the sealing cover removed, and the meal is ready for serving straight off the tray.

In Japan it has been found that a cheap space-saving way to store large quantities of rice for almost indefinite periods is to pack it in air-tight plastic bags with solid carbon dioxide and sink the whole lot, hundreds of tons at a time, to the bottom of any convenient lake.

Whisky is likely to appear on the market in plastic bottles any day now. Beer is being freeze-concentrated to a fraction of its original volume, so that it can be transported and stored more cheaply. By adding water and carbon dioxide you have instant beer out of a packet. This is not to suggest that the days of the bottle are past. In Britain a new capillary valve has been developed which, it is claimed, fills 80 per cent more bottles per minute than conventional fillers. Elsewhere a bottling machine has been built capable of filling 1000 bottles per minute. Apparatus which not only fills and seals bottles, but also makes them on the spot, is in daily use, and a totally aseptic bottling process has been perfected.

An ever-widening range of foodstuffs is being subjected to the technique of freeze-drying. But a major food company in the United States has suffered a setback in the introduction of their Frozen Milk Concentrate. In theory, concentrated frozen milk was just what the housewife wanted. It could be stored for up to 6 months in a refrigerator, and could then be made up into whole milk, light cream, or low-fat milk, by the addition of water. The market-research background revealed that over 40 per cent of families in the United States run out of milk at least once a month. Seventy per cent of people who tested it thought that concentrated frozen milk tasted just as good as whole milk, and 15 per cent liked it better. But more than 12 months after revolutionary frozen milk hit the market, its inventors had to admit that the product had hardly set the world on fire. 'Undoubtedly price is a key factor in the slow rate of sell. And the fact that it is sold from the frozen food cabinet rather than the high-traffic dairy case also contributes to a slow sales rate. Cost of production is very high because of special proces-

sing. One cannot just condense milk and freeze it–special treatment is necessary to maintain the protein structure, so that it will reconstitute properly.' As the BP researchers found with gas-oil protein, 'If people don't like it, the impact of a new food technology, however brilliant, is precisely zero.'

An alternative, extremely interesting line of approach to the problems of milk-processing is reported by the Milk Marketing Board. The technique is based upon the principles of reverse osmosis, a curious physico-chemical process that occurs widely in nature. The principles can be demonstrated by a very simple experiment.

Imagine an open-topped container divided into two sections by a thin vertical wall of cellophane. Each compartment is filled with water to the same level. A small amount of any soluble substance such as sugar is dissolved in one compartment, and a large amount of the same substance is dissolved in the other. If this apparatus is left for a time, it will be found that the level of liquid in the more concentrated side has risen, with a corresponding drop on the weaker side. What has happened is that the water has apparently passed through the membrane of cellophane in only one direction. In fact, the water molecules are able to pass in both directions, but their flow is impeded by the sugar molecules. Thus there is a net flow from the weak side to the concentrated side. The important thing, however, is that the membrane acts selectively. The sugar molecules do not pass through at all, and the weak solution therefore becomes stronger and the concentrated solution weaker.

Osmosis is the process by which liquid passes into the living cells of plants and animals where cell membranes act in much the same way as the cellophane in the experiment. In the commercial applications of osmosis, membranes based on cellulose acetate are specially made, and it has proved possible to tailor them to give particular selective properties. In the case of reverse osmosis, imagine that the sugar solution on the concentrated side is subjected to pressure. Once this pressure exceeds what is known as the 'natural osmotic pressure' of the system, the water begins to flow through the membrane in the reverse direction. Thus the concentrated solution becomes progressively more concentrated, the weak solution progressively weaker. In this way reverse osmosis provides a technique for 'squeezing' the solvent out of a solution.

Before the technique of reverse osmosis could be applied on an industrial scale, man-made membranes with controllable characteristics had to be produced, and suitable rigid but porous materials needed to support the membranes had to be made. In the last 10 years, both these requirements have been met. In practice, the feed-liquid is passed through a series of porous tubes, the tubes being lined with the membrane material. Some of the solvent passes through the walls of the tube and can then be collected.

There are two possible applications for reverse osmosis in the dairy industry. It could be used for concentrating milk itself, or liquid by-products, such as whey and skimmed milk. In this it would replace present techniques, all of which involve applying heat to drive off the water by evaporation. Alternatively, it could be used for fractionating–that is the selective extraction from milk, or milk derivatives, of useful components that are at present discarded or sold at low prices. This has been made possible by the recent development of more sophisticated membranes, some of which allow the passage of both water and certain dissolved substances, such as lactose, milk sugar, while at the same time retaining others, such as the protein fraction.

Most of the work so far done by the Milk Marketing Board researchers has been on whey, the liquid left in cheese-making after the curd is removed. It has been shown that whey can be concentrated fivefold–that is, the total of solids in it can be raised from about 6 to 30 per cent. If a fractionating membrane is used, the lactose can be removed and the whey proteins concentrated from an original ·9 to 12 per cent. Such a protein concentrate, when spray-dried, could find many uses in the

Top left At a dairy in Essex, 1-quart plastic bottles are filled with milk and sealed in a specially adapted machine. At no time is there human contact with the milk and the bottles themselves are virtually sterile at the time of filling

Bottom left The British plastic milk-bottle. It is light and tough and you throw it away when it is empty

food-manufacturing industry. The lactose solution itself can be subsequently concentrated sixfold, from an original 4·3 to 25 per cent, and then treated conventionally to yield powdered lactose.

The concentration of skimmed milk and whole milk has been shown to be possible on a laboratory scale and this technique could greatly affect the economics of the distribution of these products. Reverse-osmosis plants could be built in the areas of high milk-production, and fewer vehicles would be needed for distribution. In the end, it might even mean that the milkman will call only once a month with supplies of concentrated whole milk which, when reconstituted, will be indistinguishable from the fresh product.

At present the Milk Marketing Board's pilot plant is about to go into production. Cautiously, the scientists point out that much work will be needed before the laboratory success with these materials can be scaled up to commercial application. But there can be little doubt that the principle of reverse osmosis promises to be the most important development on the dairy industry's horizon. But whatever the new food, or the startling new science and technology that produces it, everything depends on whether *you* like it or not. For people, it seems, are choosy about what they eat. Even hungry people.

A new, whipped dessert topping that stands up longer than ordinary whipped cream. It has a sweetened, sour cream base that can be flavoured with anything from fresh fruit to French liqueurs. It has just been invented by American food technologists

Scientists at Michigan State University believe they have found a way to make blue cheese with a better flavour. By freeze-drying the cheese they produce a 'cheese crumb'; through a spray-drying method they create a 'cheese salt'. Both are combined in a sour cream base (centre) to make a new blue cheese that is said to be more chewy and nicer to eat

Britain: the Day after Tomorrow

Two-mile-high city blocks of the future. Each block would be a city in itself composed of dozens of 'village' decks. Note the area for escape-proof prison and remand homes. *Right* typical plan of a deck

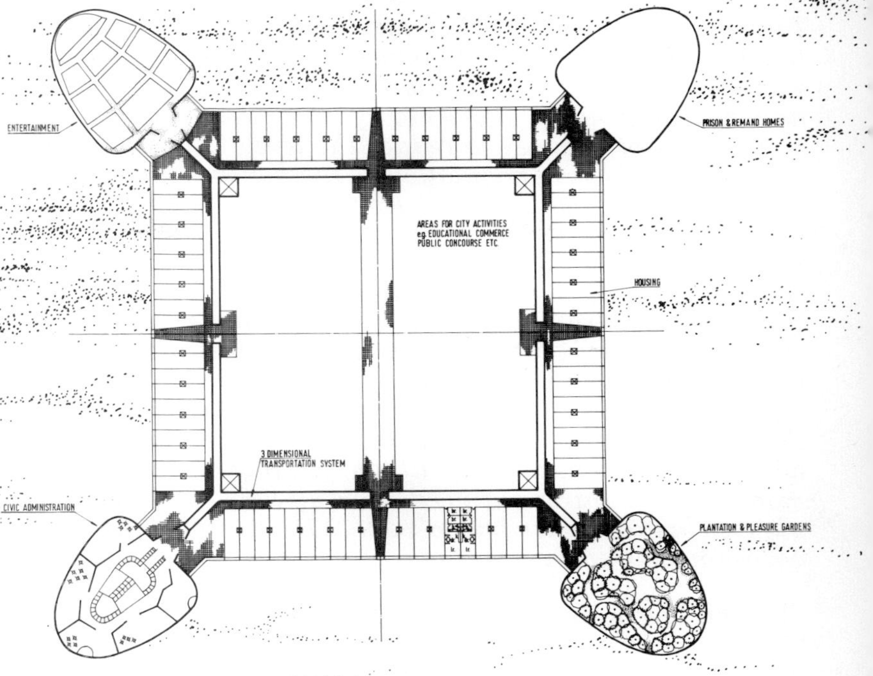

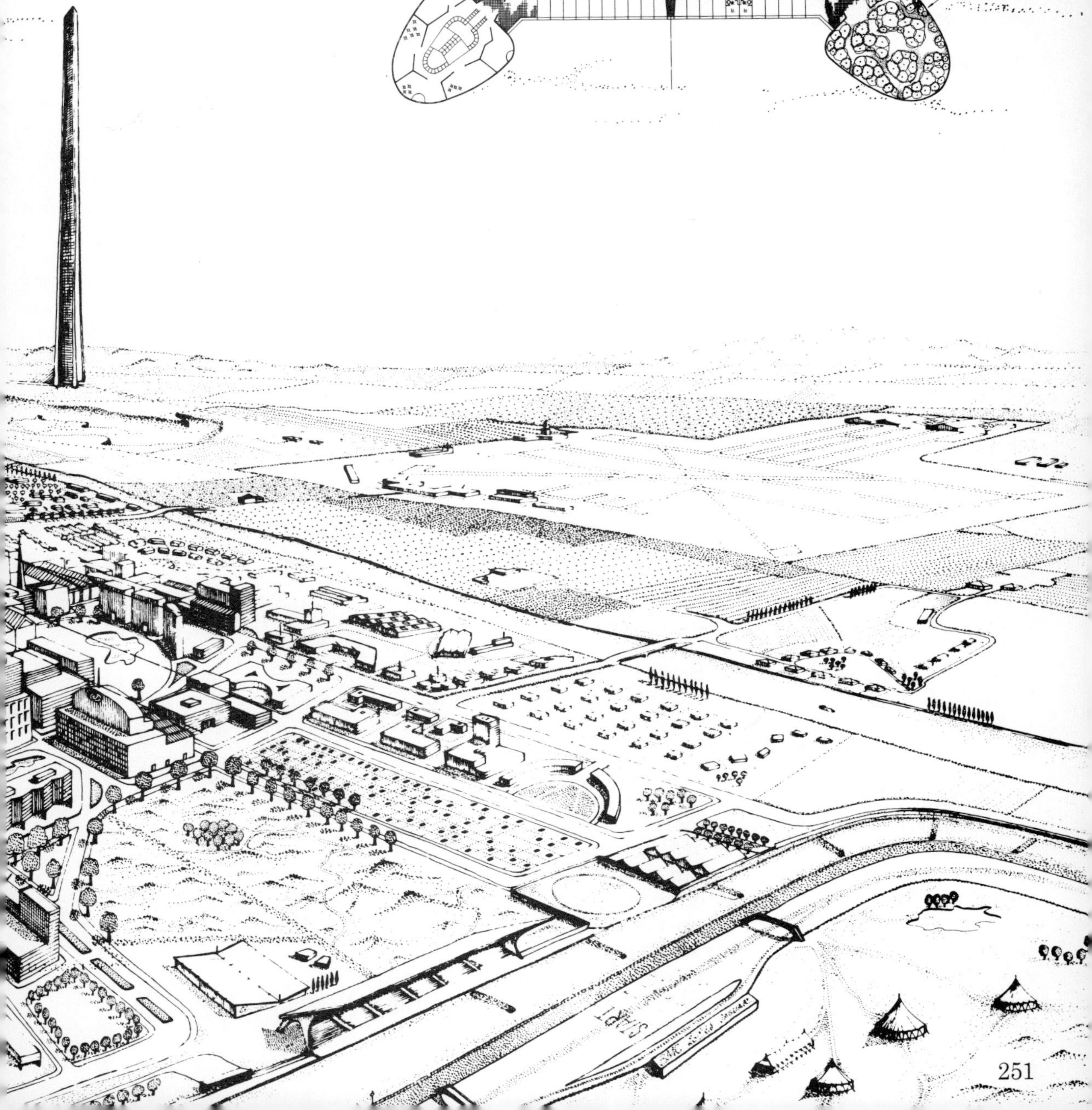

In the year AD 2120 the Secretary General of the reconstituted United Nations announces plans for a 'World Conservation Year for Human Life'. She urges member nations to become signatories to a pact whose object will be as a matter of urgency to reduce the levels of mass tranquillisation of world populations. Another immediate aim will be the creation of large open spaces alongside urban areas where people may go voluntarily to experience natural elements. 'It is a matter of deep concern that our young people no longer have the regular opportunity to experience natural phenomena such as rain, winds, and sunlight,' she says. International leaders agree and 'World Conservation Year for Human Life' is instituted by general consent. By the end of it in Britain alone there are more than thirty open spaces measuring not less than one and a half hectares each. Sixteen million children of school age, many of them untranquillised, have already been taken to glimpse 'natural phenomena' so that the memory of it will not die out within a generation.

Two-mile-high skyscraper cities, as envisaged by structural engineer Wilem Frischmann. He believes they could be one answer to land shortage due to overpopulation at the end of this century. Each city skyscraper would house 240,000 people on a plot of land only 850 feet square. The city would be made up of 4 'townships', each of which in turn would consist of ten 'villages'. A single two-mile-high building would comprise living and work accommodation, including offices, factories, shops, hospitals, prisons, churches, sewage-works and air-conditioning plants. Transport within the plastic and concrete tower would be by computer-controlled, linear motor lifts, operating vertically and horizontally. Underground tunnels would link skyscraper cities to each other

The child born in the United Kingdom one hundred and fifty years from now will grow up in an environment fundamentally different from the one we live in today. To begin with, it is more than likely that his very existence will be adding to the by now critical problem of overcrowding and from an early age he will be taught to realise it. His home will most likely be within a large complex of living units anything up to 1400 feet high. The rooms will be small by present-day standards: ceilings will be a maximum of 7 feet high, perhaps a little less in bedrooms. The whole urban complex may be constructed on artificial land at one of the several floating cities tethered to the coasts of Britain. In areas where weather control has been instituted he will still be able to see the stars at night. But if he lives anywhere else–and most people will–the chances are that his area will be enclosed in a municipal dome half a mile high and approximately 5 miles across.

Set into the upper part of the semi-opaque protective covering will be sources of artificial infra-red and ultraviolet light for use during the day and simulated moonlight for night. During some phases of the year the 24-hour-day cycle may be adjusted to 18 hours so as to allow visiting foreign dignitaries from other time zones to arrive at hours convenient to them. Suitable adjustments would be made after these visits to catch up with the calendar.

The climate will be even, comfortable and air-conditioned throughout the entire year.

A child will start his full-time schooling at the age of three and with the use of the latest techniques and electronic teaching machines will have reached what we now refer to as 'degree level' by the time he is 14. The intense nature of this educational programme will make it necessary for him to attend classes for only three hours a day, three days a week. From the age of 14 advanced education at technical universities and other institutions will be geared to practical research work rather than formal book studies, so that by the time he is 18 a young person will be qualified to enter the profession of his

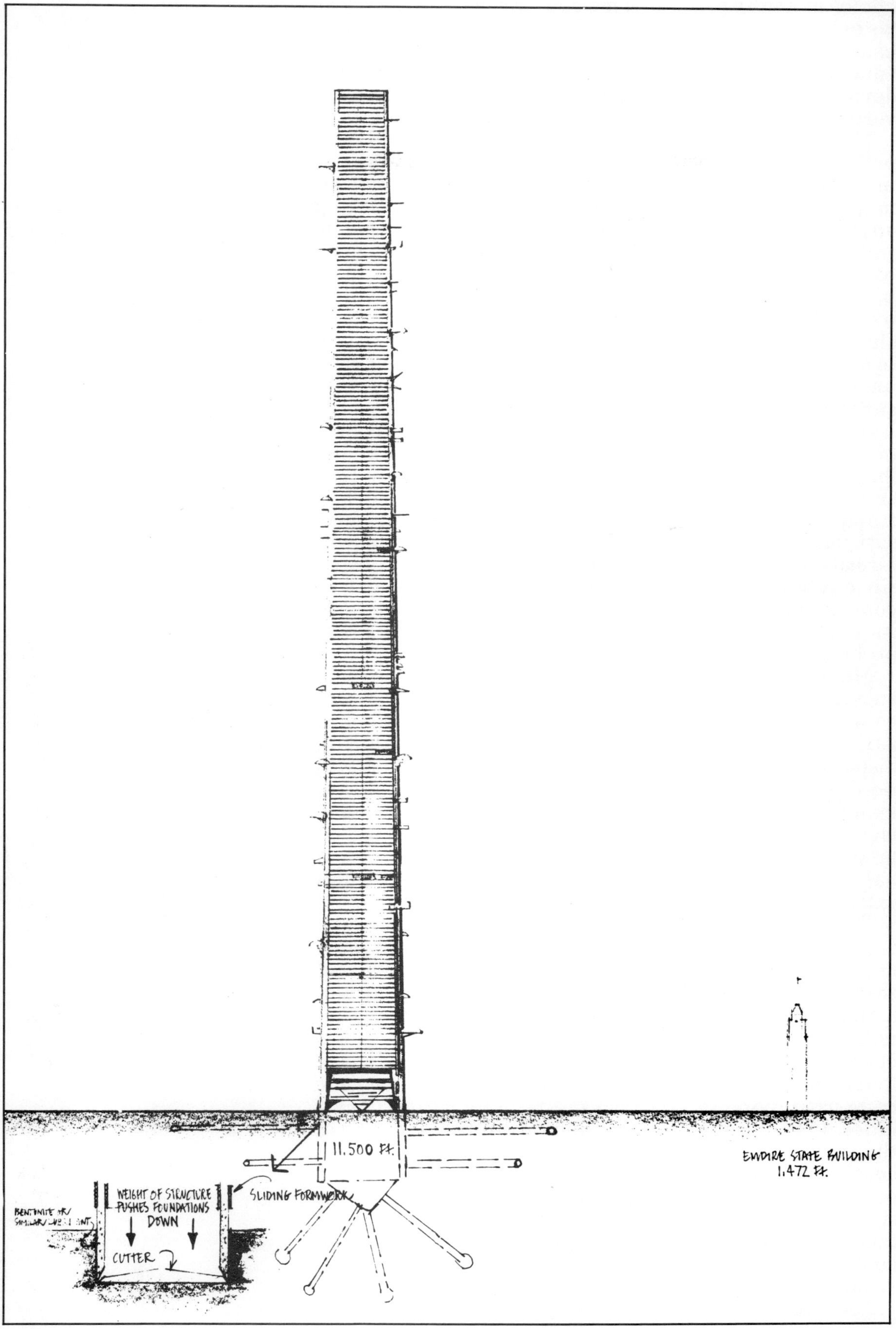
11.500 Ft.
WEIGHT OF STRUCTURE PUSHES FOUNDATIONS DOWN
CUTTER
SLIDING FORMWORK
EMPIRE STATE BUILDING
1.472 Ft.

choice as a fully practising member. Even so, illiteracy as we now understand it (the inability to read and write with ease) will be commonplace even among the highly educated because of the widespread use of electronic devices that will undertake these tasks perfectly, converting the spoken word into print and vice versa.

Meanwhile, the parents of the growing child will probably both be concerned with their own professions until they retire compulsorily at the age of 46. The mother may be a part-time computer programmer, working anything up to five hours a week. Her husband, a comptroller in an automated plant manufacturing video communications equipment, will put in a nineteen-and-a-half-hour working week and compared with many will be overworked. This typical family unit of three will have to show good reason why their number should be increased to four. If the parents want another child they will have to undergo psychiatric tests to determine whether they are suffering severe psychological damage for want of a larger family. It is unlikely that they will take this course of action, for, from birth, they will have been imbibing 'happy water' from the municipal supplies, water that contains one part per million of a harmless, tasteless tranquilliser. The effect of it will tend to make them satisfied with their way of life. Mass tranquillisation will already have been commonplace for about fifty years, since it was introduced by international agreement to control the inevitable side effects of overcrowding with its resultant increase in suicide rates and other social problems.

Much of the food available will be based on protein substitutes and, as with a book club today, a family will contract with a company to supply it with part-cooked daily menus which will be delivered once a month in disposable vacuum packs. The most complicated dish will need only a few seconds under a microwave heater to make it ready for the table. As a result of this development, modern homes will no longer have kitchens for food preparation, and the resulting saving in space on a national scale will provide room for a million and a half extra full-size living units.

The average family will take three holidays a year each lasting approximately twenty-five days. To avoid mass migrations of populations all holidays will be staggered throughout the entire year and on only one holiday will a family be permitted to travel more than one thousand miles. The purpose of these holidays will have long since been carefully rationalised by psychologists who will have found that it was necessary for people to have periodic rests from living in tall blocks, especially within domed cities. It will probably be fashionable to travel to one of the few relatively undeveloped areas left on the European continent to set up a holiday home under plastic in a kind of super urban camping site. The tents will be only one storey high and among the attractions will be that families will be able to practise the near dying art of home cooking on portable stoves powered by low radiation fuel elements.

The level of tranquilliser in camping water supplies will

be minimal and in some places, as an experimental measure, non-existent so as to allow for periods of natural aggression among the holidaymakers. This will have become possible as a result of recommendations from a 25-year-long international study into the effects on humans of extended periods in semi-tranquil states. The only acts of violence, even murder, ever to happen will occur under these conditions, but they will be rare and it will be an adequate defence in law to show that the miscreant had been drinking untranquillised water before the incident took place. The penalties for these misdemeanours will include a period of psychiatric rehabilitation together with a life ban on the use of untreated water supplies.

The most pressing social problem will undoubtedly be the superabundance of leisure time. Homo sapiens, being fundamentally a work-orientated creature, will not yet have had sufficient opportunity to evolve naturally into a more passive state and, even with the help of mass tranquillisation, there will still be many who feel that they have not enough to keep them occupied. The result will be the emergence of the most serious of all antisocial crimes, the acquisition of a second identity in order to acquire a second job illegally. This will be dealt with most harshly by the authorities because the criminals will, with premeditation, be knowingly filling a full-time 15–20-hour-week position at the expense of one of the vast army of permanently unemployed–who will tend to live in specially segregated areas in a more heavily tranquillised state.

Leisure activities will be the responsibility of a Minister for State who will be responsible directly to the chairman of the national governing body. His annual budget will be approximately 8 per cent of the gross national revenue and the money will be spent on the day-to-day leisure activities of the entire population, with special supplementary benefits for the permanently unemployed. Everyone will be catered for. For instance, young people who wish to learn to pilot spacecraft will have multi-million-pound simulators to play in; there will be well-equipped hobby societies specialising in the creation of new species of living creatures with simple intelligence. There will be Anti-Gravity Clubs and specialist organisations where women can gossip under controlled conditions. Formal further education as such will not, however, be encouraged as a leisure activity because it will have been shown that the highly educated tend to be more prone to serious psychological disorders.

The illnesses of the body that plagued man until the end of the twentieth century will, in general, now be curable. Physical medicine will, to a great extent, have replaced surgery. Indeed, the surgeon as such will have become a junior member of the medical profession. He will still have plenty of opportunity to practise his craft but the majority of his cases will be accident victims. The physician will have dominated the medical scene since the end of the century when it was shown that diseases such as cancer, arteriosclerosis and other degenerative disorders were the gross symptoms of general body deficiencies, and it was relatively useless to attempt to deal with them without treating the body mechanisms that were responsible in the first instance. Psychiatric diseases such as schizophrenia and acute depression will also be curable–indeed, serious cases will be rare. But minor behavioural disorders will be commonplace since, due to the pressures brought about by overpopulation, it will have become important to pay close attention to the merest fluctuation in the general mood and demeanour of the individual. Illnesses caused by virus will be widespread but easily treated and, while the common cold will still be prevalent, easily obtainable drugs will put a stop to the worst attack in a few hours. The bald-headed male will no longer exist since it was discovered that a simple genetic malfunction could be repaired.

It will be compulsory to report and undergo treatment for certain conditions, notably venereal disorders, and it will also be obligatory for all young people upon reaching the age of adolescence to submit to a minor operation to ensure their sterility until marriage. This will take the form of implanting long-life, microhormone capsules beneath the skin of the shoulder. Only after legal marriage will the capsules be removed under licence, and they will be replaced immediately after the birth of the first child.

All illnesses, no matter how minor, will be treated in hospital centres where patients will be obliged to remain until they are officially discharged. This measure will be designed to guard against the spread of contagious diseases in overcrowded areas. In fact, the average length of stay in hospital will be under twelve hours and per head of the population the average number of days off each year through sickness will be a mere 2.2.

The nature of a placid, semi-tranquillised society will be conducive to happy, long-lasting marriage and, though it will be a simple matter to obtain a divorce, not many marriages will end in that way. The potential breakdown of a marriage will be treated like any other behavioural disorder, with the result that specialists will be able to 'cure' matrimonial difficulties, possibly even employing an enforced stay in hospital as part of the treatment. The institution of marriage itself will have fully regained the social prestige and importance it lacked a hundred years before. One of the reasons for this will be that it will be difficult to obtain suitable living units for two without proper marriage documents–as it will also be near impossible to arrange for the removal of the microhormone implants in order to have children.

In certain circumstances marriages may be arranged from birth so as to ensure the survival of particularly rare genetic characteristics, but it will occur infrequently because to some extent it will be possible to achieve the same result using artificial techniques, though it will be a costly procedure with a risk of failure attached to it. In general, complete families will consist of only two living generations. Few grandparents will remain alive and widowhood will be almost unknown as provisions will exist to avoid it (see p. 259).

Although there will be no official religion in Britain,

the general tendency will be towards humanist thinking. The concept of a 'day of rest' will no longer have meaning, so it will be difficult for the few religious sects that still remain active to focus attention on themselves by nominating a particular day of the week as their own. The rituals once associated with birth, marriage and death, in which for centuries the Church had its social roots, is by this time fast fading. This is also reflected in the prerequisite for all priests and ministers of religion having advanced qualifications in psychology and social medicine. But despite the waning of religious influence major churches and cathedrals will be well attended by believers and non-believers alike, as they will be settings for important productions of serious music and classical drama. Admission will be by purchase of tickets at the front porch box office, the only sure way that individual churches will be able to survive financially. It is worth noting that it will have been only as a result of the combined and persistent opposition of all religious sects that proposals to institute a new approach to population-control will, so far, have proved unsuccessful. These would be based on socially acceptable programmed abnormalities, chemically induced into the unborn, in order to produce a fixed, and shorter, lifespan for certain categories of human life.

Throughout most of Europe government will be based on the concept of benevolent dictatorship. But in Britain the governing body will be an elected committee of 101 full-time members backed up by a Civil Service numbering several millions. Together they will work closely with their neighbours on the continent of Europe because their policies and political aims will be virtually identical. Throughout the rest of the world a form of amicable segregation, based upon levels of social development, will have taken place.

For by the year AD 2120 every human being on Earth will live within a class structure based loosely on degrees of intelligence. Despite this, it will no longer be thought desirable to engineer intelligence itself into Man genetically–indeed in some parts of the world it will be an illegal operation punishable by law. The reasons for this turn of events will be complex, based partly upon the need to maintain a stable world society without a top-heavy echelon of intellectuals and partly on the fully realised dangers of creating a race of super-intelligent beings.

The hallmark of an effective modern civilisation is that its weakest members are protected by the stronger ones and individuals no longer go to the wall because they cannot support themselves, either psychologically or physically. No matter who they are or what their backgrounds, the aged are assiduously cared for, the young are raised with wit and intelligence. There is a basic level of agreeable equality for all, though individuals may move on from that common level if their talents and motivations make it possible for them to do so. Today the socially aware want to believe that we are fast heading toward this state of affairs. If our grandchildren do not achieve it, then their grandchildren will, because in the final analysis it must surely be the object of human progress and endeavour. But amid all the progress, the good intentions and the new concepts that apparently will influence the world of tomorrow, are there any real pointers to show that we are heading towards a fundamentally better life for every single member of society?

Ornithologists will tell you that if you watch certain species of birds carefully you may witness a strange and terrifying spectacle. Occasionally a single member of a community becomes unacceptable to the rest. To begin with it is ignored and none of the other birds will have anything to do with it. If the outcast fails to get the message and remains with the flock, its fellows then start to harry it mercilessly. They fly close, swooping and screeching and beating their wings. They tease it and peck at it. If the poor outcast still fails to realise what is going on, it is doomed. For abruptly the teasing stops and there follows a mass and fatal onslaught. Within seconds the outcast is dead, torn to pieces by the community of which it was once a part.

It is not all that long ago in the history of our own social development that people behaved in a very similar manner. The middle-aged woman of the 15th-century English village with warts on her face and crippled by arthritis in her hands and shoulders could just as easily become outcast by her community. If she failed to move on to another village, she was mocked and occasionally assaulted. Sometimes, in the end, she would be hauled away in chains and burnt alive at the stake as a witch. Since those times men have become more civilised and humane. They no longer kill their victims, but it is worth recalling that only fifty years ago it was a common sight in Britain to see the village idiot being baited by adults as well as children. It is some indication of progress that we would now call him 'mentally retarded' and place him in a specialised institution where he could be cared for and even, to some extent, educated.

So today we may sit back and congratulate ourselves upon achieving a level of civilisation and understanding unknown to our forebears. Yet have we come as far as we sometimes think? Fundamental instincts die hard; the urge to penalise the outsider is a strong one. That capital punishment is no longer on the statute-books does not meet with universal approval, nor does the fact that we no longer beat our criminals with bundles of birch twigs until they bleed. But while the progressives may smile at the quaint foibles which still exist within a sizeable section of the population, it is worth scanning the horizon to see what might be created to satisfy the baser instincts in human make-up that cannot be totally quelled in only a few short years of enlightened leadership. Ironically, perhaps science itself may help provide an adequate diversion.

In the past few years eminent and respected biologists have been warning that we should be aware of the 'increasing pollution of our genetic stock'. By this they mean that more babies are being born into the world that would previously not have survived their full term in the womb. Because of congenital weaknesses and abnormalities they would most probably have been rejected at an early stage in their pregnancies. Yet nowadays medical skills are such that these imperfect lives can be preserved and brought to adult maturity when they themselves may reproduce with the strong chance that their offspring will inherit similar genetic weaknesses. The implications are that in years to come the people of more advanced nations will inevitably become less biologically perfect and, therefore, eventually less able to survive. Within two or three decades it may be possible to correct all genetic imperfections artificially by juggling with the genes of the unborn child or even the ovum and sperm that create him. But in a world that is fast becoming overpopulated perhaps another answer may already be finding favour first. It is easier and certainly cheaper to end the lives of imperfect infants before, or at, birth, so preserving genetic superiority while at the same time, perhaps, satisfying those deeply ingrained emotional requirements to do away with the outsiders of society. Should this course ever become widely, and officially, acceptable, it presupposes that we will have learnt to regard the value of life in a rather different light to that which has been prevalent in recent times, and to accept that some lives are more valuable than others.

Already there are signs that this conditioning process has begun. Officially registered abortion clinics in Britain–an unacceptable concept only a few years ago–now rid us of the responsibility of several hundreds of human lives a year. From time to time there is an emotional outcry against the practice but generally the nation is settling down comfortably to accept and make full use of this new facility. And there is now some evidence that the same practice could be applied to the outsiders at the other end of the age scale.

You cannot see it from the main highway because the place is screened by trees and flowering bushes. In the early spring the lawns that flank the entrance are dotted with crocuses and daffodils pushing up through the new green grass. It is a peaceful setting with the old-world charm of the late 1970s. Every ten minutes or so a large limousine draws near from the direction of Guildford, slows down, then sweeps purposefully between the entrance pillars to wind its way gently over the gravel drive between the trees and disappear from sight. Each car is driven by a chauffeur wearing identical blue uniforms, lightweight one-piece suits in the newest man-made fibres. They are young men and, somehow, they all look alike. Their passengers sitting deep in the back seats tend to be elderly people. Many of them have a single small weekend case on the bench seat beside them.

Beyond the trees the first thing you see about a quarter of a mile away is the house, standing on the crest of gradually rising land. There are lawns and flowerbeds landscaped into the scenery with infinite care and good taste, while on the skyline dark-green conifers draw the eye back to the building itself. It looks as if it were once a stately home of some well-to-do and long-forgotten gentleman. Only if you know about these things do you sense, as you draw nearer, that it is a reproduction. The proportions are pleasing and correct but somehow it does not have the weathered look of a building that has stood through 150 winters and summers. As the car pulls to a stop at the front door a couple in their forties come out to meet it. He is smiling a welcome, she is patting her hair into place. The newcomer is a man of 60. With the help of the chauffeur he eases himself from the car, shakes hands with his hosts, and together they move inside. He carries his weekend case stiffly and appears to hesitate. The woman slips her arm through his and, still smiling and talking, moves him on.

Inside, the atmosphere is reminiscent of an expensive country hotel and the newcomer is led to the arrivals desk where a young man takes over.

'It's Mr Burke, isn't it, sir?' The old man nods.

'You're most punctual. We do appreciate it. Was it a comfortable journey?'

Once more the old man nods. He is looking around as if to find his bearings, as if he were looking for something that he half expected to see.

'Now if you'll sign here, sir.' The young man offers a pen and slides a large buff form across the desk.

'Please check it if you wish. I have filled in all the relevant details. We received them by videophone this morning.' The old man signs.

'Will this evening at 8 o'clock be suitable?'

For the first time since he entered the building, the ageing visitor speaks. 'Entirely suitable.'

Four and a half hours later, after an old-fashioned meal of scrambled fresh eggs, lightly buttered wholemeal toast and tea, the old man lies waiting on his bed in his private room. He feels strangely relaxed and idly wonders about the tea that was a little stronger than he usually preferred. He smiles to himself and murmurs 'Enter' at the tap on the door. The male attendant is businesslike and swift.

'I'll be finished in a moment,' he says, as he places the micro-inoculator against the old man's forearm. The old man nods and sighs. With the vestige of a faint smile at the corners of his mouth, he closes his eyes and ceases to live.

On 1 March 1970 Professor Ian Donald, Regius Professor of Midwifery at the University of Glasgow, warned that municipal death centres might have to be set up in Britain if euthanasia were made legal. He told a meeting considering the possible legislation for euthanasia, 'It might even be suggested that municipal gas chambers should be established.' And in saying that, he perhaps foreshadowed the State-run Euthanasia Hotel of 150 years from now. For if it ever became fashionable to 'opt out', it is likely that the entire procedure would be rationalised and organised by a Ministry of Living whose function would be to see us all safely, and literally, from the cradle into the grave. One might even imagine powerful incentives being proffered to potential 'opters-out', such as insurance benefits as well as the luxury chauffeur-driven transport to your Regional Euthanasia Hotel. During slack periods the State might even offer free

Grade One funerals and cremations plus, as a very special extra, guaranteed delivery of the remains to the Garden of Remembrance of your choice. (This offer applies only within the United Kingdom.)

As unpleasant as it sounds, that Grand Guignol fantasy has the seeds of reality buried within it. On 16 May 1966 a written instruction addressed to the entire medical staff at the Neasden Hospital in North London was prepared and circulated. On the surface, it was straightforward and unambiguous, and sixteen months later it was still pinned to a notice-board in a nurses' changing-room.

When that instruction was published more widely[1] it led to acute embarrassment in many quarters. The then Minister for Health, the Rt Hon. Kenneth Robinson, issued an immediate directive to every hospital in Britain to withdraw at once any similar instructions that might exist. A special inquiry set up at Neasden Hospital reported that in spite of the apparent implications of the instruction, not one patient had ever been treated under

NEASDEN HOSPITAL

16th May, 1966

Memorandum from: PHYSICIAN SUPERINTENDENT

To: All Medical Officers (Geriatrics, Infectious Diseases, Opthalmology, Orthopaedics). Dr. James. Matron. Nursing Tutor. All Ward Sisters. M.I. Room.

Respiratory Failure and Cardiac Arrest
Resuscitation

There have been two demonstrations of the film illustrating resuscitation.

As each new batch of nurses come in for training the film will be shown.

Any staff may refresh their memory by seeing it again.

The following patients are NOT to be resuscitated:-

Very elderly, over 65 years

Malignant disease

Chronic chest disease

Chronic renal disease

Top of yellow treatment card to be marked - N.T.B.R. (i.e. Not to be resuscitated).

The following people SHOULD be resuscitated:-

Collapse as a result of diagnosis or therapeutic procedure,

e.g. Needle in Plura (even over 65 years)

Sudden unexpected collapse under 65 years,

i.e. loss of consciousness,

cessation of breathing,

no carotid pulsation.

In this case the nurse in charge should immediately institute resuscitation until the doctor arrives.

Someone else should be requested to call a doctor by dialling either of the following numbers - '60' or '0' - on the telephone, stating the ward or place where the emergency is.

[1] In *Tomorrow's World,* transmitted on 20 September 1967

its terms and there is no reason to doubt that finding. Yet at the time it was an open secret in London medical circles that the Medical Superintendent at Neasden had, in fact, probably got the idea for his directive from a similar notice at a world-famous London hospital not many miles away. In the event, it was the man from Neasden, a respected doctor with a lifetime of good medicine behind him, who took the full brunt of the row that flared over what became known as the Neasden Scandal.

Even the President of the Royal College of Physicians publicly deplored the Neasden memorandum. But other doctors went only so far as to criticise the fact that instructions had actually been committed to paper. For the probability exists that then as now medical authorities throughout the world practise a form of selective life-saving. The reasoning behind their individual life-or-death decisions is probably most carefully considered and weighed in the balance but one might justifiably suspect that frequently there are relevant factors at work that are never openly admitted to, perhaps never even fully realised. The emotional reactions of doctors themselves who are daily faced with the task of resuscitating ageing and derelict human life, coupled with the plain fact that it costs a great deal in effort and money to keep a desperately sick person alive, almost certainly play a vital part in tipping the balance when decisions concerning the termination of a life are under consideration.

In a system dominated by private medicine the doctor's decision is in some ways simpler, for the rules of that game are well understood. The patient who runs out of money to pay for his treatment is discharged and sent home to take his chances. But where there is a state-run health service, charged with providing the best possible care for the citizens it serves, any decision that conflicts with this ideal is open to serious question. Regardless of the length or the price of the treatment, we have come to expect as a social right the best that medical science can offer. But are we, even now, deluding ourselves?

While the welfare of the individual is of critical importance to himself, that view is not necessarily shared by the administrator who deals with the welfare of millions. For instance, in Britain there is at present a debate in progress over the cost-effectiveness of early diagnosis. Using techniques that are now widely available it is entirely possible to find and treat a whole variety of illnesses long before the first real symptoms show themselves. But it has been pointed out that it can cost many hundreds of pounds to find and treat, say, an early case of cervical cancer and medical accountants are not all that sure that the life of a woman in her fifties is worth the money to a nation which, in the end, pays the diagnosis and surgery bill. The lady herself is unlikely to have any doubts on that score, but the argument has already been put: why spend/waste so much money screening thousands of women to find a meagre two or three per cent who will know soon enough that they have cancer, anyway?

Once that point of view finds official acceptance an

intrinsic value is placed on life and it is not long before a bureaucrat somewhere can go on to categorise and value humanity with the precision of an accountant at the end of a financial year. If the life of a monarch is valued at £250,000 in medical fees, how much is a farm labourer's life worth? Is the potential of a child's life worth more to the nation than a woman of 45 past child-bearing age? Or, even, is an incurable mental patient worth less than a congenitally-malformed baby? Present-day taste and morality make questions like these distasteful and, for some, unnecessarily unpleasant. But even now there are signs that in tomorrow's world, perhaps only fifty years hence, we could all be conditioned to be content to accept and abide by a new moral logic which dictates that humans are valuable only providing they realise and sustain their full potential.

The irony is that this new outlook becomes a distinct possibility in direct ratio to the increased ability of scientists and doctors to extend human life beyond its 'natural' three score years and ten. When the day comes, as it surely will, and it is reported that researchers have devised a technique to regenerate all ageing human tissues, so putting off indefinitely the onset of senility, then after a short honeymoon period the moral implications of living too long will suddenly become evident. Like rickety, elderly Eskimos in the northern latitudes who used to sit outside their igloos in order to freeze to death, the indications are that our ageing sixty-year-olds, recently past their official productive lives, will be encouraged, even preconditioned, to take the necessary steps to terminate their living arrangements with a society that no longer has need of them. The criteria for living will have, finally, become directly linked to the quality of life rather than the simple instinct to survive. If you still are inclined to believe that is an unlikely prospect, consider the function of the charity committee that sits from time to time in one large American city to interview applicants for free places on the local artificial kidney machines. There are not enough free facilities to satisfy the demand. So, it is reported, the committee invites would-be users to appear before it so as to assess their relative merits. The applicants, recently discharged from local hospitals and sick near unto death with their withered organs, answer questions designed to reveal whether they are candidates with a worthwhile potential for life. The successful ones get their artificial kidneys. The losers go home to wait for the level of poison in their systems to reach fatal proportions.

Future social historians might well try to show that the tendency to revalue human life was directly linked to the moment when surgeons and doctors were able for the first time to keep patients alive whose bodily functions had disintegrated to a point where natural death was inevitable. For, once having created the machine to delay the moment of death, it was only a matter of time before a well-meaning medical man was faced with the decision to switch it off, and in so doing take an active role in terminating a life. Once that decision became a commonplace one, as it is today, human life had a tangible price on its head. Doctors found themselves asking if the quality of life at stake was worth saving. Would grandad ever recover his faculties sufficiently to appreciate living? No? Then let's, with professional regret, put grandad down.

The astute historian might go even further in his attempt to uncover the moment when 20th-century man ceased to regard life itself as sacrosanct. He might find it significant that once there was a farmer who decided that the best way to raise chickens was to grow them in boxes. If things were organised carefully enough the birds need never come out. He could feed them, water them, clean them automatically; he could light them artificially. He could dose them with tranquillisers, protect them with antibiotics, control them with hormones. Faced with this barrage of scientific care they would grow more quickly so that eventually he could kill them in large numbers and sell them at market. From chickens the idea spread to beef cattle, then to sheep-factories producing living creatures like doorhandles to be eliminated without a twinge of conscience. The historian would find that objections to these methods of food production were most often set aside on the grounds that logic was being coloured by emotion. Some psychologists even suggested that the reason people objected was that they had been brought up as children to identify too closely with nursery characters like Larry the Lamb and lovable Brer Rabbit.

Yet despite the psychologists there might be another, not altogether unemotional, factor to consider. If one still accepts that all forms of life have unique value, from the creatures in their automated farms to human beings in their city streets, the inherent risk attached to devaluing one part of that life spectrum is that it could be only a matter of time before the new and lesser value has spread right across the board, perhaps culminating one of these years in a cosy foundation-stone-laying ceremony for a Euthanasia Hotel near Guildford. The signposts that mark the road to revalued humanity are already there to be seen. In May 1970 there was an indignant outcry when the British public learnt for the first time that live foetuses from abortion clinics were being sold for medical research, and the suggestion that some doctors might actually try to grow an unborn child in the laboratory, and then slaughter it shortly before it could survive on its own, produced controlled outrage. News media transmitted the story around the world and everywhere the reaction was the same, one of muted horror. Yet it turned out to be a nine-day wonder that evaporated more quickly in the public mind than the aftermath of the World Cup. For already, without realising it, we had been sensitised against a total rejection of the idea because it was pointed out that in the long term this kind of research could actually benefit mankind. It might lead, said doctors, to the understanding and eventual cure of heart disease. Research on unborn human specimens might even help conquer cancer, that last most desperate fear of all Western societies.

So, if we are true to form, perhaps the next time we hear

the news that work has been going on with live foetuses, it will be greeted with wonder and awe and there may even be a Nobel Prize in it for somebody. For life itself is no longer a medieval miracle. Even the most amateur scientist knows enough to believe that it is just a rather complicated chemical recipe that will eventually be conjured out of a test-tube. He is probably correct. But the speed with which it is accomplished, and the later development work that will result in the ability to control all human life with fine accuracy, will signal the beginning of a new era. When the mystery is gone, when the female of the species no longer is obliged to conceive in love and give birth in pain, there will be neatly ordered categories of human life each with its own predetermined value, perhaps even its own pre-set span.

Meanwhile, if there is still an instinctive horror of the concept lurking anywhere, it is worth remembering that it may be only because we are not yet fully conditioned to accept the prospect of a revalued existence. But primitive techniques already exist to help us overcome our irrational weaknesses. Given time and the incentive to develop them, psychologists may be able to devise operant conditioning methods that will make it possible for every last one of us to accept happily the inevitable, when it finally arrives. Though it is no longer fashionable to shout 'Die for your country!' in the time of war, perhaps one day not so very far off it could be most acceptable to murmur 'Die for your society' when we have been shown that we are no longer active and productive members of it, when each of us in turn becomes one of the new 'outsiders'.

Old Burkster's Almanac

The secrets of the future are revealed by the team most closely in touch with Tomorrow's World

1972 Orbiting Space Laboratory launched.

1973 Automatic Mercury-bound spacecraft misses target.
1000 dead in Jumbo Jet collision over urban area.

1974 Major revision of world air-traffic procedures.
Space shuttle launched.
Automatic spacecraft orbits Jupiter and identifies red spot as ferrous cloud.

1975 New York's first moving sidewalk switched on.

1976 Automatic spacecraft lands on Mars and finds a form of life.

1977 First module of Spacebase in orbit with US President on board.

1978 Speech typewriter on market.
Deep-freeze multi-organ transplant banks set up.
Worldwide ban on professional boxing.
Launch of automatic spacecraft to extreme edge of solar system.

1979 Birth-control pill for men perfected.
Plastic glue available for instant repairs on human bones.
Total control of obesity with drugs.

1980 Computer terminals in the home.
Insurance companies insist on automatic burglar-alarms in all houses as condition of cover.
Channel Tunnel opens.
Artificial ageing of wines, spirits and liqueurs perfected.
Ban on use of hormones and growth-promoting substances for rearing livestock.
Abortion pill perfected.
World satellite-navigation system completed.
Large-screen, thin TV sets on market.

1981 Televising of Parliament begins.
Techniques widely available to select sex of children at moment of conception.
Campaign for conservation of insect life.

1982 London's first moving pavement switched on.
Electronic touchlines installed at Wimbledon.

1983 Insurance companies demand automatic fire-extinguishing system in every household as condition of cover.

1984 US manned landing on Mars postponed until 1990.
Simple surgery devised for permanent contact-lens implants to replace spectacles.
Water pollution controlled.
British wren extinct.

1985 Crash programme launched to preserve British songbirds.
Driverless, long-distance trains in service.
Last British foxhound pack disbanded through lack of game and country.
Pocket language-translation machines on sale in department stores.
Animal languages decoded by Man.
Videotape from security camera accepted in evidence at Old Bailey.
Worldwide ban on petrol-driven automobiles.
Nuclear-powered submarine tug in transatlantic cargo service.

1986 First Homo sapiens grown to full term artificially in a laboratory.
Channel Tunnel closed for installation of magnetic accelerator/hover system.
Automatic spacecraft (launched 1978) leaves solar system and is overtaken by advanced satellite (launched 1983).

1987 Control of arteriosclerosis and heart disease.
Soviet manned landing on Mars.

1988 US manned landing on Mars.
Air pollution controlled.
First ape taught to communicate in English.
Worldwide telephone dialling.

1989 Sea-bed farming commercially viable.
Socially acceptable narcotic drugs widely used.
Surgical grafting of permanent, mechanical-limb replacements achieved.

1990 Use of drugs to increase intelligence widespread.
3-D colour laser TV on sale.
Cure for cancer discovered.
Last national morning newspaper closes down.

1991 Hybrid soft fruit perfected that ripens in the open during European winter months.
Amalgamation of British Army, Navy and Air Force into one defence organisation with same uniforms.

1992 Computerised polling leads to frequent national referenda.
Virus diseases, including the common cold, curable.
Genetic adjustment replaces cosmetic plastic surgery.

1993 Average national working week: 22·33 hours.
Mini-helicopters for urban transport outlawed.

1994 Establishment of world computer-information bank.

1995 The end of tooth decay.
Domestic video-phone available.
First use of tactical nuclear missiles in localised war.
Creation of facsimile fresh fruit and vegetables in US laboratory.

1996 Chemical creation of living cells.
Legislation to abolish the use of coal and oil as fuels for heating.
Thermo-nuclear fusion controlled and economically viable.

1998 World's first international city completed under South Polar ice (population ½ million).

1999 Creation of intelligent artificial life achieved.

2000 Semi-voluntary euthanasia at the age of 60 encouraged by Department of Health.

2001 Manufacture of elements to order.

2008 Bank of England withdraws cash and notes in favour of credit-card economy.

2010 Compulsory birth control instituted.

2014 First three floating cities completed round coasts of Britain.

2016 2000th edition of *Tomorrow's World*.

2020 Supersonic, commuter, land-transport system opened.

2030 Average age of puberty reduced to 9 years old.

2100 Global weather-control programme launched.

Picture Credits

Acknowledgment is due to the following for permission to reproduce illustrations on the pages listed:

8–9, Kirsty Sørensen; 12, *Design* Magazine; 13, BP Chemicals (UK) Ltd; 14, Mowbot, Inc.; 15 bottom, Ken Hoskin; 22, Syndication International; 24, Council of Industrial Design; 26–7, Features International; 30–33, 35, University of California; 42, US Navy; 43 bottom, Windak Ltd; 44, 46, Westinghouse Corporation; 50 top, Elsam, Mann & Cooper Ltd; 53, General Dynamics, New York; 55, Fox Photos Ltd; 56, 57 left, Vosper Thornycroft Ltd; 64, GE Creators Ltd; 67, John Stone; 70–71, Ken Palmer; 73, Garnade; 74, Hovercraft Development Ltd; 82, Rolls-Royce Ltd; 90, Hawker Siddeley; 91, Park Studios; 94–5, The Post Office; 99, Pilkington Brothers Ltd; 100, STL Ltd; 102–3, Vickers Ltd; 105, Crown Copyright. National Physical Laboratory; 106 top, G. D. Searle & Co.; 108, Studio Toivo Steen; 112 top, Studio Pine Ltd; 112 bottom, The Marconi Co Ltd; 115, Michael Hackett, McKindoe Burns Centre; 116, *Northampton Chronicle & Echo*; 117, Mecanaids Ltd; 118, Apex Photos Ltd; 119, Associated Iliffe Press; 120, US Navy official photograph; 122, D. P. Choyce, M.S., F.R.C.S.; 125 left, Michael Hackett, McKindoe Burns Centre; 125 right, National Institute for Medical Research; 126–41, NASA; 146–7, ILC Industries, Inc.; 149–63, NASA; 167, Royal Observatory, Edinburgh; 168, Apex Photos Ltd; 169, The Council of Industrial Design; 172 top, A'Court Photographs Ltd; 172 bottom, Art-Wood Photography; 173, Reeve Photography; 175, IBM; 176, The Rank Organisation Ltd; 177 left, Fox-Waterman Photography Ltd; 177 right, Logan Photographers; 178, John Lawrence; 180 top, Parker P.R. Associates Ltd; 184–5, Energy Conversion Devices, Inc.; 190, The English Electric Co. Ltd; 191, Netherton Works Metallurgical Laboratory; 193 top, NASA; 193 bottom, Standard Telecommunication Laboratories Ltd; 196–7, NASA; 200, Gordon McLeish; 204, Stewart Bale Ltd; 205, British Insulated Callender's Cables Ltd; 207, GE Research & Development Cpr.; 208 left, Freddie Squires Ltd; 208 right, John I. Thornycroft & Co. Ltd; 215 bottom, Crown Copyright. British official photograph; 216, E.N.G. Clarke Ltd; 219–20, The Glacier Metal Co. Ltd; 222–3, ICI Ltd; 224, Rolls-Royce Ltd; 230–32, General Mills, Inc.; 235, Wamboldt-Waterfield Photography Ltd; 237, NASA; 238, 240, The British Petroleum Co. Ltd; 242–3, NASA; 246, Freddie Squires Ltd; 248–9, Michigan State University; 253, Wilem Frischmann.